Introduction to Marine Chemistry

J. P. RILEY

and

R. CHESTER

Department of Oceanography
The University of Liverpool, England

1971

London and New York

ACADEMIC PRESS INC. (LONDON) LTD
24–28 Oval Road
London, NW1

ACADEMIC PRESS INC.
111 Fifth Avenue
New York, New York 10003

Library of Congress Catalog Card Number: 71–129791
ISBN 0–12–588750–7

Printed in Great Britain by
J. W. ARROWSMITH LTD.
BRISTOL

Preface

During the past decade there has been a great upsurge in the study of the oceans as a whole. This has naturally extended into the realms of chemistry, which occupies a key place in oceanography, a science that is by its very nature interdisciplinary. As a consequence of this, there has been a considerable expansion in the teaching of chemical oceanography, both as an individual subject in its own right, and as an adjunct to courses in marine biology, geology and geophysics. The desirability of treating the oceans as a chemically unified system, embracing the hydrosphere, the biosphere and the geosphere, was recognized nearly thirty years ago by Sverdrup, Fleming and Johnson in their classical volume on oceanography. However, despite its obvious advantages this approach has rarely been adopted by other authors. Although this topic is covered in one or two advanced works, there is a dearth of modern textbooks aimed primarily at undergraduate and postgraduate students in these various disciplines. In view of this, it was felt that there is a real need for a textbook treating the subject as a whole and incorporating many of the recent advances. The present book is an attempt to fulfil this need. It has been written to provide undergraduate and postgraduate students in all the above disciplines with an up-to-date and balanced account of marine chemistry. Only a modest knowledge of chemistry is assumed. An effort has been made to cover the subject comprehensively; limitations of space have necessitated a certain degree of selectivity to obtain the required detail. Progress in chemical oceanography is being made at a very rapid rate and the text is so designed as to provide a basic framework in which it will be possible to incoporate future advances.

We particularly wish to thank all those people who have assisted us in the writing of this book. We are especially indebted to Dr. J. D. Burton, and Dr. P. leB. Williams, University of Southampton, Dr. S. E. Calvert, University of Edinburgh, and Dr. G. Skirrow of this University for reading the original drafts and for making many helpful suggestions. Helpful suggestions were also made at the proof stage by Professors Ph. H. Kuenen, R. Pytkowicz and K. H. Wedepohl. In addition, we are very grateful to Mr. D. J. Slinn of the Marine Biological Laboratory, Port Erin, and to Dr. J. P. Barlow, Cornell University, for their ready advice and pertinent criticisms of Chapter 9. We also record our thanks to Dr. H. Elderfield, Dr. G. Skirrow and Mrs. D. Gardner for proof reading. Without their assistance many errors would

v

have undoubtedly gone undetected. Our most sincere thanks are also
extended to Mr. J. Murphy, for his most valuable and willing help with
the production of diagrams. We are grateful to the various publishers
and authors for permission to use diagrams, figures and photographs
which are their copyright. Finally we would like to thank Academic
Press for their understanding, co-operation and great patience.

Liverpool   J. P. RILEY
February 1971   R. CHESTER

Contents

Chapters 1 to 9 by J. P. Riley

Chapter 1

Introduction

Chapter 2

Salinity, Chlorinity and the Physical Properties of Sea Water

Chapter 3

The Oceans and the Physical Processes Occurring in Them

Chapter 4

The Major and Minor Elements in Sea Water

Chapter 5

The Dissolved Gases in Sea Water.
Part I. Gases other than Carbon Dioxide

Chapter 6 ✓

The Dissolved Gases in Sea Water.
Part 2. Carbon Dioxide

Chapter 7

Micronutrient Elements

Chapter 8

Dissolved and Particulate Organic Compounds in the Sea

Chapter 9

Primary and Secondary Production in the Marine Environment

Chapters 10 to 13 by R. Chester

Chapter 10

Marine Sediments

Chapter 11

Processes Involved in the Formation of Deep-sea Sediments

Chapter 12

The Components of Deep-sea Sediments

Chapter 13

The Geochemistry of Deep-sea Sediments

Symbols used in the text.

Concentration and activity. Several systems for the expression of concentration are in common use. The more important of these are the molarity scale (g molecules/l of solution) designated by c_i and the molality scale (g molecules/1000 g of water) designated by m_i, where the subscript i indicates the solute species. When i is an ion, charges are not included in the subscript except where confusion is possible. Other means of indicating concentration to be found in the text are: g or mg/kg of solution (for major components), μg or ng/l of solution (for minor elements or nutrients), and μg–at/l of solution (for nutrients). Factors for conversion of μg to μg–at are to be found in the Appendix (Table 3). When an activity or activity coefficient is associated with a species, the symbols a_i and γ_i are used respectively regardless of the method of expressing concentration, where the subscript i has the significance indicated above.

UNITS

LENGTH

μm	= micrometre	= 10^{-6}m	= 10^{-3}mm
mm	= millimetre	= 10^{-3}m	
cm	= centimetre	= 10^{-2}m	
m	= metre		
km	= kilometre	= 10^{3}m	

WEIGHT

ng	= nanogram	= 10^{-9}g
μg	= microgram	= 10^{-6}g
mg	= milligram	= 10^{-3}g
g	= gram	
kg	= kilogram	= 10^{3}g
ton	= 10^{3}kg	= 10^{6}g

VOLUME

ml	= millilitre
l	= litre

CONCENTRATION

p.p.m.	= parts per million (μg/g or mg/l)
μg–at/l	= μg atom/l = μg ÷ atomic weight/l

RESISTANCE

Ω	= ohm

LIGHT FLUX

ly	= langley (1 ly min^{-1} = 1 g cal cm^{-2}min^{-1})

xiii

GENERAL SYMBOLS

a_i	the activity of species i
A.O.U.	apparent oxygen utilization (see p. 117)
c_i	the molar concentration of species i
CA	carbonate alkalinity
$Cl^o/_{oo}$	chlorinity (g/kg)
d.p.m.	radioactive disintegrations per minute
E_h	redox potential
$E_{1cm}^{1\%}$	specific extinction coefficient (extinction for a 1 cm light path of a 1% solution)
F	Faraday equivalent of electric charge (96.500 coulombs)
G	Gibbs' free energy
m_i	molal concentration of the species i (gm mol/kg)
pE	redox potential (see p. 75)
pH	$-\log a_H$ (see p. 123)
P_{100}, P_{10}, P_1	productivities at the surface, and at depths where the light intensities are 10% an 1% of the surface values respectively
R	Gas constant and electrical resistance
RQ	respiratory quotient
$S^o/_{oo}$	salinity (g/kg)
t	temperature in °C and time
T	temperature in °K
z	depth

GREEK SYMBOLS

γ_i	activity of species i
ΣCO_2	total carbon dioxide concentration of a solution
κ	electrical conductivity
λ	radioactive decay constant
μ	ionic strength
σ	density (σ_t at temperature t°C) relative to distilled water at 4°C
τ	half-life of radioactive nuclide, or residence time in the sea

Chapter 1

Introduction

I. CHEMICAL OCEANOGRAPHY

The science of oceanography is concerned with the oceans and the processes occurring in them. It embraces a range of scientific disciplines including marine biology, geography, geology, physics and chemistry and is to a great extent inter-disciplinary since, in solving a particular problem, it is frequently necessary to invoke the assistance of several sciences. Chemistry occupies a central position in this respect. In the field of physical oceanography our knowledge of the water masses of the oceans, and their origins, is largely based on measurements of chemical parameters, such as salinity and oxygen content. Determinations of the essential micro-nutrient elements nitrogen and phosphorus are of great value to marine biologists concerned with the fertility of the sea. Chemical studies are also of importance to geologists in providing clues to the modes of formation of the mysterious manganese and phosphorite nodules which grow on the ocean floor. It is the purpose of this book to give an account of the chemistry of sea water and its bearing on marine biology and sedimentary geochemistry. Physical oceanography, as such, is dealt with only at sufficient depth to assist an understanding of the chemistry. For a detailed description of the physical processes occurring in the sea the reader is referred to one of the excellent recent texts on physical oceanography such as those by Neumann and Pierson (1966) and von Arx (1962).

II. HISTORICAL

Although analyses of sea water had been carried out as early as the beginning of the nineteenth century, it was not until the voyage of the *Challenger* (Fig. 1.1) from 1873 to 1876 that marine chemistry was placed on a scientific basis. During the voyage water samples were collected at various depths down to 1,500 m. The specific gravities and carbon dioxide contents of these samples were measured at sea. Samples of

2

water and of the dissolved gases from the water were taken back to Glasgow where they were analysed by Dittmar (1884). These analyses, which were models of precision, and compare well with recent values, suggested that, within narrow limits, the ratios of the concentrations

H.M.S. Challenger.

Fig. 1.1 H.M.S. *Challenger* (Murray, 1895). By permission of H.M. Stationery Office, London.

of the major ions are constant. Dittmar suggested that it would therefore be possible to estimate the salinity of a sea water by determining one of these major constituents, and for this purpose suggested the estimation of chloride (+bromide). In 1899 Knudsen developed a highly precise titrimetric method for carrying out this determination and also investigated the relationship between this quantity and both the salinity and density of sea water (Forch *et al.*, 1902).

The occurrence of oxygen in sea water attracted attention before the middle of the nineteenth century, but it was not until Winkler (1888) developed a simple method for its determination (see p. 108) that routine studies of the distribution of the element in the sea were made. These showed that dissolved oxygen measurements could be used, like temperature-salinity data, to characterize water masses. It was soon realized that changes in the oxygen contents of waters out of contact with the atmosphere were intimately related to the biological processes of photosynthesis and respiration occurring in them. These processes also, of course, influence the dissolved carbon dioxide content of the waters. Unlike the other atmospheric gases, carbon dioxide reacts with water. The problem of the carbon dioxide equilibria in sea water is a difficult one, but was largely solved by the work of the Finnish chemist Buch between 1915 and 1933 (see Chapter 6).

It had been realized for a long time by analogy with land-plants that marine plants and phytoplankton would require the micronutrient

Fig. 1.2. R.R.S. *Discovery II*. By permission of the National Institute of Oceanography.

elements nitrogen and phosphorus for their growth. However, it was not until the development of rapid photometric methods of analysis for nitrate and phosphate in 1920–27 that it was possible to explore the close relationship between the fertility of sea water and its content of these micronutrients in detail. Analyses for these nutrients were included in the chemical programmes in the pioneering cruises between 1925 and 1927 of the *Meteor* to the South Atlantic and the *Discovery II* (Fig. 1.2) to the Southern Ocean. Data obtained on these voyages has done much to elucidate the movements of the water masses in these oceans and to demonstrate that the high fertility found in certain regions is associated with upwelling. More recently, the development of the carbon-14 method of determining primary productivity (Steeman-Nielsen, 1952) has given a fresh impetus to the study of the fertility of sea water (see p. 233). During the last decade it has been increasingly realized that the dissolved organic components of sea water, such as vitamin B_{12} play an important part in determining the succession of the phytoplankton species.

When the history of marine chemistry is considered, it is apparent that new developments have coincided with the introduction of new techniques in analytical chemistry. The last few years have been particularly productive of highly sensitive physico-chemical methods of analysis, such as mass spectrometry and gas-liquid chromatography, which have still to make their main impact on the subject. The tedium of analysing the large numbers of samples collected on research cruises is already being reduced by the adoption of automatic methods of nutrient analysis. The next logical step is the *in situ* estimation of chemical parameters. In the last five years probes have become available which provide a continuous precision record of temperature and salinity down to a depth of 3,000 m (see p. 24); these have revealed that the structure of the water column is far more complex than could have been deduced from samples collected conventionally at various depths. Electrode systems are already available for the *in situ* measurement of pH (p. 125), the partial pressures of oxygen and carbon dioxide (p. 109) and fluoride ion activity, and it is likely that other ion-selective electrodes will be developed before long for use *in situ*.

III. The Sea as a Source of Raw Materials

As the world's population increases at an ever growing rate it is necessary to seek fresh sources of food and raw materials. Man is turning increasingly to the sea to satisfy his needs. At present most of the exploitation of the sea both by fisheries and extraction of minerals has been carried out by the more industrialized nations. However, for many of the developing nations, fisheries will play a key role both in nutrition and as a source of fertilizer. Although many of the fishing grounds of the northern hemisphere are in danger of being over-fished, there are others (e.g. that off the west coast of Africa) which could be further developed for the benefit of the adjacent countries. However, it is unlikely that fish-farming will make a worth-while contribution to fisheries for a long time as, at present, practical difficulties make it barely economic to produce even luxury fish, e.g. plaice and lobsters. Further consideration of fishery science and fish culture is outside the scope of this volume and for information on these subjects specialized works such as those by Rounsefell and Everhart (1953), Iverson (1968) and Hickling (1968) should be consulted. The seaweeds which grow prolifically in many coastal waters are valuable sources of organic chemicals, particularly for the food and pharmaceutical industries. Several millions of tons are already harvested each year for the production of alginic acid, laminarin and agar-agar and the market for these and other algal products is continually expanding (Chapman, 1950; Schulter, 1961; Firth, 1969).

In the last few years the demand for water has begun to outstrip the natural supply from rainfall, particularly in the arid countries where irrigation is essential for the expansion of agriculture. In many areas desalination of sea water is the only solution to the problem and, although saline-water conversion is only in its infancy, the total production of potable water by this method exceeds 200,000 tons/day. Energy is required to bring about the separation of water or salts from sea water, and it can be shown thermodynamically that the *minimum* energy required for this process for a water of salinity $35^0/_{00}$ is ca 0.74 kW-hr per ton of product at $25°C$. In practice, it appears unlikely that it will be possible to attain an efficiency of greater than 25% for the process. Existing processes have appreciably lower efficiencies than this, and the cost of desalinated water is therefore relatively high. However, in some areas its price is even now competitive with that of potable water from other sources, and the differential will improve if cheaper power becomes available, or when efficient solar stills are developed.

Several different processes can be used for the desalination of water (see Spiegler, 1966). In most of the existing plants the separation is carried out by distillation. Multistage units are used in which the heat liberated by condensing the steam from the first stage is used to boil the water under somewhat reduced pressure in the second stage, and so on through as many as 12 stages. Distillation units with outputs of over 20,000 tons per day are now in use. Prototype plants have been constructed for the production of potable water by freezing. They are based on the fact that the ice formed by the freezing of sea water is of low salinity. The purification is carried out in several stages, sea water is refrigerated so as to produce a mush of ice crystals, after which the ice crystals are separated and washed with water of low salinity to free them from adhering brine. In the final stage the crystals are melted in a heat exchanger. During the last decade several other methods of desalination have been investigated, such as solvent extraction, electrodialysis and reverse osmosis. In the last of these, which is the most promising the sea water is forced under pressure through a membrane which is more permeable to water than to ions.

Although all the naturally occurring elements occur in the sea, the only chemicals extracted from sea water at present are those derived from its major components. Several millions of tons of sodium chloride are produced each year from the sea by solar evaporation, especially in hot countries. Magnesium and potassium can be recovered from the mother liquor remaining.

During the Second World War several plants were set up both in Great Britain and the U.S.A. for the production of magnesium metal from sea water, and practically the whole of the U.S. production of the metal is now derived from sea water. In the largest of these plants at Freeport, Texas, sea water (0·13% Mg) is treated with a suspension of lime (produced by calcining calcareous shells which are available locally). The precipitated magnesium hydroxide is separated by filtration and converted to magnesium chloride, which after purification and drying, is reduced electrolytically to magnesium (Tressler and Lemon 1951; Gross, 1970). The overall recovery of magnesium in this plant is 85–90%. Large amounts of magnesium compounds are also obtained from sea water using a similar initial precipitation stage.

Sea water is the world's principal source of bromine (average bromine content 66 g/ton). Using sulphuric acid, sea water is acidified to pH 3·5 and treated with a slight excess of chlorine. Bromine liberated according to the reaction $2Br^- + Cl_2 \rightleftharpoons Br_2 + 2Cl^-$ is stripped from the water with a current of air and then reacted in the gas phase with sulphur dioxide. Hydrogen bromide formed by the reaction $Br_2 + SO_2 + H_2O \rightarrow 2HBr + H_2SO_4$ is absorbed in a relatively small volume of water. The resultant hydrobromic acid is treated with chlorine and bromine is distilled off with steam (Yaron, 1966). This process is used in several plants with outputs exceeding 2×10^4 tons/year, such as those at Amlwch, Anglesey and Freeport, Texas.

There are enormous reserves of most elements present in the sea; for example, there would be ca $1·4 \times 10^8$ tons of an element such as silver occurring at a concentration of 0·1 $\mu g/l$. McIlhenny and Ballard (1963) have shown that it would not be economic to extract elements with a lower abundance than boron (4·6 mg/l.) with a plant through which the water is pumped. However, it is possible that in future it may be feasible to employ ion exchange resins to concentrate several valuable trace elements at once (e.g. U. Au, Ag), obviating pumping by the use of tidal fall or by using the cooling water from coastal power stations.

The sea bed contains vast mineral deposits of economic importance which are only just beginning to be exploited. At present, it is only the continental shelf regions lying under not more than 200 m of water which are being mined. These regions are generally submarine continuations of the adjacent land masses, and their rocks and mineral deposits are similar to those on the mainland. They are usually covered with detrital sediments such as sand and mud, but in some areas vast deposits of characteristic marine minerals, such as phosphorites (p. 353ff.), may be present. Among the minerals being commercially recovered from the shelf areas are oil, coal, natural gas and sulphur together with alluvial diamonds, gold and tin from drowned river valleys. The deep ocean floor, particularly that of the Pacific, is strewn with ferromanganese nodules (see p. 360ff.) which are richer in valuable metals such as copper, nickel and zinc than many of the ores now worked. The total weight of these deposits in the Pacific Ocean alone has been estimated to be more than 17×10^{11} tons (Mero, 1965). Since the average depth of water in which they lie is ca 4,500 m, their recovery is not at present a commercial proposition. For a fuller account of the mineral resources of the sea the reader is referred to the monograph by Mero (1965), and the review edited by Terry (1966).

IV. POLLUTION OF THE SEA

The sea is becoming increasingly polluted as a result of man's activities, either incidentally, or as a direct result of its use for the disposal of waste products (see, for example, Sibthorp, 1969 and Føyn, 1965). It is frequently argued that since the volume of the oceans is very large, dilution, and perhaps bacterial degradation will soon render any pollutant innocuous. While this line of reasoning is in part true, it fails to take account of the damage which may occur before the pollutant is dispersed in the oceans, or through its concentration by the biota. Harmful levels of pollution are frequently reached in bodies of water which have only limited exchange with the sea, e.g. estuaries, and fjords, and even in coastal waters with more open circulation, discharge of effluent or oil may have severe effects.

The introduction of sewage, with its high content of organic matter and nitrogen and phosphorus compounds, into water produces two effects. Bacterial oxidation of its organic content may cause the oxygen content of the water to decrease to such a low level that it will not support life. In extreme instances the oxygen may be completely exhausted and formation of toxic hydrogen sulphide will then occur. The abundance of micronutrient elements, arising from the sewage, is supplemented by further supplies of phosphorus from any detergents present in domestic waste waters and encourages the explosive proliferation of plankton in the waters (eutrophication). The rapid decay of these organisms at the end of the bloom further decreases the oxygen content of the water and may render it toxic. Pollution by sewage and effluents can produce rapid and far-reaching effects. Thus, the waters at the centre of the Baltic Sea, which 10 years ago were oxygenated practically from top to bottom, now contain hydrogen sulphide in all but the upper 75 m. Similar effects are observed in many of the American Great Lakes. Deoxygenation of water could be largely prevented by discharging sewage and organic industrial wastes only after purification by the normal process of screening and bacterial digestion of organic matter. However, such methods of treatment would not remove the nutrients which cause eutrophication, and there is an urgent need for the development of processes for the removal of phosphate from waste waters.

Great quantities of toxic substances enter the sea as waste products from industry or from agriculture. Many of these may be toxic to some forms of marine life even at concentrations below 0·2 ppm (e.g. detergents and chlorine-containing pesticides) (see Butler, 1970). Many toxic chemicals are concentrated as they pass through the various stages of the marine food chain to such an extent that there may be some hazard in eating sea-foods from certain areas. The persistent insecticide DDT is strongly concentrated in this way, and Butler (1968) has cited the occurrence of up to 15 ppm of it in fish and 800 ppm in seal blubber. The concentration of mercury in waters of the Baltic is now relatively high as a result of the run-off of organo-mercury fungicides used in the forests of Sweden. The tissues of fish living in these waters contain so much of the element that they may be toxic if eaten as a staple part of the diet.

The sea has been contaminated by artificially produced radionuclides originating both from fall-out from atomic and hydrogen bomb tests and from the discharge of radioactive effluent from nuclear power plants. While these radio-isotopes occur at concentrations too low for them to be a direct risk to health, many of them are extremely strongly concentrated by marine plants and animals, which, perhaps, may then present some risk if eaten. For example, many molluscs have concentration factors of 10^7 or more for ruthenium-106—the principal nuclide discharged from certain types of atomic power stations. For further information on the pollution of the sea by radio-isotopes and its effects, the reader is referred to the U.S. National Academy of Sciences Monographs (1957, 1959), and to the reviews by Polikarpov (1966), Mauchline and Templeton (1964) and Mauchline (1970).

In conclusion, it is evident that the pollution of the sea by man's activities may in the future have far reaching consequences. At present, only estuaries and land-locked seas are being affected, but as the scale of the pollution grows it will not be long before coastal areas become badly contaminated. It is probably impossible to prevent accidental contamination, such as that which occurred when oil from the wreck of the tanker Torrey Canyon was washed up on the beaches of south-west England in 1966 (Cooper, 1968; Smith, 1968). However, it is feasible to prevent the pollution of estuaries and coastal waters by treatment of sewage and industrial effluents before they are discharged. There is an urgent need for the development of new purification processes, and for national and international legislation to limit the discharge of waste products into the sea. Such laws should also control the dumping of radioactive and other noxious substances on the deep ocean floor, as they may represent a health hazard to future generations. In the present section it has only been possible to draw the reader's attention to the very real dangers arising from the pollution of the sea; a much fuller treatment of the subject is provided in the review edited by Hood (1970).

REFERENCES

Butler, P. A., (1968). In "Proceedings of Marsh and Estuary Management Symposium". Louisiana State Univ., July 1967, Thos. Morans Sons Inc., Louisiana.

Butler, P. A. (1970). In "Encyclopedia of Marine Resources" (F. E. Firth, ed.). Van Nostrand, New York.

Chapman, V. J. (1950). "Sea Weeds and Their Uses."

Cooper, L. H. N. (1968). Helgoländer. Wiss. Meeresunters. 17, 340.

Cox, R. A., Culkin, F. and Riley, J. P. (1967). Deep Sea Res. 14, 203.

Dittmar, W. (1884). "Report on the Scientific Results of the Exploring Voyage of H.M.S. Challenger. Physics and Chemistry", Vol. 1. H.M. Stationery Office, London.

Firth, F. E. (1969). "Encyclopedia of Marine Resources." Van Nostrand, New York.

Forch, C., Jacobsen, J. P., Knudsen, M. and Sørensen, S. P. L. (1902). K. danske. vidensk. Selsk, 12, 1.

Føyn, E. (1965). Oceanogr. mar. Biol. Ann. Rev. 3, 95.

Gross, W. H. (1970). In "Encyclopedia of Marine Resources" (F. E. Firth, ed.). Van Nostrand, New York.

Hickling, C. F. (1968). "The farming of Fish." Pergamon, Oxford.

Hood, D. W. (1970). "Impingement of Man on the Oceans." Wiley, New York.

Iverson, E. S. (1968). "Farming the Edge of the Sea." Fishing News (Books) Ltd., London.

Mauchline, J. and Templeton, W. L. (1964). Ocean. mar. Biol. Ann. Rev. 2, 229.

Mauchline, J. (1970). In "Encyclopedia of Marine Resources" (F. E. Firth, ed.). Van Nostrand, New York.

McIlhenny, W. F. and Ballard, D. A. (1963). Symposium on Economic Importance of Chemicals from the Sea. Am. Chem. Soc. Div. Chem. Marketing Econ. Washington D.C., pp. 122–131.

Mero, J. L. (1965). "The Mineral Resources of the Sea." Elsevier, Amsterdam.

Murray, J. (1895). "Report on the Scientific Results of the Exploring Voyage of H.M.S. Challenger." Summary of Scientific results Part 1, H.M. Stationery Office, London.

National Academy of Sciences—National Research Council (1957). "The Effects of Atomic Radiation on Oceanography and Fisheries", Publication No. 551, 137 pp. National Academy of Sciences—National Research Council, Washington.

National Academy of Sciences—National Research Council (1959). "Radioactive Waste Disposal into Atlantic and Gulf Coastal Waters", Publication No. 655, 37 pp. National Academy of Sciences—National Research Council, Washington.

Neumann, G. and Pierson, W. J. (1966). "Principles of Physical Oceanography." Prentice Hall, New Jersey.

Polikarpov, G. G. (1966). "Radioecology of Aquatic Organisms." North-Holland Publishing Co. Amsterdam.

Rounsefell, G. A. and Everhart, W. H. (1953). "Fishery Science, its Methods and Application." Wiley, New York.

Schulter, M. (1961). Z. Fischerei, 10, 221.

Sibthorp, M. M. (1969). "Oceanic Pollution, a Survey and Some Suggestions for Control." David Davies Memorial Inst. for International Studies.

Smith, J. E. (Ed.) (1968). "Torrey Canyon pollution and marine life." A report by the Plymouth Laboratory of the Marine Biological Association of the United Kingdom, Cambridge University Press, London, 196 pp.

Spiegler, K. S. (1966). "Principles of Desalination." Academic Press, Inc., New York.

Steeman Nielsen, E. (1952). J. Cons. perm. int. Explor. Mer. 18, 117.

Terry, R. D. (1966). "Ocean Engineering", Vol. IV, Part 1, Mineral Exploration. National Security Industrial Association, Washington, D.C.

Tressler, D. K. and Lemon, J. M. (1951). "Marine Products of Commerce." Reinhold Publishing Corp., New York.

von Arx, W. S. (1962). "Introduction to Physical Oceanography." Addison-Wesley, Reading Mass.

Winkler, L. W. (1888). Ber. dtsch. chem. Ges. 21, 2843.

Yaron, F. (1966). In "Bromine and its Compounds" (Z. E. Jolles, ed.). Ernest Benn, London.

Chapter 2

Salinity, Chlorinity and the Physical Properties of Sea Water

I. The Structure of Water and Sea Water

Before we pass on to discuss those physical properties of sea water which are of interest to the chemical oceanographer it is perhaps useful to review briefly the structures of water and ionic solutions. It should be stated at the outset that our knowledge of the liquid state is, in general, far less complete than that of either solid or gaseous states. This is particularly true of water, which is probably the most complex liquid known.

Water exists in the vapour phase mainly as monomeric H_2O, but at low temperatures it is partially associated to polymeric forms, e.g. H_4O_2 (ca 10% at 100°C and 1 atm) because of hydrogen bonding. Infra red spectrometry has shown that the H_2O molecule is L-shaped with an angle of ca 105°. X-ray and neutron diffraction methods have been used to investigate the structure of the common hexagonal form of ice (Ice I_h), and have shown (see Owston, 1958) that it is constructed of 8 identical L-shaped H_2O units arranged to make the unit cell shown in Fig. 2.1. Two hydrogens are always located adjacent to each oxygen, with only one hydrogen atom between each neighbouring oxygen atom (along each hydrogen bond connecting neighbouring water molecules). Liquid water is composed of the same elementary units, but hydrogen bonding between them enhances the dipolar interaction, and causes many of its physical properties to be anomalous (e.g. high specific and latent heats, high dipole moment, high boiling point and freezing point, great solvent ability). Kavanau (1964) and Eisenberg and Kauzman (1969) have reviewed the many structures which have been suggested for water over the last 40 years in order to explain its unusual properties. Although none of them is satisfactory in every respect, the evidence points very strongly to the simultaneous existence of more than one water species at any temperature. Frank (1961) has suggested that clusters of water molecules form and decompose with a half-life of ca 10^{-11} sec, as a result of the London or dipolar forces which are produced when redistribution of charge takes place during the dynamic formation

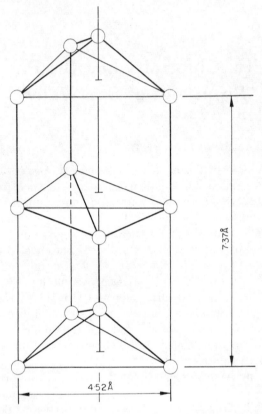

Fig. 2.1. Unit cell of Ice-I_h, omitting hydrogen atoms (Drost-Hansen, 1967).

of hydrogen bonds. Drost-Hansen (1967) considers this mechanism to be a likely one, but favours clathrate cage structures rather than clusters.

The theoretical treatment of solutions of electrolytes is a matter of extreme difficulty, because of the various complex and poorly understood ionic and molecular interactions which occur in such solutions. Present theories (see, e.g. Falkenhalgen and Kelbg, 1959) are refinements of that of Debye and Hückel (1923) and are valid only for dilute solutions. In order to make treatment of the problem possible many simplifications have to be made. The system is considered to be composed of rigid ions surrounded by a spherical symmetrical distribution of ions distributed in a continuous medium; ion-solvent and intramolecular forces are ignored. Even when these idealizations have been made, the resultant equations are difficult to solve and are applicable only to solutions having ionic strengths less than $\frac{1}{10}$ of that of sea water. Some success in the treatment of solutions having relatively high ionic strengths has been achieved by the use of thermodynamic arguments based on ion solvation (Robinson and Stokes, 1959) and by the use of formal statistical methods (see, e.g. Falkenhalgen and Kelbg, 1959). However, none of the treatments at present available can be applied to solutions having an ionic strength similar to that of sea water because of the extremely complex interaction involved (see Monk, 1961). A treatment of the various theories of electrolyte solutions is beyond the scope of this book, and the reader is referred to the book by Samoilov (1965) and the article by Horne (1968) for further information.

II. SALINITY AND CHLORINITY

It is difficult to determine the salinity of sea water by evaporation, drying of the salts and weighing. During the later stages of the evaporation, hydrolysis of magnesium chloride gives rise to hydrogen chloride, which, together with carbon dioxide from the carbonate system, is lost as the evaporation proceeds. A high temperature is necessary to remove the last traces of water and under these conditions the organic matter is oxidized. Sørensen (Forch et al., 1902) has developed a method for the determination of salinity in which compensation was made for loss of hydrogen chloride. On the basis of this work, Knudsen (1901) has defined salinity as the weight in grammes of the dissolved inorganic matter in 1 kg of sea water after all bromide and iodide have been replaced by the equivalent amount of chloride, and all carbonate converted to oxide. The total salt content of sea water is about 0·45% greater than its salinity, defined in this way.

Since salinity is difficult to determine directly, an International Commission was set up in 1899 to investigate alternative methods for its determination. The determination of halogens by precipitation as silver halides can be carried out with considerable precision. The Commission, under the direction of M. Knudsen, therefore examined the relationship between density, salinity and chloride+bromide content in a number of sea water samples. They found that there was a linear relationship between the salinity and precipitable halide concentration. They expressed the latter in terms of a new unit—chlorinity ($Cl^0/_{00}$)— which they defined as the mass in grammes of chlorine equivalent to the mass of halogens contained in 1 kg of sea water. Measurements of chlorinity and salinity on 9 samples of surface sea water gave the linear relationship

$$S^0/_{00} = 0{\cdot}03 + 1{\cdot}8050 \, Cl^0/_{00}$$

Although only two of these samples were typical ocean waters (the others were Red Sea or Baltic samples or mixtures of Baltic and North Sea waters) there was a close correlation between salinity and chlorinity; the standard and maximum deviation from the above relationship being $\pm 0{\cdot}01 \, S^0/_{00}$ and $0{\cdot}022 \, S^0/_{00}$ respectively. In the light of the results of recent work on the constancy of the ionic ratios in sea water (see p. 80), it is probable that if deep-water samples had been included in the survey the deviations would have been appreciably greater. Since chlorinity could be determined more simply than salinity the Commission recommended that it would be best to standardize on this and treat salinity as being of theoretical interest only.

For the determination of chlorinity the Commission recommended the adoption of a precise argentometric titration procedure developed by Knudsen. It is essential that the chlorinity results from all laboratories should be intercomparable, and a Standard Sea Water Service was therefore established to provide sea water of accurately known chlorinity which could be used for standardization of the silver nitrate used in the titration. In the Laboratory of the Service, bulk samples of water from the Atlantic are diluted with distilled water until their chlorinities lie close to $19{\cdot}374^0/_{00}$. After standardization by means of the Volhard method, the water is sealed in 200 ml glass ampoules and labelled with its chlorinity. When a new basic standard sea water was in preparation in 1937, the opportunity was taken to redefine chlorinity in order to make it independent of changes in the accepted atomic weights of silver and chlorine. Chlorinity was therefore redefined as the mass in grammes of pure silver necessary to precipitate the halogens in $328{\cdot}5233$ grammes of the sea water.* Although this definition preserves the relationship which Knudsen found between salinity and density it implies that, on the basis of 1963 atomic weights, the chloride+the chloride equivalent to the bromide is greater than the chlorinity by a factor of $1{\cdot}00043$.

The increasing adoption of precise conductimetric methods for the determination of the salinity of sea water (p. 18) has led to a new definition of salinity in terms of electrical conductivity. This definition is based on an extensive survey of the conductivity-chlorinity relationships of water samples from all parts of the oceans (Cox *et al.*, 1967). Chlorinity is related to the newly defined salinity by the equation

* The halogen concentration of sea water is occasionally expressed in terms of *chlorosity*, which is the number of grammes of chloride+chloride equivalent to the bromide in *one litre* of sea water at 20°C.

$S^0/_{00} = 1\cdot80655\ Cl^0/_{00}$ (Wooster *et al.*, 1969). This expression gives identical results to the older one at salinity $35^0/_{00}$. However, as the salinity decreases, the chlorinity, as calculated from the new relationship becomes greater than that calculated from Knudsen's equation and *vice versa*. The deviation is $0\cdot0026^0/_{00}$ at salinities of 32 and $38^0/_{00}$.

A. DETERMINATION OF CHLORINITY

The procedure developed by Knudsen for the precise determination of chlorinity is a refinement of Mohr's method in which the halide is titrated with a standard solution of silver nitrate. Potassium chromate is used as the indicator.

$$Cl^- + Ag^+ \rightarrow AgCl \downarrow$$

$$Br^- + Ag^+ \rightarrow AgBr \downarrow$$

$$2Ag^+ + CrO_4^{2-} \rightleftharpoons \underset{\text{red}}{Ag_2CrO_4}$$

The sea water sample is measured into a beaker by means of a 15 ml pipette having a three-way tap instead of a mark; the water is sucked up through one arm and the liquid is released by turning the tap through $180°$. The titration is carried out by means of a bulb burette (Fig. 2.2) at the top of which is a three-way tap which serves as the upper mark. Most of the capacity of the burette is in the bulb, and the stem which is normally calibrated in units of 2 ml is graduated from ca 16 to 21·5 units and permits the titration of samples having chlorinities lying between these values. After addition of a few drops of 10% potassium chromate solution, the sea water sample is stirred magnetically and silver nitrate solution (36·75 g/l.) is added from the burette by means of a stopcock with a fine jet. The titration is concluded when the precipitate acquires a slight red colour, which persists for 30 seconds. The silver nitrate solution is standardized in the same manner, but with the substitution of Standard Sea Water for the sample. Since the same ples are measured out by volume, and chlorinity is defined in terms of weight of halide per unit *weight* of sea water, the chlorinity of the samples cannot be calculated directly but must be evaluated by use of special tables. The precision of the method is limited by errors associated with the pipetting of the samples, the drainage and reading of the Knudsen burette, and with the difficulty of observing the end-point of the titration. In expert hands the method is capable of giving a precision of ca $\pm0\cdot01\ Cl^0/_{00}$. A precision of $\pm0\cdot001\ Cl^0/_{00}$ can be attained if weighed samples are titrated potentiometrically using a weight burette for the addition of the silver nitrate solution.

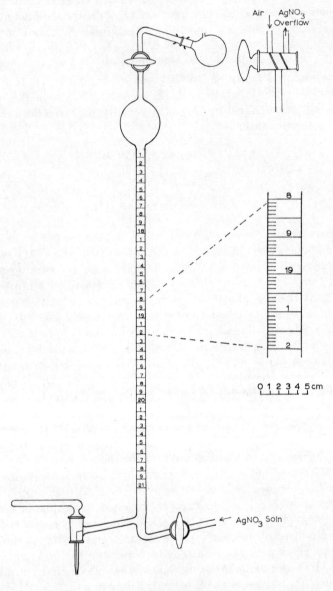

Fig. 2.2. Knudsen burette for the determination of chlorinity.

B. PHYSICAL PROCEDURES FOR THE DETERMINATION OF SALINITY

The accurate determination of chlorinity makes considerable demands on the skill and time of the analyst, and even under favourable circumstances one analyst can only carry out about 50 determinations per day. During hydrographic cruises when it is not unusual to collect over 100 samples per day, the possibility of using physical methods for the determination of salinity offers the prospect of speeding up the analyses and of using less skilled labour. Physical methods also offer the obviously desirable prospect of making *in situ* measurements of salinity.

Since chlorinity is a measure of the concentration of chloride+ equivalent bromide, it can strictly only be used in the computation of the salinity and density of sea water if the major ion/chlorinity ratios in the latter are constant. Physical determinations of salinity, however, generally give a measure of the total ionic concentration in the water. They therefore give a more reliable estimate of the density of the water since the effects of small variations in the relative proportions of the major ions are compensated.

Physical oceanographers generally do not need to know the salinity or chlorinity, as such, but require a means of assessing the *in situ* density of sea water. The precise measurement of density in the laboratory is a difficult and tedious task. However, since the density of sea water is a function of temperature and salinity, if we can measure, at constant temperature, any physical parameter, which varies with salinity, then we can relate this directly to density. Many of the physical properties of sea water do depend upon salinity, but few of them can be measured with sufficient precision for them to be used to assess the density of water for the purposes of dynamical oceanography. In fact, only refractive index, and electrical conductivity have as yet been used for this purpose. Refractive index, like density, is a bulk property of the water as a whole and as a consequence alters only relatively little as the salinity changes. As a consequence, methods using it must be capable of detecting differences of 1 in 10^5 if salinity is to be measured to $\pm 0 \cdot 01^o/_{oo}$. However, the electrical conductivity of sea water is determined by the total ionic strength of the medium and, at constant temperature, is roughly proportional to salinity. It should be stressed that since physical procedures only measure salinity indirectly, they require calibration with waters of known salinity. For a general review of the subject of salinity measurement the reader should consult Johnston (1969).

(i) Density measurements

Knudsen used a pycnometer for the determination of density in his definitive work on the relationship between the chlorinity and density of sea water, the results of which are embodied in his hydrographic tables (Forch et al., 1902). Although the pycnometer can give results of high accuracy, it cannot be used at sea and is slow in operation. Stem hydrometers have insufficient accuracy except for use with estuarine waters where large changes in salinity occur. Immersion hydrometers suspended from a torsion balance by means of a fine wire have been used but are affected by the greasiness of the water surface. The free floating total immersion hydrometer designed by Cox (1957) is free from this disadvantage; it is based on the principle of the Cartesian diver. A small, slightly compressible float, which is slightly less dense than water, is immersed in the sample and pressure is applied until the float is neutrally buoyant. A mercury manometer is used to measure the applied pressure, which is a measure of the density. This instrument is capable of a precision of $\pm 0.02^o/_{oo}$S, but suffers from the disadvantage that it cannot be used at sea since the ship's motion causes a pumping motion in the mercury manometer. Cox et al. (1970) have recently carried out a reinvestigation of the relationship between salinity and density, in which the density was measured to 2 parts in 10^7 by weighing a silica float immersed in both sea water and pure water.

(ii) Refractive index measurements

The variation of refractive index with salinity is only small, an increase in salinity of $1^o/_{oo}$ increasing the refractive index by ca 0.0002. Since it is normally desired to measure salinity to $\pm 0.01^o/_{oo}$ the use of Abbé and Pulfricht refractometers is ruled out, and it is necessary to compare the refractive index of the sample with that of a standard by means of an interferometer, which may have a precision of better than 5×10^{-7}. Variation in temperature produces only a small change in refractive index and the temperature coefficient varies only slightly with salinity. Accurate temperature control is therefore not required, provided that the sample and the standard with which it is being compared are at the *same* temperature. The interferometer provides an intrinsically simple and rapid means of measuring salinity to better than $\pm 0.01^o/_{oo}$ (see Rusby, 1967) but has not been widely adopted, although it has been used extensively aboard Russian research vessels.

(iii) Conductivity measurements

The use of conductivity measurements for the determination of salinity was suggested by Knudsen over 65 years ago. However, it is only in recent years, with the development of reliable and precise conductivity bridges, that methods depending on electrical conductivity have largely displaced the chemical method for the determination of salinity. The development of *in situ* conductivity salinometers now allows the fine salinity-depth structure of the sea to be examined.

The electrical conductivity of sea water is roughly proportional to salinity, and if salinity is to be measured to within $\pm 0.001^0/_{00}$ the conductivity must be determined to 1 in 40,000. The conductivity of sea water has a rather large temperature coefficient which varies with the salinity. At a salinity of $35^0/_{00}$, the temperature must be maintained to $\pm 0.001\,^\circ$C if the salinity is to be measured to $0.001^0/_{00}$. In a precision conductivity salinometer it is therefore necessary to have close temperature control and/or some form of compensation for temperature variations.

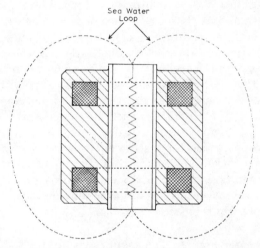

Fig. 2.3. Diagrammatic representation of inductive conductivity cell.

Alternating current is used exclusively in conductimetric salinometers in order to minimize the effects of electrolytic action on the sea water sample, and also because it is easier to amplify. However, owing to capacitance effects, it is difficult to bring alternating current signals up long cables without voltage loss and this poses a limitation on its direct use in *in situ* salinometers.

Basically, all salinometers are composed of the following parts: the conductivity cell and the bridge with its associated oscillator, amplifier and null point detector.

(*a*) *The conductivity cell.* Conductivity cells are of two principal types: (I) those having two or more electrodes, usually composed of platinum coated with platinum black, rigidly mounted in a tube of glass or insulating material, which is filled with the sample, (II) the inductive type, in which the sea water in made to form a conductive loop between the primary and secondary coils of a toroidal transformer (shown diagrammatically in Fig. 2.3). In the first type, difficulties are encountered through fouling of the electrodes with organic mucillages from the water and from polarization of the electrodes. The inductive cell is free from both of these defects and is used in most of the more recent types of precision salinometers.

It would be difficult to control the temperature of the salinometer cell to the $\pm 0.001\,°C$ necessary for the salinity to be measured to $\pm 0.001^o/_{oo}$. For this reason it is usual to control the temperature of the cell much less finely and to provide temperature compensation. Thus, if a similar cell filled with standard sea water is immersed in the same thermostat bath and used as a reference resistance it will be necessary to control the bath temperature to only $ca \pm 0.1\,°C$, since the effects of temperature variations on the conductivities of the standard and reference samples will cancel out. In most modern bench instruments the cell is not thermostatically controlled, and temperature compensation is provided by means of a resistance thermometer or thermistor which is coupled to a circuit designed to compensate for changes of conductivity with temperature, provided that they are within $\pm 3\,°C$ of the temperature at which the instrument was standardized. However, since the temperature coefficient of conductivity changes considerably with both temperature and salinity, such compensation is efficient only over a limited salinity range, and corrections must be applied if the working temperature is extreme.

(b) *The conductivity bridge.* The conductivity of the sea water sample contained within the cell is usually determined by means of a bridge circuit. In this, the resistance is compared with that of the standard sea water or a reference resistance using an alternating current (free from harmonics), with frequency of from 1,000 to 10,000 c/s. The balance point of the bridge is indicated by means of a null deflection meter or cathode ray tube coupled with an amplifier. The bridge circuit may be based on the classical Wheatstone's network in which balance is attained by variation of a resistor in one of the arms. However for precise work it is preferable to use a transformer bridge in which the voltage is varied while the resistance is kept constant. Such bridges have considerable advantages over conventional resistance bridges since they do not require precision resistors, do not change with age and are unaffected by temperature variations. In addition, capacitative leakage to earth, which is a major problem with resistance bridges, causes no difficulty.

The principles of operation of a modern salinometer with a transformer bridge may be conveniently illustrated by reference to the inductive salinometer developed by Brown and Hamon (1961). The measuring head, shown diagrammatically in Fig. 2.4, is used with the

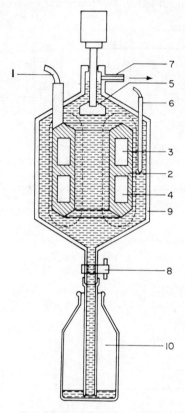

Fig. 2.4. Simplified diagram of measuring head of inductive salinometer. 1, leads from toroidal unit; 2, toroidal unit; 3, core of voltage transformer T_1; 4, core of current transformer T_2; 5, stirrer blade; 6, thermistor for temperature compensation (R_T); 7, outlet to suction pump; 8, Stopcock, controlling filling of cell; 9, clear perspex jacket; 10, sample bottle (reproduced, with permission from Brown and Hamon, 1961).

circuit shown in simplified form in Fig. 2.5. The toroidal high permeability cores of the voltage and current transformers (T_1 and T_2) are wound identically and encased together in epoxy resin; the primary winding A of T_1 is fed with 10 kc/sec from an oscillator. The cell assembly is housed inside a perspex case which is filled with the water sample by means of suction. The sample is brought to an even temperature by vigorous stirring and is equivalent to a conductive link between the magnetic circuits of both T_1 and T_2 which are separated by a magnetic shield. The resistance (R_w) of this link is determined by the geometry of the cell and varies inversely with the salinity of the sample; it is ca 60 Ω at $S = 35^0/_{00}$. In addition to the linkage through the sample, T_1

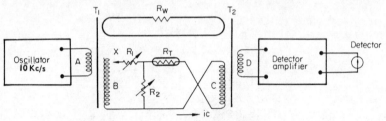

Fig. 2.5. Simplified circuit diagram for inductive salinometer. T_1, voltage transformer; T_2, current transformer; R_w, resistance of water path linking T_1 and T_2; R_1 standardizing resistor; R_2, adjustment for temperature compensation; R_T, thermistor; X, main balancing control, reading in terms of conductivity ratio (reproduced, with permission from Brown and Hamon, 1961).

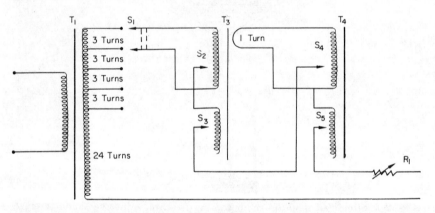

Fig. 2.6. Details of the tapped transformer ratio adjustment. S_1, 4 position switch (ratios 0·8, 0·9, 1·0, 1·1); S_2-S_5 decade ratios switches. The windings on T_3 and T_4 have the following numbers of turns; S_2 and S_4 10 × 10 turns, S_3 and S_5 10 × 1 turn (reproduced, with permission, from Brown and Hamon, 1961).

and T_2 are also linked through the windings B and C and their associated network of resistors R_1, R_2 and R_T. The effect of these windings is such that the magnetizing effect of the current i_c in the current transformer T_2 opposes the current through the water path. The bridge is balanced by adjusting the voltage output from the winding B of a bank of inter-tapped transformers, until the null deflection meter indicates that the flux in T_2 is zero. These transformers which are indicated for clarity in

Fig. 2.5 by a single variable tap X, are shown diagrammatically in Fig. 2.6. Their tappings are altered by rotary switches with dials which indicate the ratio of conductances.

A thermistor R_T is used to measure the temperature of the sample, and the associated circuitry (simplified to R_2) compensates for temperature changes by varying the current in the windings C.

Before use, the sample cell is filled with Standard Sea Water ($S = 35\cdot00\pm0\cdot01^0/_{00}$)* and the readings of the dials of the transformer bridge are set to the conductance ratio (salinity of Standard Sea Water)/35·000. After temperature equilibrium has been attained, the temperature of the sample is measured and R_2 is adjusted to give correct compensation. The transformer bridge itself is then balanced by adjustment of R_1. The apparatus is now ready for measurement of the salinities of unknown samples, provided that their temperatures do not differ by more than $\pm3°C$ from the temperature at which it was standardized. From the readings on the ratio dials, the ratio of the conductivity of the sample to that of sea water of $S = 35^0/_{00}$ is obtained. The salinity of the sample can then be read from the International Oceanographic Tables (UNESCO, 1966). These Tables also give corrections which must be applied to the conductivity ratios if the instrument is used at temperatures differing appreciably from 20°C. With this instrument, versions of which are produced commercially in Australia and the U.S.A. (Fig. 2.7), an experienced operator can analyse samples at a rate of ca 40 per hour. An accuracy of ca $\pm0\cdot003^0/_{00}S$ can be attained provided that substandards are analysed at hourly intervals to check for drift.

(c) *Effect of variation of chemical parameters on salinometer accuracy.* The agreement between conductimetric salinity measurements and those calculated from chlorinity depends on the constancy of the relative ionic composition of sea water. Thus, deep waters, with their relatively high Ca/Cl ratios (p. 81) give salinities up to 0·01% higher than those calculated from chlorinity data. Biological processes, which influence the concentration of carbon dioxide, and therefore those of bicarbonate, carbonate and hydrogen ions may also produce errors in extreme instances (Park, 1965).

* Standard Sea Water is at present standardized only in terms of chlorinity. In the future it is also proposed to standardize it in terms of its absolute electrical conductivity. This will have the advantage that changes in the major element/chlorinity ratios in the water used as standard will not affect the accuracy of determinations. It will also enable density to be computed directly from conductivity without the necessity of using salinity as an intermediate step. International Tables relating density to absolute conductivity are in course of preparation.

(d) In situ *salinometers*. In the study of internal waves and many other oceanic phenomena there is a growing need for a detailed knowledge of the salinity and temperature structure of the sea. The amassing of such data by conventional water sampling techniques is very tedious, and

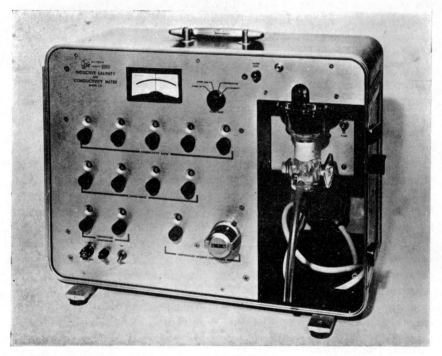

Fig. 2.7. Bench type "Hytech" inductive salinometer (photograph by courtesy of The Plessey Co., Ltd.).

unless the sampling bottles are spaced at close intervals, details of fine structure will be missed. However, the recent development of *in situ* salinity-temperature-depth recorders has now made it possible to obtain continuous plots of salinity and temperature profiles in the oceans.

The construction of *in situ* salinometers poses a number of problems not encountered in bench salinometers; the most important of these are (i) the transmission of the results to the surface, (ii) difficulties in temperature compensation caused by the appreciable time constants of thermistors and resistance thermometers, and by the large range of temperatures which may be encountered in a vertical profile, (iii) compensation for the effect of pressure on the electrical conductivity of sea water.

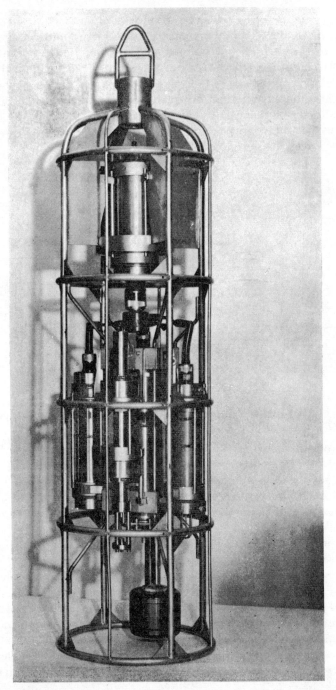

Fig. 2.8. Bissett-Berman salinity-temperature-depth probe. The instrument shown also includes a sensor for measurement of the velocity of sound. (Photograph by courtesy of The Plessey Co. Ltd.).

Salinity data can be telemetered in two separate ways. In one, the data from conductivity, temperature and pressure sensors are transmitted to the surface and salinity is then computed from these variables. In the other, the outputs of these sensors are combined in the bridge circuit and the resultant signal, which is a function of salinity variations alone, is transmitted to the surface. Much smaller demands are made on the telemetric link in instruments of this type since its precision has to be only ca 1 in 500 if the salinity is to be measured to $\pm0.02^{0}/_{00}$ over the normal oceanic range of $30-40^{0}/_{00}$.

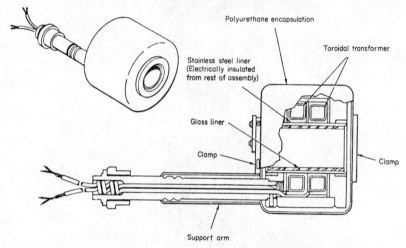

Fig. 2.9. Conductivity sensor of *in situ* salinity-temperature-depth probe (by kind permission of N. L. Brown).

A successful instrument based on this principle is the Bissett-Berman "Hytech" *in situ* salinometer developed by N. L. Brown. This instrument consists of a submersible probe unit and shipboard recorder module linked by a single core armoured cable which serves both for the suspension of the probe and as a means of transmitting the signals from it.

The probe unit (Fig. 2.8) is encased in a protective steel cage and carries sensors for the measurement of temperature, depth and salinity and their associated bridge circuits. Temperature is measured to $\pm0.02^{\circ}C$ by means of a platinum resistance thermometer. Depths down to 3,000 m are measured with a precision of ±6 m with a strain gauge pressure sensor. The sensor for salinity measurements is an inductive cell of the type shown schematically in Fig. 2.9.

The complex interrelationship between the salinity, electrical conductivity, temperature and pressure of sea water necessitates the use of highly complex compensating circuits in the salinity bridge. The circuitry of the latter which is shown in much simplified form in Fig. 2.10 consists of a series of interlocking double bridge compensating circuits for temperature and pressure effects.

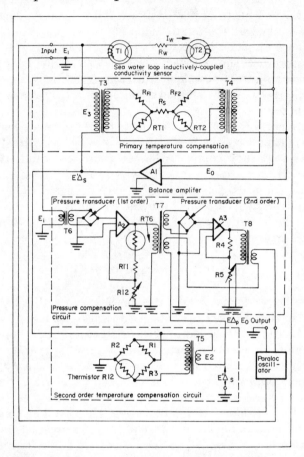

Fig. 2.10. Simplified block diagram of circuitry used with *in situ* salinity-temperature-depth probe (by kind permission of N. L. Brown).

The relationship between temperature (t) and conductivity of sea water (Λ) is described by an empirical quadratic equation of the type

$$\Lambda = \Lambda_0(1 + at + bt^2)$$

where Λ_0 is the conductivity at $0\,^\circ\mathrm{C}$, and a and b are constants. Since the conductivity of sea water varies by ca $4\cdot5\%/\,^\circ\mathrm{C}$ the temperature compensation in the salinometer must be efficient. The necessary compensation is achieved by means of the primary temperature compensation bridge which is controlled by two platinum resistance thermometers (R_{T1} and R_{T2}) in equilibrium with the sea water. By careful choice of the resistors R_{F1}, R_{F2} and R_S, the temperature coefficient of the bridge is made to match that of sea water closely, even though the temperature coefficients of the resistance thermometers are about $\frac{1}{6}$ of that of sea water and of opposite sign.

The balance amplifier applies an additional voltage $E_{\Delta S}^1$ to the input of the temperature compensation bridge, so as to drive the bridge to balance. This voltage is to a first order approximation proportional to the difference between the measured salinity and that at salinity $35^0/_{00}$, when $E_{\Delta S}^1$ is zero.

An additional correction must be made to $E_{\Delta S}^1$ since the temperature coefficient of conductivity of sea water varies slightly with salinity. The primary temperature compensation is therefore valid at $35^0/_{00}$ S. Compensation for variation in salinity is provided by a second order temperature compensation circuit. The output signal ($E_{\Delta S}^{11}$) from this circuit is independent of temperature at all salinities.

The electrical conductivity of sea water varies with pressure according to the equation

$$\frac{\Delta \Lambda}{\Lambda} = (aP + bP^2)(K_1 e^{c/t} + K_2)$$

where Λ is the conductivity at atmospheric pressure, P is the pressure, and a, b, c, K_1 and K_2 are constants. Temperature changes have a large effect on $\Delta \Lambda / \Lambda$. However, since the temperature in the depths of the oceans is more or less constant at 2–3°C, only comparatively crude temperature compensation is required. The effect of salinity on $\Delta \Lambda / \Lambda$ is quite small, and since the salinity of deep ocean water lies close to $35^0/_{00}$ no compensation for salinity is necessary in the pressure compensation circuits. The pressure compensation is achieved by means of two pressure transducers linked to a pair of bridge networks; temperature compensation is provided by the thermistor R_{T6}. The outputs of the high gain amplifiers A_2 and A_3 are proportional to $P(K_1 e^{c/t} + K_2)$ and $P^2(K_1 e^{c/t} + K_2)$ and are added together using isolated windings on the coupling transformers T_7 and T_8. By careful design of the circuitry, changes in conductivity induced by pressure can be accurately balanced out at any temperature.

The potential changes from the pressure, temperature and salinity sensing circuits are each passed through separate phase shift oscillators tuned to different standard frequencies and transformed into small changes from these frequencies. These signals are transmitted to the surface up the same single core cable. Aboard ship, the signals are passed into a filter network forming an integral part of a phase shift oscillator designed to operate over specific frequency ranges corresponding to the three parameters being measured. The separated signals are each fed into discriminators to convert them back into potentials and then into an x_1, x_2, y recorder by means of which temperature and salinity are plotted as a function of depth. Alternatively, the signals may be fed into a digital logger and the stored data may be processed later by means of a computer.

The Hytech *in situ* salinometer can be used at depths down to 3,000 m and shows precisions for depth, temperature and salinity of $\pm 0.2\%$, $\pm 0.02\,^\circ\text{C}$ and $\pm 0.03^0/_{00}$ respectively.

III. PHYSICAL PROPERTIES OF SEA WATER

In the space available in this book, it is only possible to give a brief account of the physical properties of sea water. For a more detailed treatment of the subject and for extensive tables of physical and physico-chemical properties of sea water, Horne (1969) should be consulted.

A. DENSITY AND COMPRESSIBILITY OF SEA WATER

As yet, no accurate values for the absolute density (g/cm^3) of sea water have been published. Only specific gravities are available, which are defined as the ratio of the weight of a sea water at a specified temperature to that of the same volume of a pure water at $4\,^\circ\text{C}$. The specific gravity of a water of temperature t° is usually expressed in terms of either specific gravity anomaly (specific gravity -1), or $\sigma_t(= (\text{specific gravity at } t\,^\circ\text{C}-1) \times 1,000)$. The relationships between the chlorinity, temperature and specific gravity of sea water were investigated by Forch *et al.* (1902). Their results are embodied in Knudsen's Hydrographic Tables and are shown in abridged form in Table 2.1.*

Sea water has a small, but finite compressibility which varies with both temperature and salinity. Detailed tables have been compiled by Ekman (1910) and Matthews (1938) showing the variation of specific gravity and specific volume (the reciprocal of specific gravity) of sea water with pressure, salinity and temperature. Table 2.2 shows the specific gravity and compression (percentage reduction in volume relative to the volume at atmospheric pressure) of sea water of salinity $35^0/_{00}$ and temperature $0\,^\circ\text{C}$, similar to those of most oceanic deep waters,

---

* Unfortunately, at the time that this work was carried out it was not appreciated that repeated distillation could alter the density of pure water by as much as 2 parts in 10^5 (owing to fractionation of the isotopes of both hydrogen and oxygen). The uncertainty of the absolute values of the specific gravities given in the Hydrographic Tables is probably $\text{ca} \pm 2\text{--}3$ parts in 10^5 since nothing is known about the origin of the pure water used as standard. Recent work by Cox *et al.* (1970) using an isotopically defined water suggests that Knudsen's tables are low by an average of 0.006 σ_t.

TABLE 2.1

Specific gravity anomaly $\times 10^5$ of sea water

$S^0/_{00}$	Temperature (°C)									
	0	2	4	6	8	10	15	20	25	30
0	−13	−3	0	−5	−16	−32	−87	−177	−293	−433
5	397	403	402	394	381	362	301	207	87	−57
10	801	804	799	788	772	750	685	586	462	315
15	1,204	1,204	1,195	1,181	1,162	1,138	1,067	964	836	686
20	1,607	1,603	1,589	1,573	1,551	1,532	1,450	1,342	1,210	1,057
25	2,008	2,001	1,988	1,970	1,947	1,920	1,832	1,720	1,585	1,428
30	2,410	2,400	2,384	2,363	2,340	2,308	2,215	2,098	1,960	1,801
32	2,571	2,560	2,543	2,521	2,494	2,464	2,364	2,250	2,110	1,950
34	2,732	2,719	2,701	2,678	2,651	2,619	2,522	2,402	2,261	2,100
36	2,893	2,879	2,860	2,836	2,808	2,775	2,676	2,554	2,412	2,250
38	3,055	3,040	3,019	2,994	2,965	2,931	2,830	2,707	2,563	2,400
40	3,216	3,200	3,179	3,153	3,122	3,088	2,985	2,860	2,714	2,550
42	3,377	3,361	3,337	3,310	3,279	3,243	3,138	3,011	2,864	2,700

TABLE 2.2

*Specific gravity and volume reduction of sea water under pressure**

Pressure (db)	Specific gravity	% decrease in volume
0	1·02813	0·000
100	1·02860	0·046
200	1·02908	0·093
500	1·03050	0·231
1,000	1·03285	0·460
2,000	1·03747	0·909
3,000	1·04199	1·349
4,000	1·04640	1·778
5,000	1·05071	2·197
6,000	1·05494	2·609
7,000	1·05908	3·011
8,000	1·06314	3·406
9,000	1·06713	3·794
10,000	1·07104	4·175

* Salinity, $35 \cdot 0^0/_{00}$; temperature 0°C.

at pressures of up to 1,000 bars. Data on the compression of sea water of various salinities and temperatures under a pressure of 100 bars (corresponding to a depth of ca 1,000 m) are given in Table 2.3.

TABLE 2.3

Percentage reduction in volume of sea water under a pressure of 1,000 db at various temperatures and salinities

$S^0/_{00}$	Temperature (°C)			
	0	10	20	30
0	0·500	0·470	0·451	0·440
10	0·486	0·459	0·442	0·432
20	0·474	0·448	0·432	0·423
30	0·462	0·438	0·424	0·415
35	0·457	0·433	0·419	0·411
40	0·450	0·428	0·415	0·407

TABLE 2.4

Relative conductivity of sea water (UNESCO, 1966)

$S^0/_{00}$	15°C	20°C
30	0·87100	0·8713
31	0·89705	0·8973
32	0·92296	0·9232
33	0·94876	0·9489
34	0·97443	0·9745
35	1·00000	1·0000
36	1·02545	1·0254
37	1·05079	1·0506
38	1·07601	1·0758
39	1·10112	1·1008
40	1·12613	1·1257

B. ELECTRICAL CONDUCTIVITY OF SEA WATER

A knowledge of the precise relationship between the electrical conductivity and salinity of sea water is necessary on account of the growing use of salinometers. In practice, these instruments are usually standardized with Standard Sea Water (salinity 35·00±0·01%) and the ratio of its conductivity to that of the sample is measured. Comprehensive tables are available (UNESCO, 1966) relating the conductivity ratio at both 15° and 20° to salinity (cf. Table 2.4). This compilation also contains tables of corrections to be added to the conductivity ratios, by means of non-thermostatic salinometers, at other temperatures, to convert their readings to those at the standard temperatures. The difficult task of measuring the *absolute* conductivity of sea water with an accuracy of 1 in 10^5 is at present being undertaken at the British National Institute of Oceanography.

The electrical conductivity of sea water increases under the influence

TABLE 2.5

Effect of pressure on the conductivity of sea water*

Temp.	Pressure (db)	$S^0/_{00}$ 31	$S^0/_{00}$ 35	$S^0/_{00}$ 39	Temp.	$S^0/_{00}$ 31	$S^0/_{00}$ 35	$S^0/_{00}$ 39
0°C	1,000	1·599	1·556	1·512	15°C	1·032	1·008	0·985
	2,000	3·089	3·006	2·922		1·996	1·951	1·906
	3,000	4·475	4·354	4·233		2·895	2·830	2·764
	4,000	5·759	5·603	5·448		3·731	3·646	3·562
	5,000	6·944	6·757	6·569		4·506	4·403	4·301
	6,000	8·034	7·817	7·599		5·221	5·102	4·984
	7,000	9·031	8·787	8·543		5·879	5·745	5·612
	8,000	9·939	9·670	9·401		6·481	6·334	6·187
	9,000	10·761	10·469	10·178		7·031	6·871	6·711
	10,000	11·499	11·188	10·877		7·529	7·358	7·187
5°C	1,000	1·368	1·333	1·298	20°C	0·907	0·888	0·868
	2,000	2·646	2·578	2·510		1·755	1·718	1·680
	3,000	3·835	3·737	3·639		2·546	2·492	2·438
	4,000	4·939	4·813	4·686		3·282	3·212	3·142
	5,000	5·960	5·807	5·655		3·964	3·879	3·795
	6,000	6·901	6·724	6·547		4·594	4·496	4·399
	7,000	7·764	7·565	7·366		5·174	5·064	4·954
	8,000	8·552	8·333	8·114		5·706	5·585	5·464
	9,000	9·269	9·031	8·794		6·192	6·060	5·929
	10,000	9·915	9·661	9·408		6·633	6·492	6·351
10°C	1,000	1·183	1·154	1·125	25°C	0·799	0·783	0·767
	2,000	2·287	2·232	2·177		1·547	1·516	1·485
	3,000	3·317	3·237	3·157		2·245	2·200	2·156
	4,000	4·273	4·170	4·067		2·895	2·837	2·780
	5,000	5·159	5·034	4·910		3·498	3·429	3·359
	6,000	5·976	5·832	5·688		4·056	3·976	3·896
	7,000	6·728	6·565	6·402		4·571	4·481	4·390
	8,000	7·415	7·236	7·057		5·045	4·945	4·845
	9,000	8·041	7·847	7·652		5·478	5·369	5·261
	10,000	8·608	8·400	8·192		5·872	5·756	5·640

* Percentage increase compared with the conductivity at one atmosphere.

of hydrostatic pressure and correction must be made for this effect in salinometers constructed to work *in situ* (cf. p. 28). Bradshaw and Schleicher (1965) have measured the percentage increase in conductivity produced by pressures of up to 1,000 bars (Table 2.5). They have found that whereas the effect of salinity variations on the pressure coefficient of conductivity is slight, the effect of temperature is relatively large.

C. REFRACTIVE INDEX OF SEA WATER

The refractive index (for the sodium D doublet) of sea water at various temperatures and salinities has been measured by Utterback *et al.* (1934), whose results have been recomputed by Cox (1965) and are presented in Table 2.6. If the refractive index is to be used for the determination of the salinity of sea water, normal refractometers do not provide sufficient accuracy. For this purpose it is necessary to use an interferometer in which the refractive index of the sample is compared with that of Copenhagen Standard Sea Water (see Rusby, 1967 and Mehu and Johannin-Gilles, 1969).

TABLE 2.6

*Refractive index of sea water**

$S^0/_{00}$	Temperature (°C)					
	0	5	10	15	20	25
0	1·3 3395	1·3 3385	1·3 3370	1·3 3340	1·3 3300	1·3 3250
5	3500	3485	3465	3435	3395	3345
10	3600	3585	3565	3530	3485	3435
15	3700	3685	3660	3625	3580	3525
20	3795	3780	3750	3715	3670	3620
25	3895	3875	3845	3805	3760	3710
30	3991	3966	3935	3898	3851	3798
31	4011	3985	3954	3916	3869	3816
32	4030	4004	3973	3934	3886	3834
33	4049	4023	3992	3953	3904	3851
34	4068	4042	4011	3971	3922	3868
35	4088	4061	4030	3990	3940	3886
36	4107	4080	4049	4008	3958	3904
37	4127	4099	4068	4026	3976	3922
38	4146	4118	4086	4044	3994	3940
39	4166	4139	4105	4062	4012	3958
40	(4185)	(4157)	(4124)	(4080)	(4031)	(3976)
41	(4204)	(4176)	(4143)	(4098)	(4049)	(3994)

* For sodium D light.

D. TRANSPARENCY OF SEA WATER

The transparency of sea water to visible radiation is a major factor in determining the depth of the euphotic zone. Only part of the radiation reaching the surface enters the sea and the remainder is reflected. Under direct illumination from the sun at an elevation of 45°, ca 98% of the incident light is transmitted. However, at solar elevations of less than 30° the losses by reflection increase progressively, e.g. at an elevation of 10° only ca 65% of the light enters the surface. With diffuse radiation from the sky the reflection loss is ca 7%. If the surface of the sea is rough the losses by reflection may be increased as much as 50%.

Light entering the sea is attenuated in four principal ways:

(i) absorption by the pure sea water,
(ii) scattering by dissolved ions and molecules and by particles of colloidal dimensions (Rayleigh scattering),
(iii) scattering, reflection and absorption by suspended inorganic and organic particles,
(iv) absorption by dissolved organic materials.

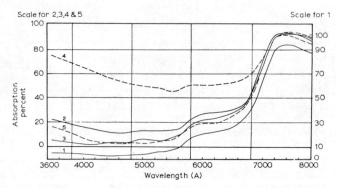

Fig. 2.11. Percent absorption of light of various wavelengths by water (1 m light path). 1, Distilled water (note different scale); 2, Sargasso Sea water unfiltered; 3, Sargasso Sea water, filtered through fine filter; 4, Coastal water, unfiltered; 5, the same coastal water filtered through a fine filter (after Clarke and James, 1939).

The attenuation of light is generally expressed in terms of the extinction $E = \log I_0/I$, where I_0 and I are the respective intensities of the light before and after attenuation by passage through 1 m of water. Thus, the value of E takes account of attenuation by both absorption and scattering. Over the wavelength range 3,500–8,000 Å, the absorption spectrum of micro-filtered deep ocean water is almost identical to that of distilled water (Fig. 2.11, Curve 3), see also Visser

(1967). However, the extinction values of near-shore waters are considerably greater, mainly because of scattering by organic and inorganic particulate matter* (Fig. 2.11, curve 4). Even after this particulate matter has been removed by micro-filtration, these waters may show enhanced extinction values for blue light (Fig. 2.11, curve 5) owing to the presence of dissolved plant pigments and humic substances (*Gelbstoff*, see p. 190).

For a more extensive treatment of the optical properties of sea water, the reader should consult the monograph by Jerlov (1968). The attenuation of solar radiation in sea water with reference to marine photosynthesis has been reviewed by Strickland (1958).

E. COLLIGATIVE PROPERTIES OF SEA WATER

Some recent values for the freezing point of sea water are presented in Table 2.7 (private communication from Miss L. A. Mayneord and Dr. C. N. Murray). They are estimated to be accurate to ca $\pm 0.004\,°C$. The osmotic pressure of sea water does not appear to have been measured directly, and the most reliable data available for this property have been calculated from the lowering of the vapour pressure (Table 2.8; Robinson, 1954).

TABLE 2.7
Freezing point of sea water

Salinity ⁰/₀₀	5	10	15	20	25	30	31
Freezing point (°C)	−0·268	−0·535	−0·801	−1·068	−1·341	−1·621	−1·678
Salinity ⁰/₀₀	32	33	34	35	36	37	38
Freezing point (°C)	−1·734	−1·791	−1·849	−1·906	−1·964	−2·018	−2·079
Salinity ⁰/₀₀	39	40					
Freezing point (°C)	−2·138	−2·196					

TABLE 2.8
Osmotic pressure of sea water at 25°C

Chlorinity ⁰/₀₀	12	13	14	15	16	17	18	19	20
Osmotic pressure (atm)	15·51	16·85	18·19	19·55	20·91	22·28	23·66	25·06	26·47

* This may amount to 1–2 mg/l in the waters of the English Channel.

REFERENCES

Bradshaw, A. and Schleicher, K. E. (1965). *Deep Sea Res.* **12**, 151.

Brown, N. L. and Hamon, B. V. (1961). *Deep Sea Res.* **8**, 65.

Clarke, G. L. and James, H. R. (1939). *J. Opt. Soc. Am.* **29**, 43.

Cox, R. (1957). *J. Cons. perm. int. Explor. Mer.* **22**, 38.

Cox, R. (1965). "The physical properties of sea water" *in* "Chemical Oceanography" (J. P. Riley and G. Skirrow eds) Vol. 1. Academic Press, London.

Cox, R., Culkin, F. and Riley, J. P. (1967). *Deep Sea Res.* **14**, 203.

Cox, R., McCartney, M. J. and Culkin, F. (1970). *Deep Sea Res.* **17**, 679.

Debye, P. and Hückel, E. (1923). *Phys. Z.*, **24**, 185 and 334.

Drost-Hansen, W. (1967). *J. Colloid Interf. Sci.*, **25**, 131.

Eisenberg, D. and Kauzman, W. (1969). "The Structure and Properties of Water" Oxford University Press, Oxford.

Ekman, V. W. (1910). *Publ. Circ. Cons. Explor. Mer.* **43**, 47.

Falkenhalgen, H. and Kelbg, G. (1959). *In* "Modern Aspects of Electrochemistry" (J. O'M. Bockris ed.) Vol. II. Academic Press, New York.

Forch, C., Knudsen, M. and Sørensen, S. P. L. (1902). *K. danske Vidensk Selsk.* **12**, 1.

Frank, H. S. (1961). "Some Questions about Water Structure." Desalination Res. Conf. Proc. U.S. *Nat. Acad. Sci. Nat. Res. Counc.* No. 942.

Horne, R. N. (1968). *Surv. Prog. Chem.* **4**, 1.

Horne, R. A. (1969). "Marine Chemistry." Wiley-Interscience, New York.

Jerlov, N. G. (1968). "Optical Oceanography." Elsevier, Amsterdam.

Johnston, R. (1969). *Oceanogr. mar. biol. Ann. Rev.* **7**, 31.

Kavanau, J. L. (1964). "Water and Solute-water Interactions." Holden-Day Inc., San Francisco, California.

Knudsen, M. (1901). 2me Conference International pour L'Exploration de la Mer, Report, Supplement 9.

Knudsen, M. (1903). *Pub. Circ. Cons. Perm. int. Explor. Mer.* **5**, 3 pp.

Matthews, D. J. (1938). Tables for the specific volume of sea water under pressure. *Cons. Perm. int. Explor. Mer.* Copenhagen.

Mehu, A. and Johannin-Gilles, A. (1969). *Deep Sea Res.* **16**, 605.

Monk, C. B. (1961). "Electrolytic Dissociation." Academic Press, New York.

Owston, P. G. (1958). *Adv. Phys.* **7**, 172 and 181.

Park, K. (1965). *J. Oceanogr. Soc. Japan*, **21**, 124.

Robinson, R. A. (1954). *J. mar. biol. Ass. U.K.* **33**, 449.

Robinson, R. A. and Stokes, R. H. (1959). "Electrolyte Solutions", 2nd ed. Academic Press, New York.

Rusby, J. S. M. (1967). *Deep Sea Res.* **14**, 427.

Samoilov, O. Y. (1965). "Structure of Aqueous Electrolyte Solutions and Hydration of Ions." Consultants Bureau, New York.

Strickland, J. D. H. (1958). *J. Fish. Res. Bd Can.* **15**, 453.

UNESCO (1966). "International Oceanographic Tables." National Institute of Oceanography. Wormley, England.

Utterback, C. L., Thompson, T. G. and Thomas, B. D. (1934). *J. Cons. perm. int. Explor. Mer.* **9**, 35.

Visser, M. P. (1967). *NATO Committee on Oceanographic Research*, Tech. Rep. **39**.

Wooster, W. S., Lee A. J. and Dietrich, G. (1969). *Deep Sea Res.* **16**, 321.

Chapter 3

The Oceans and the Physical Processes Occurring in Them

I. Introduction

It is helpful to present some data on the oceans before considering the physical processes which influence the distribution of dissolved and particulate materials in the sea. Approximately 361×10^6 km² (71%) of the earth's total surface of 510×10^6 km² is covered with water. The

TABLE 3.1

Areas, volumes and depths of oceans and seas (after Defant, 1961)

Sea or Ocean	Area (10⁶ km²)	Volume (10⁶ km³)	Mean depth (m)	Greatest depth (m)
Oceans incl. adjacent seas				
Atlantic Ocean	106·2	353·5	3,331	8,526*
Indian Ocean	74·9	291·9	3,897	7,450†
Pacific Ocean	179·7	723·7	4,028	11,034‡
All Oceans	361·1	1,370·3	3,795	—
Seas				
Mediterranean and Black Sea	2·97	4·32	1,458	4,404
Hudson Bay	1·23	0·16	128	229
Baltic	0·42	0·02	55	463
North Sea	0·58	0·05	94	665
English Channel+Irish Sea	0·18	0·01	58	263
Red Sea	0·44	0·22	491	2,359
Persian Gulf	0·24	0·01	25	84
Japan Sea	1·00	1·36	1,350	3,712

* Puerto Rico Trough north of Puerto Rico.
† Java Trench south of Java.
‡ Mariana Trench (11° N. 143° E.) according to sounding by Russian research ship Vityaz.

land areas are concentrated mainly in the northern hemisphere, and land predominates over sea only between 45° and 70° N. In contrast, land surface amounts to only 2·5% of the total area between 35° S. and 65° S. The earth's surface is divided by the continents into the three major oceans—the Atlantic, Pacific and Indian Oceans. From a geographical standpoint it is convenient to regard the North Polar Sea as part of the Atlantic Ocean and to apportion the Antarctic Ocean between the three oceans. However, the oceanographic features of these polar water masses differ so markedly from those of the major oceans that they are best considered separately. In addition to the oceans as such, there are a number of smaller, more or less land-locked seas, including the Mediterranean, the North Sea and the Baltic Sea, each with its own characteristic features. The areas, volumes and mean depths of the oceans (including their adjoining seas) and the more important of these seas are given in Table 3.1.

II. The Topography of the Oceans

The sea floor can be divided into a number of topographical regions (Fig. 3.1). At the edge of the land, lying between the high and low water

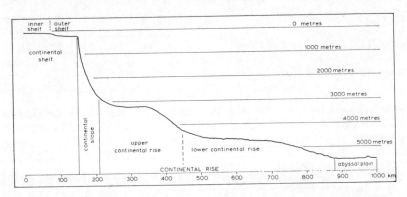

Fig. 3.1. Typical profile of the continental margin.

marks there may be a beach. Seawards of this is a shallow submarine terrace of average slope 0·1° which is termed the continental shelf. This may vary in width from a few miles to several hundreds of miles, and is terminated, usually at a depth of 150–200 m, by the upper edge of the continental slope. It has an area of ca 7·6% of that of the ocean floor and in most regions is a submerged extension of the continental blocks. The enormous deposits of minerals, oil and natural gas which lie under the continental shelf represent a great potential wealth which is only just beginning to be exploited. The waters overlying some parts of the shelf are highly fertile, and it is in these areas that the principal sea fisheries are found.

The continental slope has an average slope angle of ca 4° and in some regions, particularly around the Pacific, the angle scarcely decreases until the greatest depths are reached. In other areas the slope gradually levels out to one or more continental rises before the deep ocean floor is attained.

Much of the deep sea floor is occupied by flat and featureless abyssal plains from which rise hills on which may be numerous sea mounts. The greatest depths in the oceans are found in the deep trenches, the majority of which are in the Pacific. They frequently occur on the convex sides of arc shaped groups of islands. The most prominent feature of the ocean floors is the submarine mountain chain which runs continuously down the length of the North and South Atlantic, into the Indian Ocean and then across the Pacific. This ridge system exerts a considerable influence on the circulation of the deep waters of the oceans. Thus, the deep waters on the eastern and western sides of the Atlantic are separated from one another by the Mid-Atlantic Ridge, and exchange between them only takes place through gaps in the ridge lying close to the equator. The flow of deep water is further limited by ridges running roughly at right angles to the main ridge system. For further information on the topography of the ocean floor the reader should consult Shepard (1967), and Heezen and Menard (1963).

III. COLLECTION OF DATA

Our knowledge of the structure of the oceans has been largely built up by the examination of analytical data for water samples collected from various depths at hydrographic stations. Although it is now possible to measure salinity and a few other chemical parameters *in situ*, most components of sea water must still be determined in the laboratory. It is essential that when they reach the laboratory the water samples should be representative of the water at the sampling depth. This necessitates careful design of the sampling bottle to ensure that it flushes out rapidly when lowered, and seals perfectly when closed. It should also be resistant to corrosion and not contaminate the sample. The 1.3 l water bottle designed by the British National Institute of Oceanography (Fig. 3.2) fulfils all these requirements and is easy to operate. It consists of a stout polypropylene tube which can be closed by two spring-loaded hinged rubber caps. It is lowered open on a wire to the appropriate depth and a brass weight, called a messenger, is dropped down the cable. This actuates a mechanism which causes the caps to shut and also inverts a frame bearing reversing thermometers, setting the thermometers. The temperature and depth at the instant of sampling can be estimated from the readings of the protected and unprotected thermometers respectively (for the mode of action of these thermometers see Sverdrup *et al.*, 1942). Generally several bottles (up to 20) are attached to the wire at predetermined intervals, and each one is closed successively by a messenger released by the closing of the bottle immediately above it.

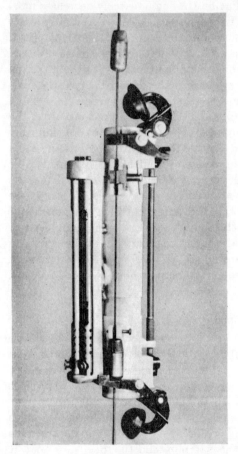

When the bottles have been brought back on board, the sample is run off through a stopcock and should be analysed as soon as possible to minimize errors which may arise from biological processes or evaporation. The parameters usually determined in physical oceanographic surveys are salinity, dissolved oxygen and silicate. In biological work certain micronutrients such as nitrate and phosphate are also commonly estimated and in addition, many other more specialized measurements such as plant pigment concentrations and primary productivity may be required.

In many areas the structure of the upper layers of the sea varies with time. Diurnal variations, associated with solar heating and biochemical processes do not extend to depths

Fig. 3.2. National Institute of Oceanography pattern water sampling bottle, shown open.

greater than 20 m whereas changes associated with seasonal variations and internal waves may occur down to depths of 200 m or more. To study such phenomena it is necessary to repeat the sampling programme at regular intervals. Graphs are then drawn relating the depth with temperature, salinity, etc., at each station. If a series of stations in a line has been worked the data may be plotted to give a vertical

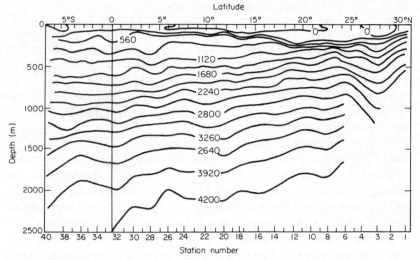

Fig. 3.3. Vertical profile of distribution of silicon (as μg Si/1) in the eastern Equatorial Pacific Ocean, May-June, 1952 (from Wooster and Cromwell, 1958). (Originally published by the University of California Press; reprinted by permission of the Regents of the University of California).

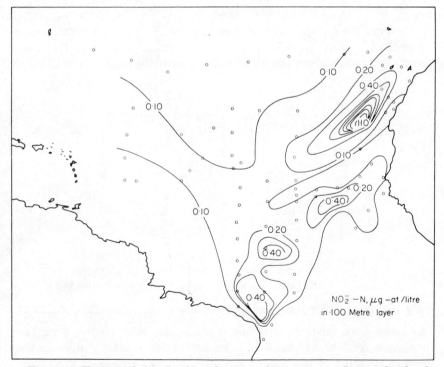

Fig. 3.4. Horizontal distribution of nitrite-nitrogen, μg–at/l at a depth of 100 m in the tropical Atlantic Ocean (1 μg–at $NO_2 = N/l = 14$ μg NO_2-N/l). (Vaccaro, 1965).

profile of the water properties (see, e.g., Fig. 3.3). When data from extensive surveys are available, it may be possible to show the horizontal distribution of a property at a particular depth by means of a contoured map (see, e.g. Fig. 3.4). Micro-nutrient elements tend to be distributed very patchily in the upper layers of the sea because of their utilization by phytoplankton colonies, and it is difficult to map their horizontal distribution in a meaningful way.

By studying these graphs and maps the oceanographer must try to build up a model of the sea which will account for the observed distributions of temperature, salinity, dissolved oxygen, micro-nutrients, etc. Among the processes which contribute to this distribution may be horizontal transport, mixing, eddy diffusion, evaporation, exchange with the atmosphere and biological processes. The remainder of this chapter will be devoted to an elementary description of the physical processes which lead to the transport and mixing of water in the sea. For a more detailed coverage the reader should consult Sverdrup *et al.* (1942) and Neumann (1968). The influence of living organisms on the chemical composition of sea water will be discussed in later chapters.

IV. TEMPERATURE DISTRIBUTION IN THE SEA

Geographically, the surface temperature of the open ocean varies roughly latitudinally, with the isotherms running more or less east-west. The temperature ranges from 28° C near the equator to −2° C in

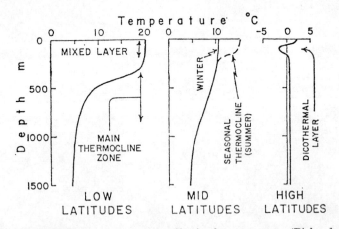

Fig. 3.5. Typical mean temperature profiles in the open ocean (Pickard, 1964).

the polar seas. Inshore, if ocean currents run close to the land, the isotherms tend to be parallel to the coast. The coastal waters at the eastern margins of the oceans are often cooler because of upwelling.

The oceans can be divided into three vertical temperature zones. (Fig. 3.5). The uppermost of these, with a thickness of 50–200 m, usually consists of well mixed water having a temperature similar to that of the surface. Beneath this lies a zone in which the temperature decreases rapidly with depth; this is known as the permanent thermocline. It occurs at a depth between ca 200 and 1,000 m in temperate and subtropical waters, but is absent from polar waters because there is a net loss of heat through the surface. Below the thermocline is a deep zone in which temperature falls more gradually.

In summer, when the water temperature at middle latitudes rises, a temperature gradient is set up in the upper layer at a depth of 30–50 m. In autumn, loss of heat from the surface causes the setting up of convection currents which destroy this seasonal thermocline.

V. WATER CIRCULATION IN THE OCEANS

The most important factor in determining the structure of the oceans is the circulation pattern. The movement of water is maintained by two principal driving forces—density differences and wind—for which solar heating is directly or indirectly the energy source. Tidal currents, although present at all depths in the oceans, have little influence on water circulation as they produce only slight net transport of water.

When wind blows over the surface of the sea, fluid friction causes the water to move. In the open ocean the direction of movement of the water differs from that of the wind, because of the rotation of the earth. This produces a geostrophic or Coriolis effect which causes the surface currents to tend to flow in a direction at 45° to the right of the wind in the northern hemisphere, and 45° to the left of it in the southern hemisphere. The force producing this effect varies as the sine of the latitude and is zero at the equator. Wind is the driving force producing the surface circulation in the oceans. Its influence extends generally to only a few hundreds of metres, being least in the equatorial regions where it may reach down to a depth of only 20–200 m. Wind-driven currents reach their greatest thicknesses in the western boundary currents of temperate latitudes, e.g. the Gulf Stream which has a depth of 1,000 m.

Wind stress has a further important effect on the oceanic circulation. In regions of low latitude, the trade winds blowing along the coast towards the equator, assisted by the geostrophic effect, cause the surface water to move away from the coast on the eastern side of the ocean. Its place is taken by water which rises from depths of 200–400 m. This cool upwelling water contains relatively high concentrations of nutrients, such as nitrate and phosphate (Fig. 3.6). As a result, these areas are frequently highly fertile and many of them support thriving fisheries. This is particularly true of the region of upwelling off the Atlantic coast of South Africa, and in the Pacific along the coast of Peru. For a largely theoretical account of upwelling the reader should consult the review by Smith (1968).

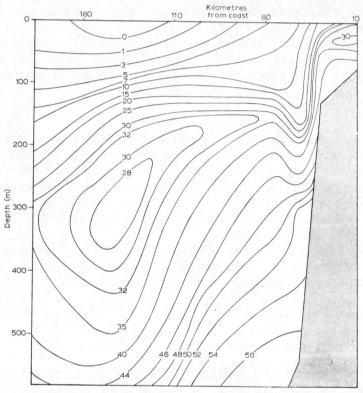

Fig. 3.6. Upwelling of phosphate-rich water off the West African Coast (section along Lat. 22° N. on 13–14 April 1969). (as $\mu g \ PO_4^{3-} - P/l$)

In contrast, the circulation of the lower layers of the sea is maintained mainly by thermodynamic processes depending on density differences in the water. Such density differences are produced by the physical conditions prevailing at the surface which cause changes in the temperature and/or salinity of the water. The temperature is increased by absorption of solar radiation and lowered by evaporation or conduction to the atmosphere. The salinity is raised by evaporation and freezing (since the ice is almost pure water) and is reduced by rainfall, run-off from land and the melting of ice. At high latitudes the density of the water is increased as a result of the intense cooling, and in the Antarctic this effect is augmented by the increase in salinity caused by the formation of ice. Convection currents are then set up, and the dense water will gradually sink until it encounters water of slightly greater density when it will spread out horizontally. In the tropics the rate of evaporation is high, but the saline water thus produced remains at the surface as it has a low density because of its high temperature. However, in the Red Sea and the Mediterranean, the rate of evaporation is sufficiently high to produce water of unusually high salinity. This sinks to intermediate depths and flows over the shallow sills at their mouths and out into the ocean, and is replaced by a density current of less saline water entering in the opposite direction at the surface. These outflows of high salinity water form intermediate layers in the North Atlantic and Indian Oceans which can be recognized over wide areas. (see e.g. Fig. 3.9)

Our knowledge of the current systems of the oceans is most detailed for the upper ten metres or so, since direct measurements can be made by comparing the actual courses of ships with those calculated from their speeds and directions. Much less is known about the movements of deep water, and most of our present knowledge has been inferred from the distribution of temperature and chemical parameters, such as salinity and dissolved oxygen. In recent years the development of the neutral buoyancy float by Swallow (1955) has made it possible to make direct measurements of such currents, and much detailed information about a number of currents has been obtained by their use. For the present purposes only an elementary account of oceanic circulation is required, and the reader is referred to the monograph by Neumann (1968) for a more detailed and theoretical treatment.

A. SURFACE CURRENTS

The patterns of the surface currents of the Atlantic, Pacific and southern Indian Oceans are broadly similar (Fig. 3.7). However, the influence of the shapes of the adjacent coast-lines produces differences in detail. In the North Pacific and North Atlantic the surface circulation is in a clockwise direction, and in the South Pacific, South Atlantic and Indian Oceans the current flow is in an anticlockwise direction. Because of the geostrophic effect the currents on the western side of the oceans (i.e. the Gulf Stream, the Kuroshio and the Brazil and Agulhas currents) are narrower and faster flowing than the return currents on the eastern sides. In all these oceans, at low latitudes north and south of the equator, there are westward flowing currents which in some regions may be separated by an eastward travelling countercurrent.

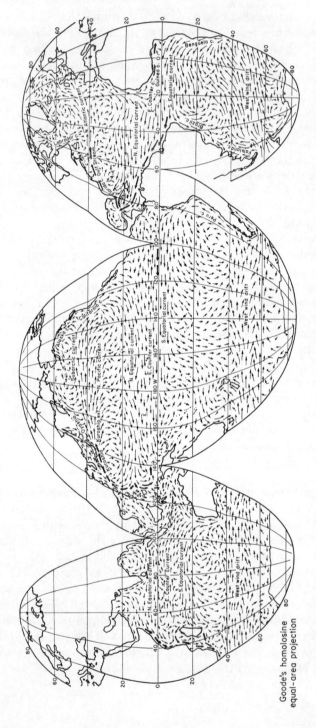

Fig. 3.7. Surface currents of the oceans in February-March (Sverdrup, 1945). By permission of the Goode Base Map Series, Department of Geography, University of Chicago (publisher). Copyright by the University of Chicago.

(i) Atlantic Ocean

The North Equatorial current is driven westwards by the north-east trade winds towards the West Indies and *en route* is joined by part of the South Equatorial current. The combined current then divides into two parts; one entering the Caribbean Sea. Easterly winds drive this water into the Gulf of Mexico, from which it escapes to the North Atlantic by way of the Gulf of Florida as the Florida current. It is then rejoined by the other part which has passed to the east of the Antilles Arc. The combined current—the Gulf Stream—flows north-eastwards along the American coast. Recent work (summarized by Stommel, 1965) has shown that the course and width of the Gulf Stream are continually changing; the average width is ca 150 km, but this may decrease to 50 km. The average surface speed in the neighbourhood of Chesapeake Bay is about 150 cm/sec (3kt) which corresponds to a water transport of ca 70×10^6 tons/sec. At the latitude of Cape Hatteras the current turns eastwards and is driven across the Atlantic by the prevailing south-west winds as the North Atlantic current, which as it is broad, has a velocity only $\frac{1}{10}$ to $\frac{1}{30}$ of that of the Gulf Stream. When it nears Europe this current divides into two branches one of which turns north-east and enters the Norwegian Sea, and the other turns south towards the Canary Islands to complete the circulation.

The surface circulation of the South Atlantic is more or less a mirror image of that of the North Atlantic. The south-east trade winds, acting between the equator and 10–15° S., drive the South Equatorial Current westwards. A minor part of this water passes across the equator into the North Atlantic and the remainder is deflected along the South American coast as the warm and saline Brazil Current. At about 40° S. the Current turns east and merges with the West Wind Drift which is driven by the prevailing west winds. This current is bounded on the north by the Subtropical Convergence and turns northwards up the coast of Africa and becomes the Benguela Current.

Between the North and South Equatorial Currents, at about 7° N. where winds are light, a current occurs which flows in the opposite direction. This Equatorial Countercurrent is formed from westward turning branches of the two Equatorial Currents and is best developed on the eastern side of the Ocean.

(ii) Pacific Ocean

The surface water circulation of the Pacific Ocean resembles that of the Atlantic. Trade winds drive the North and South Equatorial Currents in a westerly direction across the width of the Ocean at speeds of 25–30 and 50–60 cm/sec respectively. Since the wind systems show a lack of symmetry these currents are not distributed symmetrically about the equator, but are displaced ca 7° to the north. When the North Equatorial Current reaches the Philippines it divides. Some of the water turns back southwards to feed the Equatorial Countercurrent (see below), but the major part runs north-east, close to the coast of Japan as the Kuroshio. This warm current closely resembles the Florida current, being deep, narrow and swift. At about 35° N. the current turns east and becomes broader; it is then called the Kuroshio Extension and is the counterpart of the Gulf Stream, transporting water at a similar rate (30–60 $\times 10^6$ m^3/sec). At about 170° E. it is joined by the cold Oyashio Current coming from the Bering Sea. The combined current, which is now known as the North Pacific Current, flows eastward until it nears North America. It then divides into two parts, of which one turns north through the Gulf of Alaska to the Bering Sea, while the other flows southwards as the California Current to complete the North Pacific gyre.

The surface circulation of the South Pacific is anticlockwise. The South Equatorial Current after crossing the ocean turns southwards along the east coast of Australia and then merges with the West Wind Drift. The circulation is completed when it turns northwards as the Peru Current. As will be seen from Fig. 3.7 the main circulatory pattern of both the Northern and Southern Pacific is complicated by the presence of enormous eddies or gyrals.

The countercurrent system of the Pacific is much better developed than that of the Atlantic, and the narrow North Equatorial Countercurrent can be traced at 8° N., between the North and South Equatorial Currents for a distance of ca 12,000 km. A further, rather poorly developed, countercurrent occurs at about 10° S. These current systems are fed from backward turning branches of the North and South Equatorial Currents respectively.

(iii) Indian Ocean

The surface circulation of the Northern Indian Ocean is greatly influenced by the surrounding land masses. That of the southern Indian Ocean resembles those of the other southern oceans in having an anticlockwise gyre which is bounded on the south by the West Wind Drift. Between November and March the north-east trade winds produce an equatorial current system similar to that in the Pacific. However, between April and September the south-west monsoons cause the westward North Equatorial Current to reverse its direction and the Equatorial Countercurrent to disappear, and this leads to the establishment of the Somali current.

(iv) Antarctic Ocean

The Antarctic or Southern Ocean is best considered separately from the three other Oceans which open into it, as its waters have special characteristics. The principal surface current is the easterly Circumpolar Current, which is called the West Wind Drift since it is produced by the stress of the prevailing westerly wind. The circulation is mainly a closed one, but the geostrophic effect does tend to produce a northerly movement of the surface water, resulting in water entering the Atlantic and Pacific as the Falkland and Peru Currents respectively. Although the Circumpolar Current is comparatively slow, it is deep and has the greatest transport rate of any of the ocean currents, this being, at its maximum, more than 140×10^6 m³/sec.

B. DEEP CIRCULATION

We have seen above (p. 45) that the movements of deep water are produced by the sinking from the surface of water which has become denser through being cooled or increased in salinity. This water then

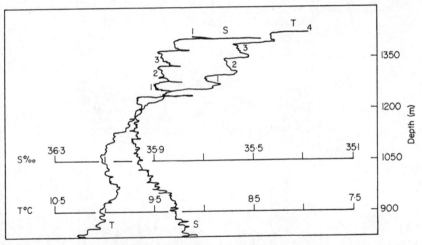

Fig. 3.8. Part of a record taken with a Bissett-Berman temperature-salinity-depth probe at a station in the Atlantic 500 km to the west of Gibraltar. The record shows the complex thermohaline structure of the water beneath the outflowing warm saline water from the Mediterranean. Note the step structure and unique inversions (1–4); after Howe and Tait (1970).

spreads out when it reaches the appropriate density surface. At mid and low latitudes it is often possible to distinguish five layers: surface, upper, intermediate, deep and bottom, each of which has a greater density than the one lying above it, and is prevented from mixing with the adjacent layers by the density gradient. At high latitudes the upper and deep waters are often similar, and the intermediate layer is frequently missing. The use of *in situ* temperature-salinity-depth probes (p. 24) reveals that these layers often have extremely complex structures (Fig. 3.8 and Pingree, 1969). Seasonal variations affect only the surface layer which may have a thickness of 100–200 m. However,

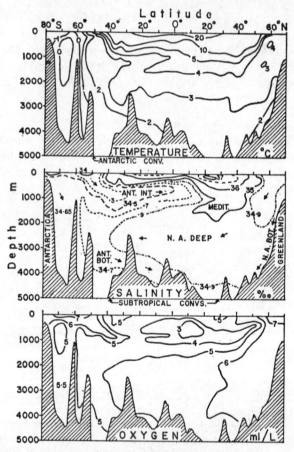

Fig. 3.9. South-north longitudinal sections of water parameters along the western trough of the Atlantic Ocean (Pickard, 1964).

water movements produced by wind stress influence both the surface and upper layers, but die out as the base of the latter is reached at depths of up to 800 m.

The existence of these various layers is clearly demonstrated in Fig. 3.9 which shows the temperature, salinity and dissolved oxygen distribution along a north-west section of the West Atlantic. As would be expected, it is only in the surface and upper layers that considerable horizontal or vertical temperature gradients are found. Freezing ice over the continental shelf of Antarctica, particularly in the Weddell Sea, produces water having a density ($\sigma_t = 27\cdot89$), slightly greater than that of the adjacent circumpolar water ($\sigma_t = 27\cdot84$). This water therefore sinks and forms the northward moving Antarctic Bottom Water. Antarctic Intermediate Water is formed by the sinking of water of salinity ca $33\cdot8^0/_{00}$ and temperature $2\cdot2$°C at the Antarctic Convergence at 50° S. It can be readily traced north of the equator as a salinity minimum at a depth of ca 1,000 m. Beneath it, flowing southwards at a depth of 1,500–4,000 m, is the North Atlantic Deep Water. This highly oxygenated water is formed by the rapid sinking of chilled surface water during the winter off the south of Greenland. There is some evidence that this water is underlain by even denser North Atlantic Bottom Water flowing from the north over the Greenland-Iceland Ridge. Between 20° N. and 40° N. the salinity section shows evidence of the intrusion of high salinity Mediterranean water (see p. 45). The central water masses of the Atlantic and other oceans result from the sinking of water in winter at Subtropical Convergences at 35–40° N. and 35–40° S.

The deep water circulation pattern of the Pacific is more complex than that of the Atlantic and is characterized by the fact that only very limited exchange of water occurs between the northern and southern parts of the oceans. At depths between 100 and 800 m lie the North and South Central Pacific Water masses. These are separated by the higher salinity water of the Pacific Equatorial Water Mass that extends across the entire width of the ocean, and which is separated from the surface waters by a strongly developed thermocline. The North Pacific Central Water extends to 40° N. and northwards of it lies the Pacific Subarctic Water Mass, having a low temperature (2–4°C) and low salinity ($33\cdot5$–$34\cdot5^0/_{00}$). This is formed by sinking dense water at the Subarctic Convergence, which results from the mixing of warm saline water of the Kuroshio Extension with cold, less saline waters of the Oyashio current coming from the Bering Sea. The resultant Subarctic Current flows eastward until it nears the American coast, where it turns south-east and its density is decreased by heating and mixing until it merges with the North Equatorial Current. An Intermediate Water which circulates in a clockwise direction lies beneath the North Pacific Central Water Mass. This may have been formed by vertical mixing in the Subarctic region.

The South Pacific Central Water Masses are bounded on the south by the Subtropical Convergence at 40° S. Beneath them lies an Intermediate Water Mass which has been formed by sub-surface mixing in the region of the Subantarctic Convergence. The northward flow of this water is barred by the Equatorial Water Mass, and it therefore mixes with the water lying above and below it.

Unlike the deep waters of the Atlantic, those of the Pacific are not formed by sinking in the Arctic and Antarctic regions. They originate from an offshoot of the Antarctic Circumpolar Current, which spreads north between 160° and 180° W., and can be traced into the North Pacific (see Fig. 3.10), where it must eventually rise into the upper

Fig. 3.10. Deep water circulation in the oceans according to Stommel (1958). By permission of Pergamon Press.

layers. The deep and bottom waters of the Pacific with their extremely slow circulation therefore act as a sink for the waters of the other oceans and this is reflected in their unusually high content of trace metals.

C. CURRENTS IN INSHORE AREAS

The currents produced by wind are sufficiently strong to obscure the rather feeble permanent currents in many shallow inshore waters. In such waters the directions of the wind and the current which it produces are more nearly parallel than they are in the open ocean. The wind factor (the ratio of the current velocity to wind velocity) is generally greater, and is influenced by the geography of the coastline and the direction of the wind. It is greatest if the movement of water is not restricted in the direction of the wind, as for example with a southwest wind blowing into the English Channel. The velocities of these currents may be further increased if their passage is partially restricted, and velocities of 150 cm/sec can be reached in the Straits of Dover when strong winds are blowing, because of the funnelling effect of the coastline.

Although the velocities of tidal currents in coastal areas may reach 150 cm/sec, and in narrow straits can attain three times that value, in fact they bring about little net transport of water. However, their interaction with an uneven sea bed or the sides of a channel frequently gives rise to turbulence which may cause both horizontal and vertical mixing.

D. ESTUARINE CIRCULATION

In estuaries considerable changes in the salinity take place in an upstream direction. In some estuaries, where there is a strong tidal flow, turbulence over a rough bottom will keep the water well mixed. In others, where there is little turbulence, the almost non-saline river water forms an upper layer over the more saline water with which it mixes and moves seawards. This produces a density current which flows in the opposite direction near the bottom (Figs. 3.11a and 3.11b).

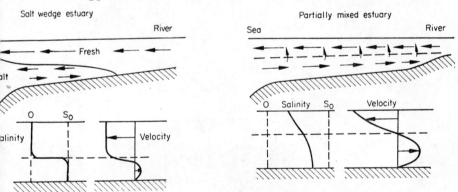

Fig. 3.11. (a) Salt wedge estuary. (b) Two-layer flow with entrainment. Upper diagrams show a section along the estuary. Lower diagrams show typical salinity and current velocity profiles (from Bowden, 1967).

Superimposed on this circulation are much faster tidal currents which move the water bodily up and down stream with the ebb and flow. The extent to which this type of estuarine circulation is developed varies considerably from one river to another, and for any particular river depends on several factors such as the river flow, tidal currents and the amount of vertical mixing which occurs. When the less dense water

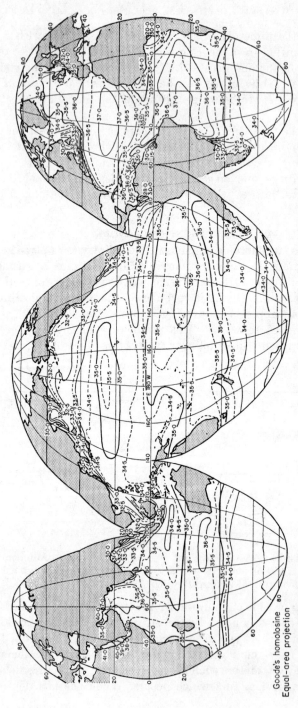

Goode's homolosine
Equal-area projection

Fig. 3.12. Surface salinity of the oceans in the northern summer (Sverdrup, 1945). By permission of George Allen and Unwin Ltd. (Projection by courtesy of the University of Chicago).

reaches the sea the Coriolis force (p. 43) causes it to deflect towards the right in the northern hemisphere. The estuarine water therefore tends to flow along the coast, extraining and mixing with the sea water as it does so. For a more detailed account of estuarine processes the reader should consult the review by Bowden (1967).

VI. Salinity Distribution in the Oceans

The principal natural processes which lead to changes in the salinity of sea water are those which bring about the removal or addition of fresh water. Increase in salinity is caused by evaporation and by the removal of almost pure water, as ice, during freezing. Decrease in salinity results from atmospheric precipitation, run-off from land and the melting of ice. Thus, these mechanisms affect only the surface layer of the sea, and once a body of water has left the surface its salinity can only be changed by mixing with adjacent water masses having different salinities.

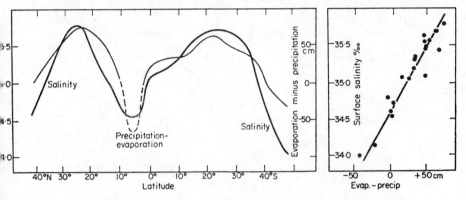

Fig. 3.13. *Left:* Mean values of surface salinity and evaporation minus precipitation for all oceans plotted against latitude. *Right:* Plot of surface salinity versus evaporation-precipitation, (Sverdrup, 1945; by permission of George Allen and Unwin).

Since the factors affecting the salinity of surface waters are basically climatic ones, the salinity distribution in these waters tends to be zonal and ranges in the open ocean from 32–37·5⁰/₀₀. The salinities in all the oceans show the same pattern (see Fig. 3.12). Maximum values occur in tropical latitudes at ca 20–30° N. and 15–20° S., where trade winds and high temperature combine to produce an excess of evaporation over precipitation (See Fig. 3.13). A minimum is found in equatorial regions because of the lighter winds and high rainfall. At higher latitudes than the tropics, the salinity decreases because precipitation exceeds evaporation, and this effect is accentuated in summer in polar seas by the melting of ice. Lower salinity values are found near the coasts of continents because of the discharge of rivers. Land-locked seas show the greatest range of salinities, ranging from the faintly brackish water of the northern Baltic Sea to higher than oceanic values in the Mediterranean Sea $(37-40^0/_{00})$ and Red Sea $(40-41^0/_{00})$, where the rate of evaporation is high. Seasonal variations in salinity in the surface layers of the ocean are usually less than 0.5%, except near ice or where there are marked seasonal variations in rainfall, e.g. the Bay of Bengal.

We have seen previously (p. 44) that the vertical structure of the oceans is determined by the densities of the various water masses. These are closely related to their temperatures, and variations in salinity are not normally great enough to override the effect of temperature. It is therefore possible to find stable situations where high salinity water overlies low salinity water.

The vertical distributions of salinity with depth in the Atlantic and Pacific Oceans are shown diagrammatically in Fig. 3.14. At low and

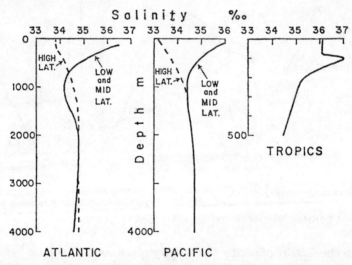

Fig. 3.14. Typical salinity profiles in the open ocean (Pickard, 1964).

mid-latitudes there is a pronounced salinity minimum at 600–1,000 m, below which the salinity increases to ca 2,000 m; in tropical waters a well defined maximum occurs at ca 100 m, near the top of the thermocline. At high latitudes, where convergence occurs this minimum is absent and the salinity increases with depth to ca 2,000 m. The salinity of all ocean waters lying deeper than 4,000 m is relatively constant at $34.6-34.9^0/_{00}$.

VII. Mixing Processes in the Sea

The discussion in this chapter has so far been concerned with the circulatory pattern of the oceans and the origins of the principal water masses. It is now necessary to consider the processes which bring about transport of chemical substances, both within the various bodies of water and from one to another. At the outset it should be made clear that transport by molecular diffusion is very much slower than by advection and eddy diffusion.* If molecular diffusion was the only mechanism for equilibration it would be expected that the concentration of a solute would increase with depth because of gravitational settling and that at equilibrium this would be balanced by upward movement caused by thermal energy. Pytkowicz (1968) has given a theoretical treatment of the subject from which he derives the expression

$$N_z = N_0 \exp\left[\frac{1}{RT}\int_0^z (M - \bar{v}\rho)gdz\right]$$

where N_z and N_0 are the number of molecules of the substance (molecular or ionic weight M and having a partial molar volume of $\bar{v}$) at depth z and the surface respectively, ρ is the density of the solution and g is the acceleration due to gravity. He has shown that for sodium chloride the gradient would amount to ca 18% per kilometre. Processes of this type are clearly not responsible for the large scale salinity structure of the oceans; however, they may perhaps play a part in establishing the microstructure. Most of the vertical and horizontal mixing in the sea must be attributed to processes produced by turbulence.

A. TURBULENT MIXING

In nature, the flow of moving bodies of water is rarely of the laminar type in which sheets of water move with a steady speed. Almost always the flow is turbulent, and random motion of smaller or larger eddies of the fluid are superimposed on the simple flow pattern. This causes interchange of water and dissolved salts between one layer and another, i.e. mixing results. Turbulent mixing can occur in both the vertical and horizontal planes. Because the oceans are much wider than they are deep, horizontal mixing effects are several thousands of times greater than those in the vertical plane. This is accentuated when a stable density gradient is present as this suppresses vertical mixing, but has negligible influence on horizontal turbulence which takes place along planes of equal density. Because of these differences horizontal and vertical mixing will be treated separately.

*In advection, large scale water movements occur, the dissolved materials being carried along with the water, e.g. in ocean currents and upwelling. In eddy diffusion, an exchange of chemical properties takes place without overall transport of water.

B. HORIZONTAL MIXING

Although horizontal turbulent mixing takes place at all depths in the oceans, the process has been studied in detail only in the surface layer. The most pronounced eddies are those associated with the major wind-driven surface currents. Thus, eddies having diameters of ca 100 km occur between the South Equatorial Current and the West Wind Drift and in the North Atlantic Current. Smaller eddies are formed from the large ones and these in their turn break down into a cascade of smaller and smaller eddies.

Small scale horizontal turbulence can be made visible by releasing the fluorescent pink dye Rhodamine B into the sea. The dye patch will be transported by eddies larger than itself, while being dispersed by eddies which are smaller than it. Progressively larger eddies become involved in the dispersion as the dye spot spreads. Quantitative aspects of turbulent mixing can be examined by making measurements of the distribution of dye across the patch at regular intervals using a fluorimeter, which can detect it at concentration levels or $1 : 10^9$ or less.

C. VERTICAL MIXING

Vertical mixing is produced by two principal mechanisms depending on turbulence produced either by wind action or by the frictional shear of currents flowing over one another or over the bottom. The scale of these processes is much smaller than those involved in horizontal mixing and the amount of mixing is correspondingly much less.

When wind blows over the surface of the sea it produces waves. If these are small and regular they produce little mixing in deep water, since the particles of water in the wave execute an almost circular motion, moving forwards at the crest of the wave and backwards at the trough. These movements fall off rapidly with depth. As the wind increases, the surface of the sea will become more disturbed and waves of various wavelengths will be superimposed. The interaction between these waves will lead to increased mixing. At even higher wind velocities, or near to the shore the waves will break, causing strong turbulence the effect of which, however, only influences the shallow surface layer.

Wind also produces vertical mixing in the sea by another mechanism. As has been mentioned on page 43, wind stress tends to cause the surface water to move at 45° to the wind. This movement is transmitted to the water below by eddy viscosity. With increasing depth the velocity of the induced subsurface current decreases uniformly and its direction deviates from that of the wind more and more, until at a certain depth the drift is at 180° to the surface current. This depth, which lies between

50 and 200 m depending *inter alia* on the wind speed, is for practical purposes the limit of the wind's influence.

Stable temperature gradients act as barriers to vertical mixing since for mixing to occur in a layer of increasing density, potential energy must be supplied by the turbulence. If the turbulence produced by the wind is not sufficient to break down this gradient (thermocline), then mixing will not take place between the waters lying above and below it.

In deep erwaters, beyond the range of wind-produced turbulence, the principal cause of mixing is frictional shear caused by currents. Eddy diffusion occurs at the boundary between two currents and results in mixing of their waters. Similarly, friction of currents passing over the bottom causes turbulence which, if the current is sufficiently strong, may extend a considerable distance from the bottom, and even to the surface. Thus, the high productivity of the water over the Grand Banks off Newfoundland is caused by the turbulence produced by the strong current over the rough bottom. This brings nutrient-rich water to the surface. Tidal currents cause top to bottom mixing in the eastern English Channel and inhibit the establishment of a thermocline.

REFERENCES

Bowden, K. F. (1967). *In* "Estuaries" (G. H. Lauff, ed.). American Association for the Advancement of Science, Washington, D.C.

Defant, A. (1961). "Physical Oceanography", Vol. 1. Pergamon Press, Oxford.

Heezen, B. C. and Menard, H. W. (1963). *In* "The Sea" (M. N. Hill, ed.), Vol. 3. Wiley, New York.

Howe, M. R. and Tait, R. I. (1970). *Deep Sea Res.* **17**, 963.

Neumann, G. (1968). "Ocean Currents." Elsevier, Amsterdam.

Pickard, G. L. (1964). "Descriptive Physical Oceanography." Pergamon Press, Oxford.

Pingree, R. D. (1969). *Deep Sea Res.* **16**, 275.

Pytkowicz, R. M. (1968). *Oceanogr. mar. Biol. Ann. Rev.* **6**, 83.

Shepard, F. P. (1967). "Submarine Geology." Harper Brothers, New York.

Smith, R. L. (1968). *Oceanogr. mar. Biol. Ann. Rev.*, **6**, 11.

Stommel, H. (1958). *Deep Sea Res.* **5**, 80.

Stommel, H. (1965). "The Gulf Stream." University of California Press, Berkeley.

Sverdrup, H. U., Johnson M. W. and Fleming, R. H. (1942). "The Oceans, their Physics, Chemistry and General Biology." Prentice-Hall, New York.

Sverdrup, H. U. (1945). "Oceanography for Meteorologists." George Allen and Unwin Ltd., London.

Swallow, J. C. (1955). *Deep Sea Res.* **3**, 74.

Vaccaro, R. F. (1965). *In* "Chemical Oceanography" (J. P. Riley and G. Skirrow, eds), Vol. 1. Academic Press, London.

Wooster, W. S. and Cromwell, T. (1958). *Bull. Scripps. Instn Oceanogr.* **7**, 1969.

Chapter 4

The Major and Minor Elements in Sea Water

I. INTRODUCTION

At the present time, over 75 elements have been detected or determined in sea water. With the development of more sensitive analytical methods, such as neutron activation, it is probable that the other naturally occurring elements will soon be determined, even though their concentrations may be less than 10^{-9} g/l. It is convenient for purposes of discussion to consider the elements present in sea water in four main classes:

  (i) the dissolved atmospheric gases;
 (ii) the micronutrient elements (e.g. ionic forms of nitrogen, phosphorus) which are essential for the growth of all marine plants;
(iii) the geochemically relatively unreactive *major* elements which occur at concentrations above 1 mg/kg;
(iv) the minor or trace elements which occur at lower concentrations but which are not covered by classes (i) or (ii).

This chapter will be devoted to a consideration of the marine chemistry of the major and minor elements and classes (i) and (ii) will be discussed in subsequent chapters.

II. GEOCHEMICAL BALANCE OF THE OCEANS

A. GENERAL

The oceans have probably been formed by reaction of the primary rocks with volatile substances which have distilled from the interior of the earth under the influence of radioactive heating. Most of the cations and a fraction of the anions in the oceans appear to have originated from the weathering of igneous silicates, whereas most of the major anions have come from the volatiles (Wedepohl, 1966).

Goldschmidt (1933) has set up a material balance for the formation of the ocean and sediments by the weathering of the primary igneous rocks. If a mass A of igneous primary material weathers to form total masses of sedimentary material B and elements dissolved in the ocean C, and if a_e, b_e and c_e are the average weight fractions of an individual element e in igneous rocks, sedimentary material and in the dissolved solids of sea water respectively, then

$$a_e A = b_e B + c_e C$$

Using computer techniques, Horne and Adams (1966) have elaborated this model and used it to study the distributions of 65 elements. They have concluded that for every 1 l of sea water formed, ca 1·2 kg of primary igneous rock has been weathered. For most elements they have found good agreement between the amount liberated by the decomposition and that found in the sea and sedimentary material. However, some elements (B, S, Cl, As, Se, Br. I) are not in balance, probably because a significant proportion of their total mass in the sedimentary and oceanic environment is of volcanic origin. Our ignorance of the origins of the relatively high concentrations of manganese found in marine sediments is reflected in the high degree of imbalance found for manganese.

The present ionic composition of sea water results from a balance between the rate at which dissolved matter is added to the ocean from the land and the atmosphere, and the rate at which it is removed from the sea by incorporation into the sediments or by being returned to the atmosphere. It is this geochemical balance which appears to have kept the relative composition of sea water fairly constant since Cambrian times and has prevented the build-up of toxic concentrations of trace elements, such as copper, arsenic and selenium.

Materials enter the oceans either by fall-out from the atmosphere or from the influx of river water. Some of these substances are incorporated in the oceans as dissolved or colloidal species. However, part of this material is in the form of non-reactive solid phases such as quartz, feldspar or clays which soon settle to the ocean floor (see Chapter 12). The pH and concentrations of the major and many of the minor dissolved cations in sea water are regulated by equilibria between the dissolved cations and aluminosilicates either authigenic, suspended, or in the sediments. Sillén (1963, 1965, 1967) has used models in terms of such reactions to explain the composition of sea water. In the most recent of these he postulates that sea water is one phase in a nine component equilibrium system consisting of the atmosphere, the sea, quartz (SiO_2), calcite ($CaCO_3$), the aluminosilicate clay minerals—kaolinite, illite, chlorite and montmorillonite—and a ninth phase which may be the authigenic mineral phillipsite (p. 369ff.), or perhaps dolomite

$(CaMg(CO_3)_2)$. With the exception of dolomite all these minerals exist in marine sediments apparently at equilibrium with sea water. Thus, the equilibrium between kaolinite and chlorite may be an important factor in maintaining the magnesium content of sea water at a constant level.

$$Al_2Si_2O_5(OH)_4 + SiO_2 + 5Mg^{2+} + 7H_2O \rightleftharpoons Mg_5Al_2Si_3O_{10}(OH)_8 + 10H^+$$
kaolinite   chlorite

Although these concepts can explain the composition of sea water qualitatively, much more must be learnt about the reactions involved before they can be used to give a quantitative explanation.

Clay minerals, mainly those of terrestrial origin delivered by the rivers, assist in two ways in maintaining the balance of the major cations in the oceans.

(i) By acting as ion exchangers (Grim, 1952),
$$\text{e.g. } Mg(R) + Ca^{2+} \rightleftharpoons Ca(R) + Mg^{2+}$$
$$Na(R) + K^+ \rightleftharpoons K(R) + Na^+$$
$$(R = \text{aluminosilicate lattice})$$

Experiments have shown that the above reactions proceed rapidly, and that the position of equilibrium lies well to the right. It is thus easy to see why sodium is more abundant in sea water than potassium $(Na/K = 27 \cdot 0)$ and why this relationship is reversed in deep sea sediments $(Na/K = 0 \cdot 42)$. Base exchange with montmorillonite assists in the maintenance of the magnesium balance.

(ii) Degraded illites and chlorites formed by terrestrial weathering are reconstituted by taking up potassium and magnesium from sea water (Holland, 1965).

Although the calcium and strontium concentrations in sea water are controlled to a large extent by cation exchange with clay minerals, equilibria with carbonate minerals are also extremely important regulating factors (see p. 82 and Holland, 1965):

$$\text{e.g. } CaCO_3 + 2H^+ \rightleftharpoons Ca^{2+} + H_2O + CO_2$$

For many trace elements, particularly transition metals, the above mechanisms are less important and the geochemical balance is maintained by adsorption onto authigenic minerals or by biological processes.

The mechanisms maintaining the geochemical balances of the anions in sea water are mostly very different from those associated with the cations. Chloride and bromide are geochemically inert, and it is probable that more than 75% of these halogens added from the volatiles is still in the oceans (Rubey, 1951). Most of the chloride reaching the sea from rivers has been simply recycled and very little is added from rock weathering.

Considerable amounts of sulphate reach the sea from the weathering of sulphide minerals; the geochemical balance is probably maintained by deposition of sulphide minerals and sulphur in the sediments via bacterial reduction (Berner, 1964).

Relative to igneous rocks, argillaceous marine sediments are considerably enriched in boron. It is probable that the geochemical balance of this element is maintained principally by adsorption onto clay minerals; subsequent diagenetic changes result in the formation of tourmaline. Biological concentration mechanisms may also play some part in the balance since this element is strongly concentrated in some siliceous organisms, such as sponges.

It has been suggested (Holland, 1965) that the partial pressure of carbon dioxide in the sea is crudely buffered by systems involving the coexistence of quartz, chlorite, dolomite and calcite (see above, p. 61).

B. RESIDENCE TIMES

Since the concentrations of the various elements in sea water approximate to a steady state, the rates at which they are supplied to the ocean are balanced by the rates at which they are removed. Barth (1952) has introduced the concept of the residence time of an element (τ), which is the average time it remains in the sea before being removed by the processes mentioned above. Thus,

$$\tau = \frac{A}{(dA/dt)}$$

Where A is the total amount of the element dissolved or in suspension in the oceans, and dA/dt is the rate at which it is added or precipitated. It is assumed that the element is completely mixed in a time appreciably shorter than the residence time, and that neither A nor dA/dt change in 3–4 times this period.

Values of dA/dt have been estimated in two ways. (i) From data on the average composition of river water and the total amount of river water delivered into the sea annually (Barth, 1952). It is necessary to make allowance for the significant amounts of elements which are transferred from the sea via the atmosphere and precipitation to the rivers (cf. IIC.1). (ii) From data on the median rate of marine sedimentation and the average composition of marine sediments (Goldberg and Arrhenius, 1958). Despite the difficulties involved in quantitatively assessing the many factors involved there is surprisingly good agreement between estimates of τ made by the two methods.

TABLE 4.1

Average abundances of the elements in the earth's crust (Ahrens, 1965), river waters (Livingstone, 1963; Durrum and Haffty, 1963 and Turekian, 1969), and ocean water (mainly after Turekian 1969, with additions and corrections); oceanic residence times (Goldberg, 1965) and probable principal dissolved species in sea water (after Sillén, 1963)

Element	Earth's crust $\mu g/g$	Average river water $\mu g/l$	Ocean water $\mu g/l$ for $S = 35^{0}/_{00}$	Oceanic residence time (years)	Probable main dissolved species
Hydrogen	$1 \cdot 119 \times 10^8$		$1 \cdot 078 \times 10^8$		H_2O
Helium			$7 \cdot 2 \times 10^{-3}$		$He(g)$
Lithium	20	3	180	$2 \cdot 0 \times 10^7$	Li^+
Beryllium	2·8	<0·1	$0 \cdot 6 \times 10^{-3}$	150	Hydroxy complexes
Boron	10	10	$4 \cdot 5 \times 10^3$		H_3BO_3, $B(OH)_4^-$
Carbon†		1,200	$2 \cdot 8 \times 10^4$		HCO_3^-
Nitrogen		250*	500*†		N_2, NO_3^-
Oxygen		$8 \cdot 8 \times 10^8$	$8 \cdot 56 \times 10^8$		H_2O
Fluorine	625	100	$1 \cdot 4 \times 10^3$		F^-, MgF^+
Neon			0·12		$Ne(g)$
Sodium	$2 \cdot 4 \times 10^4$	9,000	$11 \cdot 05 \times 10^6$	$2 \cdot 6 \times 10^8$	Na^+
Magnesium	$2 \cdot 0 \times 10^4$	4,100	$1 \cdot 326 \times 10^6$	$4 \cdot 5 \times 10^7$	Mg^{2+}
Aluminium	$8 \cdot 2 \times 10^4$	400	5†	100	$Al(OH)_3$?
Silicon	$28 \cdot 2 \times 10^4$	4,000	10^3†		$Si(OH)_4$
Phosphorus	$1 \cdot 0 \times 10^3$	20	70†	$8 \cdot 0 \times 10^3$	HPO_4^{2-}
Sulphur	260	3,700	$9 \cdot 28 \times 10^5$		SO_4^{2-}
Chlorine	130	8,000	$1 \cdot 987 \times 10^7$		Cl^-
Argon			450		$Ar(g)$
Potassium	$2 \cdot 4 \times 10^4$	2,300	$4 \cdot 16 \times 10^5$	$1 \cdot 1 \times 10^7$	K^+
Calcium	$4 \cdot 2 \times 10^4$	1,500	$4 \cdot 22 \times 10^5$	$8 \cdot 0 \times 10^6$	Ca^{2+}
Scandium	22	0·004	$1 \cdot 5 \times 10^{-3}$	$5 \cdot 6 \times 10^3$	Hydroxy complexes ?

Element					Species
Titanium	5.7×10^3	3	1	160	$Ti(OH)_4$?
Vanadium	135	1	1.5	1×10^4	$(H_2V_4O_{13})^{4-}$, HVO_4^{2-}, VO_3^- ?
Chromium	100	1	0.6†	350	Hydroxy complexes
Manganese	950	~5	2†	1,400	$Mn(OH)_{3(4)}$?
Iron	5.6×10^4	670	3†	140	$Fe(OH)_3$?
Cobalt	25	0.2	0.08†	1.8×10^4	Co^{2+}
Nickel	75	0.3	2	1.8×10^4	Ni^{2+}
Copper	55	5	3†	5×10^4	Cu^{2+}, $CuOH^+$
Zinc	70	10	5	1.8×10^5	Zn^{2+}
Gallium	15	0.1	3×10^{-2}	1.4×10^3	Hydroxy complexes ?
Germanium	1.5	~1	6×10^{-2}	7.0×10^3	$Ge(OH)_4$
Arsenic	1.8		2.3		$HAsO_4^{2-}$
Selenium	0.05	0.2	0.45		SeO_3^{2-}
Bromine	2.5	~20	6.8×10^4		Br^-
Krypton			0.21		$Kr(g)$
Rubidium	90	1	120	2.7×10^5	Rb^+
Strontium	375	50	8.5×10^3	7.5×10^3	Sr^{2+}
Yttrium	33	40	0.013		Hydroxy complexes ?
Zirconium	165	3	2.6×10^{-2}		Hydroxy complexes ?
Niobium	20		1×10^{-2}	300	
Molybdenum	1.5	1	10	5.0×10^5	MoO_4^{2-}
Technetium					
Ruthenium			7×10^{-4}		
Rhodium					
Palladium					
Silver	0.07	0.3	0.1	2.1×10^6	$AgCl_2^-$
Cadmium	0.2		5×10^{-2}	5×10^5	$CdCl^+$, Cd^{2+}
Indium	0.1		1×10^{-4}		Hydroxy complexes ?
Tin	2	0.04	0.01		Hydroxy complexes ?
Antimony	0.2	1	0.2	3.5×10^5	$Sb(OH)_6^-$?
Tellurium					
Iodine	0.5	~5	60†		IO_3^-, I^-

TABLE 4.1—continued

Element	Earth's crust $\mu g/g$	Average river water $\mu g/l$	Ocean water $\mu g/l$ for $S = 35^0/_{00}$	Oceanic residence time (years)	Probable main dissolved species
Xenon			5×10^{-2}		$Xe(g)$
Caesium	3	0·05	0·5		Cs^+
Barium	425	10	30†	$4·0 \times 10^4$	Ba^{2+}
Lanthanum	30	0·2	34×10^{-4}	$8·4 \times 10^4$	La^{3+}
Cerium	60		12×10^{-4}	440	Ce^{3+}?
Praseodymium	8·2		6×10^{-4}	80	Pr^{3+}
Neodymium	28		28×10^{-4}	320	Nd^{3+}
Promethium					
Samarium	6		45×10^{-5}	270	Sm^{3+}
Europium	1·2		13×10^{-5}	180	Eu^{3+}
Gadolinium	5·4		70×10^{-5}	300	Gd^{3+}
Terbium	0·9		14×10^{-5}	260	Tb^{3+}
Dysprosium	3		91×10^{-5}	460	Dy^{3+}
Holmium	1·2		2×10^{-4}	530	Ho^{3+}
Erbium	2·8		9×10^{-4}	690	Er^{3+}
Thulium	0·5		2×10^{-4}	1,800	Tm^{3+}
Ytterbium	3		8×10^{-4}	530	Yb^{3+}
Hafnium	3				
Tantalum	2		2×10^{-2}		
Tungsten	1·5	0·03	0·12	10^3	WO_4^{2-}
Rhenium	0·005		1×10^{-2}		ReO_4^-
Osmium					
Iridium					
Platinum					
Gold	0·004	0·002	5×10^{-3}†	$5·6 \times 10^5$	$AuCl_2^+$

Element					
Mercury	0·08	0·07	5×10^{-2}†	4.2×10^{4}	$HgCl_4^{2-}$
Thallium	0·45		1×10^{-2}		Tl^{3+}
Lead	12·5	3	3×10^{-2}†	2×10^{3}	Pb^{2+}, $PbOH^+$ $PbCl^+$
Bismuth	0·17		2×10^{-2}	4.5×10^{4}	?
Polonium			2×10^{-11}		
Astatine					
Radon	2×10^{-12}		0.6×10^{-12}†		$Rn(g)$
Francium					
Radium	4×10^{-7}		1.0×10^{-7}†		Ra^{2+}
Actinium					
Thorium	9·6	0·1	4×10^{-5}†	350	Hydroxy complexes ?
Protactinium			2×10^{-10}†		
Uranium	2·7	0·04	3·3	5×10^{5}	$UO_2(CO_3)_3^{4-}$?

* Combined nitrogen; ca 15 mg dissolved molecular nitrogen/l.
† Considerable variations occur.   ‡ Inorganic carbon.
§ The composition of river waters depends greatly on the geology of the source area.

The observed residence times range from $2 \cdot 6 \times 10^8$ years for sodium to about 100 years for aluminium (see Table 4.1) and reflect the great variations in geochemical reactivity which the elements exhibit in sea water. The alkali and alkaline earth metals of lower atomic number (except beryllium) have residence times of 10^6 years or more owing to their low geochemical reactivity. However, as the atomic number of the elements of these groups increases, the residence time decreases progressively owing to the high ion exchange affinity which these larger ions have for clay minerals.

Some elements such as aluminium, iron, chromium, titanium, beryllium and thorium have residence times of 100–1,000 years, and for them the assumption of mixing in time shorter than 3–4 times the residence time is not valid. The ions of these elements are rapidly hydrolysed at the pH of sea water and are incorporated into authigenic minerals such as ferro-manganese nodules and phillipsite. During continental weathering these elements tend to resist solution and thus pass into the sea as particulate matter, such as clays, feldspars, etc., which is soon deposited in the sediments. Because of the short residence times of such elements we may expect to find variations in their concentrations from one ocean to another. Goldberg and Koide (1962) have found indications of such a variation for thorium.

The comparatively short residence times of many transition elements (e.g. copper, 50,000 years; nickel, 15,000 years; cobalt, 18,000 years; and manganese 7,000 years) are a sign of the efficiency with which they are removed from solution during the deposition of ferromanganese minerals.

If no other removal mechanism was operative, it would be expected that the concentration of a cation in sea water would be regulated by the solubility of the least soluble compound which it forms with one of the anionic species present in the water. Krauskopf (1956) has investigated this problem by adding excess of the ions of ten metals to sea water samples maintained at pH $7 \cdot 8$–$8 \cdot 2$. After equilibrium had been attained, the precipitates were removed and the concentrations of the dissolved metal ions in the residual solution were determined. Calcium, strontium and barium were present at approximately their normal sea water concentrations. However, Pb, Ni, Co, Cu, Zn and Cd were all found to be undersaturated in sea water to an extent ranging from 10^2–10^7 fold in rough proportion to their residence times. Sillén (1963) has pointed out that most of the latter trace elements do not occur in marine sediments as discrete minerals but exist in solid solution in compounds of more abundant elements having a similar ionic radius and charge. For example, cobalt and lead may occur in solid solution in

a-goethite (as $(Fe,Co)OOH$) and manganese dioxide (as $(Mn,Pb)O_2$) respectively, both of which are common constituents of deep sea sediments. When such substitution occurs, the trace element will have an apparent solubility product many times lower than if it existed as a pure compound.

At the pH and pE (see p. 76) of sea water CoOOH would be the most stable phase existing in equilibrium with Co^{2+}, which is the most abundant species of this element in sea water. From the equilibrium constant for:

$$CoOOH + 3H^+ + e^- \rightleftharpoons Co^{2+} + 2H_2O, \log K = 29 \cdot 3 \ (25°C)$$

it can be shown that $\log c_{Co} = -6 \cdot 7$ whereas the measured value for sea water is ca $-9 \cdot 0$. The discrepancy between these values is probably accounted for by the fact that CoOOH only exists in the sediment in solid solution in FeOOH. The equilibrium concentrations of Co^{2+} must therefore be multiplied by the activity for CoOOH in the solid solution which is $\ll 1$ (Sillén, 1963). When allowance is made for these differences in activity there is reasonable agreement between the calculated and observed concentrations of many trace elements such as Pb and Co.

The presence of barite ($BaSO_4$) as a discrete mineral in marine sediments (see p. 357ff.), and the relative insolubility of barium sulphate suggests that the concentration of barium in sea water may be controlled by equilibrium with this solid phase. Chow and Goldberg (1960) have shown that although the barium concentration of sea water is somewhat variable, it lies fairly close to the saturation concentration when the effect of temperature and pressure on the solubility product of $BaSO_4$ is taken into account. In fact in some areas the *deep* water appears to contain barium at about the saturation level. Hanor (1969) has carried out a theoretical study of the system and has pointed out that the solubility of barium sulphate in sea water can be most conveniently thought of in terms of the molal equilibrium product,

$$(m_{Ba^{2+}})(m_{SO_4^{2-}}) = (Kd_{BaSO_4})(N_{BaSO_4})/(\gamma_{Ba^{2+}})(\gamma_{SO_4^{2-}})$$

where m_i is the molality of an ion i in sea water, N_{BaSO_4} is the mole fraction of the $BaSO_4$ in the barite and Kd_{BaSO_4} is its dissociation constant. When considering barite saturation it is therefore necessary to take into account, not only the formation of $BaSO_4^0$, but also the existence of other cation-sulphate pairs (see Table 4.3), which considerably reduces the free sulphate ion activity. In addition, since marine barite contains substantial amounts of celestite ($SrSO_4$) in solid solution (see p. 358) N_{BaSO_4} is decreased and this reduces the concentration of barium in equilibrium with this solid phase. Thus, sea water which is capable of dissolving barium from pure solid barium sulphate may be saturated with respect to natural barite.

III. CHEMICAL SPECIATION

In marine chemistry it is customary to define dissolved species as those which will pass a $0·5$ μm filter. Although this division between dissolved and particulate matter is convenient, it is quite arbitrary and it should be realized that "dissolved species" will also include most of the colloidal forms of the element.

A. PARTICULATE SPECIES

The suspended particulate matter in sea water is very variable in both composition and amount. Its inorganic fraction consists mainly of minerals such as feldspars, clays and quartz formed by weathering of terrestrial rocks (see Chapter 11). Other inorganic species which may be present are the siliceous and calcareous remains of dead organisms and authigenic minerals, produced by interaction of dissolved or colloidal species (see Chapter 12). The organic particulate matter in the oceans consists mainly of living organisms and their decay and metabolic products (see Chapter 8). The inorganic content of such particles may be quite high, and appreciable amounts of trace elements, such as iron and copper, may be adsorbed onto them or held in organic combination in them as chelates or in porphyrins. The waters of the open oceans away from the continental shelf and at depths of greater than 200 m contain little suspended matter (concentrations of $0·5$–250 μg/l, average ca 40 μg/l, have been recorded by Jacobs and Ewing, 1969). This material, which is predominantly in the <1 μm size range consists mainly of clay minerals. There is some evidence that the concentration of suspended matter may increase within a few hundred metres of the bottom in regions of rapid sedimentation owing to re-suspension of sediment by currents.

B. DISSOLVED SPECIES

(i) Activity coefficients

Non-ideal behaviour is encountered when attempts are made to apply the laws of chemical equilibrium to solutions of electrolytes, particularly when the ionic strength is high. The non-ideal behaviour arises because ionic solutes show powerful long-range coulombic interactions which cannot, even approximately, be ignored. In order to preserve the formal simplicity shown by thermodynamic expressions for ideal systems, the actual concentration, c, is replaced by the thermodynamic activity, a, where $a = c \cdot \gamma$, γ being termed the activity coefficient of the species in question. Both a and γ are functions of the composition and particularly of the ionic strength* and temperature. Although, in general, activities and activity coefficients can be assigned to molecular species, it is not possible to measure individual ionic activities or ionic activity coefficients, but only mean activities or activity coefficients for pairs of ions. However, in order to permit some sort of theoretical examination and systematic study of complex ionic solutions of the ionic strength and composition of sea water, values can be assigned to individual ionic activities by making certain assumptions. Thus, the activity of the hydrogen ion may be assumed to be given by applying the expression for pH (pH $= -\log_{10}a_H$) despite the fact that pH itself is only of operational significance and its meaning in solutions of high ionic strength is doubtful. If assumptions of this type are made, it is often possible to estimate other individual ionic activity coefficients. Table 4.2 shows the activity coefficients of the major ions in a sea water medium. Most of these values have been determined experimentally (mainly by potentiometric methods), but some of them have been deduced from measurements made in other media of similar ionic strength (Garrels and Thompson, 1962). Few values are available for the activity coefficients of the minor ions of sea water, and Krauskopf (1956) has suggested the use of the values 0·7 and 0·12 for mono and divalent cations respectively; for uncharged species and organic compounds a value of 1·13 should be used.

(ii) Dissolved inorganic species in sea water

The "dissolved" inorganic species present in sea water are principally electrolytes although uncharged species, such as H_3BO_3 and dissolved gases, are included in this category, together with colloidal micelles. Competitive complexing with the major anions (e.g. Cl^-, OH^-, HCO_3^-, SO_4^{2-}, F^- and perhaps organic ligands) is the principal factor controlling the nature of the inorganic species present in sea water. Until recently it has been difficult to obtain direct evidence about such species. However, the development of high sensitivity (e.g. pulse) polarographic methods now enables the speciation of the more abundant trace elements (e.g. zinc) to be studied (Branica *et al.*, 1969). Most of our knowledge of them has been derived from calculations based on measurements of stability constants of complexes in media much less complicated than sea water. The interaction between the cations and anions may range from purely electrostatic attraction, found in ion-pair aggregates,

---

* Ionic strength $(\mu) = \frac{1}{2} C_i z_i^2$ where C_i is the molal concentration of the ion and z_i is its charge.

to covalent bonding between the cationic group and donor groups (ligands). If the ligand is attached to the central atom by two or more bonds a *chelate* complex results. All these species will exist in solvated forms in which the hydration may range from simple ion-dipole interaction (e.g. with the alkali metals) to co-ordination (e.g. for

TABLE 4.2

Activity coefficients of dissolved species in sea water (temperature 25°C, chlorinity = 19⁰/₀₀ (ionic strength 0·7))

Species	Activity coeffic- ient	Reference
Na^+	0·68	Platford (1965a)
K^+	0·64	Garrels and Thompson (1962)
Mg^{2+}	$\begin{cases} 0·36 \\ 0·06 \end{cases}$	Thompson (1966) / Platford (1965b)
Ca^{2+}	0·21	Berner (1965)
SO_4^{2-}	0·12	Garrels and Thompson (1962)
CO_3^{2-}	0·022	Berner (1965)
HCO_3^-	0·56	Berner (1965)
$MgCO_3^\circ$, $CaCO_3^\circ$, $CaSO_4^\circ$, $NaHCO_3^\circ$	1·13	Garrels and Thompson (1962)
$NaCO_3^-$, $NaSO_4^-$, $CaHCO_3^+$, $MgHCO_3^+$, $\quad KSO_4^-$	0·68	Garrels and Thompson (1962)
F^-	0·39	P. G. Brewer (unpublished)

transition metals). Other ions or molecules present in the water may compete with water molecules for positions in the co-ordination sphere of the central metal atom.

The alkali and alkaline earth metals do not form strong complexes, and their tendency even to form ion-pairs (such as $NaCO_3^-$, $MgCO_3^\circ$) is rather limited. Garrels and Thompson (1962), Hanor (1969) and Kester and Pytkowicz (1969) have considered the occurrence of ion pairs of the major cations of sea water with sulphate, bicarbonate and carbonate; ion pair formation with the chloride ion is negligible. Except for the value for the Na—SO_4 pair, there is good general agreement between the results of all the authors. Sodium and potassium exist to the extent of ca 98% as simple cations (Table 4.3); only ca 1% occurring as ion pairs with sulphate. About 10% of the calcium and magnesium, 4·6% of the strontium, and 6·5% of the barium (Hanor, 1969) in sea water occur in the form of ion-pairs with sulphate. Considerable proportions of the sulphate, fluoride and carbonate in sea water occur as ion pairs mainly with magnesium, e.g. ca 50% of fluoride is present as MgF^+ (Brewer, *et al.*, 1970) and 2% is present as CaF^+ (Elgquist, 1970), and ca 20% of sulphate is paired with Mg (Fisher, 1967; Pytkowicz and Gates, 1968; see also Kester and Pytkowicz, 1969).

TABLE 4.3

Species of major cations and anions occurring at 25°C in sea water of chlorinity 19·0⁰/₀₀, pH 8·1 at 25°C under 1 atmosphere pressure (from Hanor, 1969)

| Cation | Molality | Molal percentage distribution of cations | | | |
		Free ion	Me—SO₄ pair	Me—HCO₃ pair	Me—CO₃ pair
Na^+	0·4823	99·0	1·0	0·0	0·0
K^+	0·01020	98·5	1·5	0·0	0·0
Mg^{2+}	0·05485	89·9	9·2	0·6	0·3
Ca^{2+}	0·01062	91·5	7·6	0·7	0·2
Sr^{2+}	0·00080	94·9	4·6	0·4	0·1

| Anion | Molality | Molal percentage distribution of anions | | | | | | |
		Free ion	Na—A pair	K—A pair	Mg—A pair	Ca—A pair	Sr—A pair	Ba—A pair
SO_4^{2-}	0·02909	62·9	16·4	0·5	17·4	2·8	0·0	0·0
HCO_3^-	0·00186	74·1	8·3	0·0	14·4	3·2	0·0	0·0
CO_3^{2-}	0·00011	10·2	19·4	0·0	63·2	7·1	0·0	0·0

At the pH of sea water, complexing by hydroxyl ion is important for many cations of oxidation number $+2$, or more, and is probably predominant for iron (III), titanium (IV), and zirconium (IV). Other anions compete with the hydroxyl ion, and their complexes are formed to an extent dependent on the stability constants. For a metal ion M (omitting its charge) forming a complex MA with the ligand A, the stability constant K_{MA} is given by $K_{MA} = a_{MA}/a_M \cdot a_A$. Thus, for complexing with F^-

$$\log a_{MF} a_M = \log K_{MF} + \log a_F$$
$$= \log K_{MF} - 4 \cdot 2$$

Estimated values for $\log K_{MA}$ at 25°C for some ligand-cation pairs in a non-complexing medium (such as $NaClO_4$ or KNO_3) of ionic strength similar to sea water are given in Table 4.4 (after Sillén, 1964). Where a major part of the metal is complexed by a ligand, the value given is in bold type, a value given in italics indicates somewhat weaker complexing. An asterisk shows that there is evidence for complex formation and a dash indicates that there is no evidence of complex formation. For one complex of a metal to predominate over another, for example the sulphate complex of a metal over the hydroxide form

$$\log K_{\text{MOH}} - \log K_{\text{MSO}_4} > \log a_{\text{OH}} - \log a_{\text{SO}_4} = -6 \cdot 0 - (-1 \cdot 6 - 0 \cdot 9) \approx -3 \cdot 5$$

Sulphate complexes are thus formed by several $2+$ ions such as Mg^{2+} and Co^{2+} which are complexed only weakly by hydroxyl ions. At the sulphate concentration prevailing in sea water ($\log c_{\text{SO}_4} = -1 \cdot 6$) ca 5–10% of these metals are present as sulphate complexes.

TABLE 4.4

Stability constants (expressed as log K, at 25°C) for ligand: metal ion pairs in a non-complexing medium of ionic strength similar to that of sea water (after Sillén, 1964)

Ligand	OH⁻	F⁻	Cl⁻	Br⁻	SO_4^{2-}	CO_3^{2-}
log c (ligand)	$-6 \cdot 9$	$-4 \cdot 2$	$-0 \cdot 3$	$-3 \cdot 1$	$-1 \cdot 6$	$-3 \cdot 7$
Be^{2+}	**−6**	**5·0**	1	—	0·7	*
Mg^{2+}	−12	1·3	0·2	—	*1·0*	*2*
Ca^{2+}	−13	0·5	—	—	*1·0*	*2*
Sc^{3+}	**−5**	**6·2**	*1·1*	1·2	*	*
Y^{3+}	−9	3·9	*0·4*	0·5	*2*	*
La^{3+}	−10	2·7	0		*1·5*	*7*
Th^{4+}	**−4**	**7·5**	0·3		*3·3*	*
UO_2^{2+}	**−6**	*4·5*	0	−0·3	1·8	*
Fe^{3+}	**−3**	*5·2*	*0·6*	0	2·3	
Cu^{2+}	*−8*	0·7	0	0	*1*	**5**
Ag^+	−12	0·2	**3**	*4*	0·2	
Zn^{2+}	−9	0·7	*0*	0	1	
Cd^{2+}	−10	0·5	**1·5**	1·6	*1*	*
Hg^{2+}	**−3·5**	1·0	**6·7**	**9·0**	1·3	
Al^{3+}	**−5**	*6·1*	—		*1·3*	
Pb^{2+}	*−8*	0·3	**1·0**	1·3	*1·3*	
Bi^{3+}	**−2**	*	*2·2*	2·3		

Chloride complexing is important only for a few cations which have an outer shell of 18 electrons, e.g. Au^{3+}, Ag^+, Hg^{2+}, Cd^{2+}, Zn^{2+}. These probably exist as $AgCl°$, $HgCl^+$, $HgCl_2°$, etc., but where the stability constants are sufficiently high, significant quantities of higher complexes (such as $AgCl_2^-$, $HgCl_3^-$, $HgCl_4^{2-}$ and $CdCl_2^\circ$) and even mixed complexes (e.g. $HgBrCl^\circ$) may be present. Carbonate complexing is probably only important for the uranyl ion and perhaps thorium.

The calculation of the relative proportions of the various species existing in sea water may be illustrated by reference to cadmium. Let $w = c_{CdOH^+}$; $x = c_{CdSO_4^\circ}$; $y = c_{CdCl^+}$; and $z = c_{Cd^{2+}}$. Take the stability constants for these complexes as given in Table 4.4. Assume activity coefficients of 0.7, 0.1 and 1.0 for monovalent ions, divalent ions and uncharged species. Then, since the total concentration of cadmium in sea water is 2×10^{-9} M

$$w + x + y + z = 2 \times 10^{-9} \text{ M}$$

$$K_{CdSO_4^\circ} = 10 = \frac{a_{CdSO^\circ}}{a_{Cd^{2+}} \cdot a_{SO_4^{2-}}} = \frac{x}{z(0.1)\,(2.84 \times 10^{-2})\,(0.1)}$$

$$x = 2.8 \times 10^{-3} z$$

$$K_{CdCl^+} = 32 = \frac{a_{CdCl^+}}{a_{Cd^{2+}} \cdot a_{Cl^-}} = \frac{0.7y}{z(0.1)\,(0.7)\,(0.536)}$$

$$y = 1.715z$$

$$K_{CdOH^+} = 10^{-10} = \frac{a_{CdOH^+}}{a_{Cd^{2+}} \cdot a_{OH^-}} = \frac{0.7w}{z(0.1)\,.\,10^{-6}(0.7)}$$

$$w = 10^{-17} z$$

Whence

$c_{Cd^{2+}} = 0.74 \times 10^{-9}$ M; $c_{CdSO_4^\circ} = 0.002 \times 10^{-9}$ M; $c_{CdCl^+} = 1.26 \times 10^{-9}$ M; $c_{CdOH^+} = 2 \times 10^{-26}$ M.

Thus, the $CdCl^+$ ion pair is the predominant species of cadmium in sea water; the ion pairs with sulphate and hydroxyl are insignificant. These deductions are supported by experimental evidence obtained polarographically (Barić and Branica, 1967). In contrast, the closely related element appears to exist principally as Zn^{2+} and $ZnOH^+$.

(iii) Redox potential and redox systems

For elements having more than one oxidation state the species present at equilibrium in sea water will be controlled not only by the pH but also by the redox potential of the system. If we consider the half cell reaction

$$Ox + ne^- \rightleftharpoons Red$$

$$\log K = \log a_{red} - \log a_{ox} + np\text{E}$$

Then $pE = pE° + n^{-1} \log(a_{ox}/a_{red})$ where $pE° = n^{-1} \log K$, and pE is a pure number, analogous to pH, which corresponds with the oxidizing power of the system at equilibrium (Sillén, 1965). This quantity is related to the more generally used redox potential, E_h(V), by the relationship:

$$pE = -\log[e^-] = \frac{E_h}{2 \cdot 3 \; RT/F} = \frac{E_h}{0 \cdot 05915} \text{ at } 25°C$$

However, for convenience and simplicity pE will be used in subsequent calculations.

In well buffered solutions containing a redox system which comes to equilibrium rapidly (e.g. $Fe^{2+} - Fe^{3+}$), E_h can be measured accurately using a platinum indicator electrode. However, in natural waters in which the concentrations of such redox pairs are low, the redox potential is controlled by the concentration of dissolved oxygen. Unfortunately, the reaction at the oxygen electrode is slow and is not completely reversible. As a result, traces of impurity on the electrode surface, or in the solution take over the electrode function. For this reason, measured E_h values in such systems are usually considerably lower than the theoretical values, and published E_h values should be regarded with caution (Garrels and Christ, 1965; Morris and Stumm, 1967; Whitfield, 1969). In oxygenated systems the most accurate values for E_h, and thus pE, are those calculated from the oxygen concentration, by considering the reaction $O_2(g) + 4H^+ + 4e^- \rightleftharpoons 2H_2O$ for which log $K = 83 \cdot 1$. At 20°C and pH $8 \cdot 1$ with water saturated with air (log $P_{O_2} = -0 \cdot 69$), since

$$\log K = 2 \log a_{H_2O} - \log P_{O_2} - 4 \log a_{H^+} - 4 \log a_e$$
$$83 \cdot 10 = -0 \cdot 01 \times 2 + 0 \cdot 69 + 8 \cdot 1 \times 4 + 4pE$$
$$\text{hence } pE = 12 \cdot 5$$

In normal well-oxygenated sea waters the values of pH and pE will not diverge much from the above figures. However, in poorly ventilated basins, organic matter falling from above will strip the water of oxygen and cause a *fall* in the value of pE; carbon dioxide produced during the reaction will cause the pH to fall also. If the water becomes completely deoxygenated (i.e. *anoxic*), sulphate-reducing bacteria will proliferate, and lead to the production of hydrogen sulphide, and to a further marked decrease in pE (see also p. 115).

(iv) Redox equilibria

To illustrate the principles used in redox calculations let us consider the oxidation state of thallium in sea water ignoring any possible complexes.

$$Tl^{3+} + 2e^- \rightleftharpoons Tl^+$$

the standard potential $(E°)$ for this reaction is $-1·247$ V at $25°$C. The redox potential (E) of sea water is $12·5 \times RT/F = 12·5 \times 0·05915 = 0·729$ V. From the Nernst equation,

$$E = E° + \frac{RT}{nF} \ln \frac{a_{Tl^{3+}}}{a_{Tl^+}}$$

where n is the number of electrons involved in the reaction. Thus,

$$0·729 = -1·247 + \frac{RT}{2F} \ln \frac{a_{Tl^{3+}}}{a_{Tl^+}}$$

$$\therefore \quad 1·976 = \frac{0·0591}{2} \log_{10} \frac{a_{Tl^{3+}}}{a_{Tl^+}}$$

i.e. $Tl^{3+}/Tl^+ \simeq 10^{100}$ and thallium should therefore be present entirely in the $3+$ oxidation state.

Similar calculations show that only vanishingly small concentrations of chlorate and bromate would occur at equilibrium with the chloride and bromide present in sea water. For example, in the equilibrium $ClO_3^- + 6H^+ + 6e^- \rightleftharpoons Cl^- + 3H_2O$, $\log K = 147·2$, from which it can be shown that at pH $8·1$ and pE $12·5$, $[ClO_3^-]/[Cl^-] \approx 10^{-23}$. However, for the corresponding reaction for iodine, $\log K = 110·1$, and at equilibrium iodate should be the predominant species $([IO_3^-]/[I^-] = 10^{13·5})$ (but see below p. 78).

In many instances, the position of equilibrium will be affected by solid phases present, and in the neighbourhood of the ocean floor turbulence may assist in bringing about reaction of minerals of the sediments with dissolved species. Thus, thermodynamic calculations suggest that manganese dioxide and not $MnOOH$ or $Mn(OH)_2$ should be the stable solid phase at pH $8·1$ and pE $12·5$, and this is borne out by X-ray diffraction studies of manganese nodules. Redox considerations suggest that the principal ionic forms of manganese under these conditions should be Mn^{2+} and MnO_4^-.

$MnO_2(s) + 4H^+ + 2e^- \rightleftharpoons Mn^{2+} + 2H_2O$, $\log K = 41·6$

$\log a_{Mn^{2+}} = 41·6 - 4pH - 2pE = 41·6 - 32·4 - 25·0 = -15·8$

$MnO_4^- + 4H^+ + 3e^- \rightleftharpoons MnO_2(s) + 2H_2O$, $\log K = 85·8$

$\log a_{MnO_4^-} = -85·8 + 4pH + 3pE = -85·5 + 32·4 + 37·5 = -15·6$

It would thus be expected that at equilibrium, sea water would contain approximately equal proportions of manganese in the $2+$ and $7+$ oxidation states. However, the equilibrium concentrations predicted are of the order of 10^9 times less than those actually found in the sea. It is doubtful whether such high concentrations of dissolved manganese could exist for long in the forms considered in the above calculation, and Sillén (1961) has postulated that the element is mainly present as uncharged hydroxide species, perhaps $Mn(OH)_3$ or $Mn(OH)_4$.

It is only occasionally that the form and oxidation states of trace element species in sea water can be verified. Unfortunately, in several such instances the species which predominate are not those which theory would suggest, and this casts doubt on the theoretical interpretation. Nitrogen shows the most important deviation. If equilibrium had been established in the reaction $4NO_3^- + 4H^+ \rightleftharpoons 2N_2(g) + 5O_2 + 2H_2O$, then, the major part of the nitrogen would be present in the sea as nitrate ion and not as dissolved gas. At equilibrium at pH 8·1 and pE 12·5, $p_{N_2} = 10^{-13}$ atm and $c_{NO_3^-} \approx 0.2$ M; and no other species would reach concentrations as high as 10^{-6} M (Sillén, 1965). In fact, $p_{N_2} = 0.758$ atm, and the concentration of nitrate ion never exceeds ca 30 μM. Sillén has suggested that a mechanism may exist in the upper layers of the sea by means of which nitrate is converted to molecular nitrogen. When the equilibrium $SeO_4^{2-} + 2H^+ + 2e^- \rightleftharpoons SeO_3^{2-} + H_2O$ (log $K = 29.7$) is considered, it can be shown that at pH 8·1 and pE 12·5, Se^{6+} should predominate over Se^{4+} in the ratio $1 : 10^{-11.5}$. However, recent work (Chau and Riley, 1965) has shown that the ratio is $1 : > 100$. Similarly, chromium should exist in oxygenated sea water as the CrO_4^{2-} ion, in fact, it appears to be partly present in the $3+$ oxidation state, presumably as hydroxide complexes (Fukai, 1969; see also Elderfield, 1970).

The large differences which are often found between theoretical models and actual systems may be caused by several factors.

(1) Equilibrium may not have been attained owing to the slowness of the reaction. In this connection it is interesting to note that an activation energy of 45 kcal is high enough to prevent any appreciable reaction in 10^6 years at 25°C. This energy is not large compared with the energies of bonds which may have to be broken in the rate-determining steps of reactions. For example, thermodynamic considerations suggest that potassium iodide in solution, or in solid form, should be spontaneously oxidized to iodate by oxygen, but it is in fact not oxidized at a detectable rate. The iodine species in the sea are evidently not in equilibrium since iodide represents ca $\frac{1}{3}-\frac{1}{4}$ of the total iodine present. Similar considerations may apply to the marine occurrence of chromium (III) since many of its reactions are notoriously slow.

(2) Biological activity may have led to reactions opposing the attainment of equilibrium particularly with micronutrient elements. Changes in pH and pE in the immediate vicinity of organisms and detritus may influence the equilibria with some trace elements.

(3) Photochemical processes occurring in the upper layers of the sea.

(4) Important species of the element in question having been neglected in the calculations.

(5) Equilibrium data being unreliable or incomplete.

(6) Incorrect interpretation of the analytical data for the actual system.

C. COLLOIDAL SPECIES

At the natural pH of sea water a considerable number of polyvalent ions (e.g. Fe^{3+}, Al^{3+}, Ti^{4+}, Zn^{4+}, Th^{4+}) are slowly hydrolysed and converted to complex colloidal hydroxy compounds, which may eventually coalesce and precipitate. Because of its biological importance, ferric-iron has been most widely studied, but investigations have been made difficult because of its extreme dilution. Biedermann and Chow (1966) have examined the hydrolysis equilibria of iron (III) species in 0·5 M sodium chloride medium using comparatively high iron concentrations ($> 0·001$ M). They found that the hydrolysis is a slow process and leads via a soluble species having one OH group/iron atom to precipitation of a very insoluble compound $Fe(OH)_{2·7}Cl_{0·3}$ which X-ray diffraction showed to be identical with the compound originally designated α-FeOOH (goethite), which is known to occur in marine sediments.

$$Fe^{3+} + 2·7H_2O + 0·30\ Cl^- \rightleftharpoons Fe(OH)_{2·7}Cl_{0·3} + 2·7\ H^+$$

The log solubility product of this compound is $-3·04$. They found no evidence of appreciable amounts of soluble hydrolysis products at equilibrium. Sea water which has been filtered through a 0·5 μm filter is usually found to contain ca 1–4 μg/l of iron. Since colloidal iron species dissolve only very slowly in the 0·05 N acid usually used in the determination of iron, it is likely that these traces of iron are present as the uncharged $Fe(OH)_3$ complex (Sillén, 1963).

D. ORGANICALLY COMPLEXED TRACE METALS IN SEA WATER

Ocean water contains only small amounts of organic compounds, such as humic acids and amino acids, which are capable of forming even weak chelate complexes with trace metals at pH 8. It is therefore likely that chelates are of only minor importance in speciation in oceanic waters (Williams and Solorzano, 1967). Thus, less than 0·1% of the boron present is complexed by the dissolved carbohydrate (Williams and Strack, 1966), and even ferric iron, which forms strong complexes with chelating agents, is present in organically bound form in ecologically insignificant proportions. (Williams, 1969). There is evidence to suggest that organically bound copper and zinc occur in the sea, but these may well be present as colloidal species, in combination with, for example, lipids, rather than in true solution (Williams, 1969; Longerich and Hood, 1968). In addition, porphyrin complexes remaining after the decay of marine plants and animals may have a more than transitory existence before being broken down further. Although the proportions of chelated trace metals are probably insignificant in oceanic waters, they may not be so in inshore and especially estuarine waters which frequently contain quite high concentrations of dissolved organic compounds.

Laboratory studies have shown that many marine organisms are capable of obtaining their supplies of essential trace elements by breaking down chelates formed with EDTA. In the preparation of plankton cultures it is thus helpful to add trace elements in chelated form to prevent hydrolysis and precipitation.

IV. MAJOR ELEMENTS

A. CONSTANCY OF RELATIVE IONIC COMPOSITION OF SEA WATER

As long ago as 1819, Marcet presented a paper to the Royal Society in which he suggested that "specimens of sea water contain the same ingredients all over the world, these bearing very nearly the same proportions to each other, so that they differ only as to the total amount of their saline contents". The extensive study by Dittmar (1884) of 77 samples of sea water, collected during the circumnavigational cruise of H.M.S. *Challenger* (1873–77), confirmed the general truth of this observation. However, Dittmar found some variations in the ratios of the major ions to chloride which were greater than could be accounted for by analytical errors; in particular, the calcium/chloride ratio was higher in deeper waters than in surface waters. Since Dittmar's time there has been a growing tendency to consider that the relative composition of sea water is constant.

A reinvestigation of the problem has been carried out recently by workers at the British National Institute of Oceanography and at the University of Liverpool (Cox and Culkin, 1966; Morris and Riley, 1966; Riley and Tongudai, 1967). Well over one hundred samples representative of the major seas and ocean basins were analysed by modern analytical methods. The average results for ocean waters are given in Table 4.5. If land-locked seas (such as the Baltic and Black seas) are ignored, it is found that within the limits of analytical errors, there are no significant variations of the ratios of sodium, potassium, sulphate, bromide, strontium* and boron to chlorinity; these elements are

---

* However, Mackenzie (1964) and Angino *et al*. (1966) claim to have found considerable variations in the Sr/Cl ratio in sea water. Only slight variations have been found by Andersen and Hume (1968) and Andersen *et al*. (1970) who used more reliable analytical procedures.

said to be conservative. However, the calcium/chlorinity ratio is significantly (up to 0·5%) higher in deep and intermediate waters than in surface waters. This enrichment probably arises because these waters are cold and rich in carbon dioxide and actively dissolve the calcium carbonate constituting the frustules of dead plankton falling from the surface. Significant variations also occur in the magnesium/chlorinity ratio, but further investigation is necessary before their cause can be established.

TABLE 4.5

Composition of ocean water

Ion	g/kg of water of salinity $35^0/_{00}$	g/kg/Cl $^0/_{00}$
Chloride	19·344	—
Sodium	10·773*	0·5561*
Sulphate	2·712	0·1400
Magnesium	1·294*	0·0668*
Calcium	0·412*	0·02125*
Potassium	0·399*	0·0206*
(Bicarbonate	0·142	—)
Bromide	0·0674	0·00348
Strontium	0·0079*	0·00041*
Boron	0·00445†	0·00023†
Fluoride	0·00128†	$6·67 \times 10^{-5}$†

* Average of mean results of Cox and Culkin (1966) and Riley and Tongudai (1967).
† Greenhalgh and Riley, unpublished results.

B. CONDITIONS UNDER WHICH MAJOR ELEMENTS MAY NOT BE CONSERVATIVE

Although the major elements in ocean water are usually conservative or almost so, considerable variations in their concentration ratios may be found in the marine environment, under some atypical conditions.

(i) Estuaries and land-locked seas

The composition of river water is largely controlled by the nature of the rocks in which it originates. As a consequence, considerable variations may be found in the ionic concentration ratios from one river to another, although the total ionic concentration rarely exceeds 200 mg/l. Characteristically, the SO_4 : Cl, HCO_3 : Cl, K : Na, Mg : Na and Ca : Mg ratios are generally much greater in river water than in sea water (Table 1.1). In land-locked seas, such as the Baltic and Black Seas, run-off from land may affect the composition of the water appreciably.

Thus, in the water of the central regions of the Baltic ($Cl = 3\cdot16^0/_{00}$) the SO_4 : Cl ratio may rise as high as $0\cdot1455$ (normal ratio $0\cdot1400$) (Kwiecinski, 1965). Gripenberg (1937) has been able to characterize four different regions of the Baltic according to their Ca : Cl ratios. In contrast, the Mg : Cl ratios do not differ appreciably from that of ocean waters, perhaps because the fresh water reaching the Baltic is poor in magnesium, whereas it is rich in calcium.

Variations in the concentration ratios of the major elements in estuarine waters of appreciable salinity are usually not detectable, since the quantities of dissolved solids in river water are relatively small.

(ii) Anoxic basins

In the deep waters of certain basins (e.g., the Black Sea, Cariaco Trench and some Norwegian fjords) the oxidative decomposition of organic matter raining from above exhausts the dissolved oxygen and causes the redox potential to fall drastically (see p. 76). Such conditions are ideal for the proliferation of the sulphate-reducing bacteria, which are attached to sediments. These bacteria are active in converting the sulphate present in these waters into sulphide, and this leads to a decrease in the SO_4/Cl ratio; ratios as low as $0\cdot1360$ have been found at 2,000 m in the Black Sea (Skopintsev et al., 1958). Some sulphur may be lost from these waters by deposition as pyrites (FeS_2).

(iii) Freezing

Some fractionation of the major ions takes place during the freezing of sea water. Thus, for example, sea ice retains proportionately more sulphate than chloride, and as a consequence the ratio is lowered in the water. This effect has been observed in the North Pacific where the water of the minimum temperature layer has been found to be relatively deficient in sulphate (Fukai and Shiokawa, 1955).

(iv) Precipitation and dissolution of carbonate minerals

In warm, shallow waters, where rapid chemical or biological pre-cipitation of calcium carbonate is occurring, sufficient calcium may be withdrawn from the water to affect the Ca/Cl ratio significantly. Such conditions are found on the Bahama Banks where aragonite is being precipitated in large quantities, and variations in the Ca/Cl ratio of as much as $\pm3\%$ have been reported (Traganza and Szabo, 1967). Solution of calcium carbonate from the walls of isolated rock pools, under influence of high P_{CO_2} values induced by biological respiration, leads to increase in the Ca/Cl ratio of the water. Both of these mech-anisms may also perhaps, to a lesser extent, affect the Mg/Cl and Sr/Cl ratios.

(v) Submarine vulcanism

The injection of molten volcanic magma into sea water probably has little effect on the concentration ratios of the principal ions in the water, although it greatly increases the concentration of dissolved silicon (Stefansson, 1966). However, it is quite probable that the high fluoride/chlorinity ratio ($8 \cdot 0$–$9 \cdot 0 \times 10^{-5}$; normal ratio $6 \cdot 7 \times 10^{-5}$) found in the deep waters in a few localities close to the mid-Atlantic Ridge has been produced by injection of volcanic gases (Riley, 1965).*

(vi) Admixture with geological brines

It is likely that highly saline water enters the ocean through fissures in the ocean floor in a few places, of which the Red Sea provides the best known example. Basins ca 200 m deep, containing hot, highly saline water (temperature 45–58°C; salinity 255–326⁰/₀₀) have been found, at a depth of 2,000 m, in its central region (see Degens and Ross, 1969, and Riley, 1967).† The major element/chlorinity ratios in this water are appreciably different from those in normal sea water; magnesium, sulphate, fluoride, bromide and boron being particularly deficient (Table 4.6). Although the boundary between the dense brine ($d^{45°} = 1 \cdot 19$) and the normal Red Sea Bottom Water is quite sharp, there is a narrow mixed layer between the two waters suggestive of upward transport of the dissolved salts.

TABLE 4.6

Ratios of concentrations of major elements/chlorinity of water from the Discovery Deep in the Red Sea and in normal sea water (Brewer et al., 1965)

	Na⁺	K⁺	Ca²⁺	Mg²⁺	Sr²⁺ ×10³	SO₄²⁻	Br⁻ ×10³	F⁻× 10⁵	B× 10⁴
Discovery deep water	0·5978	0·0139	0·0303	0·0052	0·29	0·0048	0·78	0·03	0·5
Normal sea water	0·5561	0·0206	0·0213	0·0668	0·41	0·1400	3·48	6·67	2·3

---

* Brewer *et al.* (1970) have suggested that the excess fluoride may be present in a colloidal form, since measurements with a fluoride-specific ion electrode have shown only the same fluoride activity as surface samples.

† The deuterium and oxygen-18 concentrations in the brine are similar to those of sea water in the vicinity of the comparatively shallow sill at the southern end of the Red Sea. Craig (1969) has postulated that in this region sea water penetrates, via the sea floor, into the thick evaporite beds which underlie the sediments on the floor of the Red Sea. Dissolution of salt from the evaporites increases the salinity of the water and the high geothermal heat flux raises its temperature. The high density of the brine, relative to normal sea-water causes it to sink and flow northwards, emerging 1,000 km further north at a depth of 2,000 m in the brine filled deeps in the central rift of the Red Sea. This hypothesis receives confirmation from data on the ratios of dissolved nitrogen and argon. It has been suggested (Craig, 1969) that the low SO_4^{2-} : Cl ratio in brine results from removal of much of the sulphate of the original sea water by selective membrane filtration through clay minerals (see p. 86).

(vii) Evaporation of sea water in isolated basins

If sea water becomes trapped in a basin separated from the sea by a very shallow bar and if the climate is hot and arid, the conditions are favourable for evaporation and the formation of beds of evaporite minerals. Although, at present, such combinations of conditions are found only in a few regions, they must have been common in the past, as witness the vast evaporite beds found in many parts of the world (e.g. Stassfurt Deposits).

The equilibria attained during the evaporation of sea water were first studied by van't Hoff (1912) and have been extensively reviewed by Borchert (1965). The crystallization of salts during the progressive evaporation of sea water can be divided into four principal stages (Table 4.7). When the crystallization is carried out in the laboratory, Stage IV leads first to the precipitation of Na—K—Mg sulphates and finally K—Mg chlorides crystallize from the residual brine; the actual minerals formed depend on the temperature used in the evaporation. In nature, the concentrated brine becomes deoxygenated by the organic matter which it contains. This leads to proliferation of sulphate-reducing bacteria and thence to loss of sulphate from the brine by conversion to volatile hydrogen sulphide. As a consequence, natural marine evaporite beds are almost devoid of sulphate minerals, and the most important solid phase deposited in Stage IV is carnallite $KCl . MgCl_2 . 6H_2O$.

TABLE 4.7

Phases formed during the progressive evaporation of sea water

Stage No.	Density of brine	Weight % of liquid remaining	Principal solid phases deposited	% of total dissolved solids
	1·026	100		
I	1·140	50	Calcium carbonate+dolomite	1
II	1·214	10	Gypsum ($CaSO_4 . 2H_2O$)	3
III	1·236	3·9	Halite (NaCl)	70
IV	—	—	Sodium-magnesium-potassium sulphates and chlorides	26

(viii) Exchange of major ions between sea and atmosphere or sediments

It is of interest to examine the chemical processes occurring when sea water is dispersed into the atmosphere or when it forms the interstitial water of marine sediments. Both of these interactions lead to fractionation of major elements, and, presumably also of minor elements, although the latter have not yet been investigated in any detail.

(a) *Exchange of ions between the sea and the atmosphere.* Bubbles bursting at the surface of the sea disperse ca 1×10^9 tons of the dissolved ions into the atmosphere each year (Eriksson, 1959). These ions are eventually returned to the sea either directly or indirectly via precipitation, together with a contribution from the land. Considerable fractionation of the ions occurs during their transfer from sea to atmosphere (see e.g. Sugawara, 1965), probably because of surface effects involving organic matter. When bubbles are formed in the sea they tend to gather dissolved and particulate organic matter from the water and carry it to the surface where it forms a film of a very complex nature. This surface-active organic rich layer takes up certain ions selectively. Under the action of even moderate winds white-cap bubbles will be formed and when these burst they will disperse both the ions and the organic matter* into the atmosphere in the form of an aerosol. Depending on the humidity, the areosol exists in the atmosphere either as solid particles or droplets in the size range $0 \cdot 1$–$20 \ \mu m$. These *salt particles* may play an important part in the nucleation of rain clouds. Their principal component ions are sodium, chloride and sulphate. However, the concentration ratios of these and their other component ions differ from those in sea water. Ions such as calcium, potassium, magnesium and sulphate are enriched (relative to sodium) in the atmosphere, whereas chloride and bromide are slightly impoverished. It is likely that this mechanism will lead to the concentration of transition metals into the atmospheric phase because of the strong chelates which many of them form with organic matter (see also p. 210).

The atmospheric chemistry of the halogens poses a number of unsolved problems. The Br/Cl ratio is only slightly higher in the air than in the sea and it is thought that both elements are injected into the atmosphere via the bubble breaking process (Duce et al., 1965). Secondary processes evidently occur, since considerable proportions of both elements exist in the atmosphere as gases. The gaseous form of chlorine may be hydrogen chloride, formed following reaction of the salt particles with sulphur trioxide or nitrogen dioxide. Photochemical oxidation of bromide in the salt particles may give rise to bromine (Duce et al., 1965). The I/Cl ratio in the atmosphere is many times greater than in the sea. Most of the atmospheric iodine occurs in gaseous form, probably as the free element formed either by photochemical oxidation of iodide at the

---

* Surface active organic matter has been observed in the aerosol on the windward side of breaking surf (Blanchard, 1964; see also Barger and Garrett, 1970).

sea surface or by decomposition of an iodine-rich organic film at the surface. The rate of loss of iodine from the ocean by the former mechanism has been estimated to be 4×10^{11} g/year (Miyake and Tsunogai, 1963).

Boric acid is appreciably volatile, and it is probable that the relative enrichment of boron in rainwater results from distillation of this compound from the surface of the sea.

(b) *Interstitial waters in bottom sediments.* The concentration ratios and absolute concentrations of a number of ions in the interstitial waters of marine sediments (see also p. 405) may differ appreciably from those in sea water. The chlorinity of these waters usually lies within $\pm 1^0/_{00}$ of the overlying sea water, but significant variations occur, both down the sediment column and from one locality to another. It is thought (Siever *et al.*, 1965) that these differences are not effects of paleosalinity, but are caused by the montmorillonite and other clays constituting the sediment, acting under pressure as semi-permeable membranes. By this mechanism ions are excluded while water itself is allowed to pass upwards.

Changes in the element/chlorinity ratios in the interstitial waters may be produced by interaction with the minerals of the sediments. Magnesium has been found to be slightly depleted owing to uptake by chlorite and perhaps to reaction with calcium carbonate to form dolomite. In contrast, potassium tends to be enriched because of the hydrolysis of feldspars. The waters tend to be either at equilibrium or slightly undersaturated with respect to calcite, unlike surface sea water.

V. Variations of the Hydrogen and Oxygen Isotope Ratios in the Water of the Oceans

In addition to the principal isotopes ^1H and ^{16}O, naturally occurring hydrogen and oxygen contain small proportions of a number of heavier isotopes, e.g. ^2H, ^3H, ^{17}O, ^{18}O. Deuterium (^2H,D), oxygen-17 and oxygen-18 are present to the extent of $0 \cdot 0156\%$, $0 \cdot 037\%$ and $0 \cdot 20\%$ respectively. The main molecular species of water are therefore $H_2{}^{16}O$, HDO and $H_2{}^{18}O$ which are present at equilibrium in the approximate molar ratio $10^6 : 320 : 2,000$. The differences in physical properties of these species leads to fractionation in the marine environment. This fractionation is much more pronounced for deuterium than for oxygen-18.

Several investigations have been made with a view to using these variations to trace water masses since, unlike salinity, the isotopic composition is a tracer for the *water* component of the sea (see Yoshio and Nobuko, 1968). Craig and Gordon (1965, 1966) have used such data in conjunction with salinity measurements to confirm and augment classical theories of the origins of deep water masses. Thus, they were able to show that the North Atlantic Deep Water was of convective origin and to demonstrate the role of freezing in the formation of the bottom water of the Weddell Sea. The deep waters of both the Pacific and Indian Oceans differ in isotope-salinity relationships from their surface waters; they are probably derived from mixing of North Atlantic Deep Water and Antarctic Bottom Water with Intermediate Water from the Weddell Sea.

Mass spectrometry provides the only satisfactory means of determining deuterium and oxygen-18 in water. Before the deuterium determination is carried out, the hydrogen is liberated from the sample by reaction with uranium. Oxygen-18 is determined by examination of the $C^{16}O^{18}O$ peak of carbon dioxide which has been equilibrated with the sample. It is not possible, at present, to determine the absolute isotope abundances, but measurements must be made relative to a standard. Owing to the variations in the D/H and $^{18}O/^{16}O$ ratios of the tap waters used as standards it is difficult to compare the results of earlier workers. For this reason it is now customary to use as standard the National Bureau of Standards reference distilled water sample NBS–1. In oceanographic work it is usual to express the deuterium or ^{18}O per mille enrichment (δ) relative to "standard mean ocean water" (SMOW) and Where the R values are the respective isotopic ratios.

$$\delta = \left[\left(\frac{R_{sample}}{R_{SMOW}} - 1 \right) \times 1,000 \right]$$

The (SMOW) has been defined (Craig, 1961a) in terms of NBS–1 so that

$$D/H(SMOW) = 1 \cdot 050 \, D/H(NBS–1)$$
$$^{18}O/^{16}O(SMOW) = 1 \cdot 008 \, {}^{18}O/^{16}O(NBS–1)$$

A. PROCESSES AFFECTING THE ISOTOPIC COMPOSITION OF SEA WATER

The processes producing isotopic variations in surface waters are those which also lead to salinity variations, namely mixing, precipitation, evaporation and, to a lesser extent, freezing. Since isotopic species differ in their vapour pressures and freezing points they are transported at different rates through the water cycle. This means that the isotopic composition of sea water is not a constant function of salinity in all parts of the oceans; regions of similar salinity may, in fact, have quite different values. However, very similar correlations of δ and salinity may be produced in two regions of the sea by quite different processes, e.g. by evaporation and precipitation in equatorial regions and freezing in the Antarctic. There is a good linear correlation between the δD and ^{18}O values in the sea, precipitation and polar melt waters (Craig, 1961b).

During the freezing of water, deuterium and ^{18}O become enriched in the ice to the extent of $20^0/_{00}$ and $2^0/_{00}$ at equilibrium. However, the effect of freezing on the isotope ratios in the sea appears to be only local. Even in the Antarctic the effect of precipitation hides the isotope fractionation caused by meltwater (Craig and Gordon, 1965).

The saturation vapour pressures of both $HD^{16}O$ and $H_2^{18}O$ are lower than that of $H_2^{16}O$, and in precipitation these species concentrate in the liquid phase, and the atmospheric water vapour becomes progressively enriched with the lighter isotopes. It is found that the mean precipitation over the sea in tropical and sub-tropical regions with high rainfall has $\delta^{18}O$ values in the range 0 to $-5^0/_{00}$. In temperate regions, the values range from -5 to $-15^0/_{00}$ and decrease to even lower values $(< -20^0/_{00})$ in polar areas.

Evaporation probably exerts a stronger influence over the isotope ratios than precipitation, and results in the concentration of the heavier, less volatile isotopes in the surface layer. Relative humidity of the atmosphere is an important factor controlling the isotopic enrichment particularly at high salinities, high humidity tending to reduce the ratios (Lloyd, 1966). The mechanism of concentration of deuterium and oxygen-18 during evaporation is complex, and for a detailed account of the marine chemistry of these isotopes the reader should consult the monograph by Craig and Gordon (1965).

VI. MINOR ELEMENTS

Unlike the major elements, many of the minor elements are not conservative in sea water. This is partly because of their greater geochemical and biological reactivity, and partly because processes involving adsorption and biological uptake will produce relatively much more effect on elements occurring at low concentrations. In addition, unusual localized concentrations of these elements may be produced by run-off from land and by volcanic action.

Much of the data on the occurrence of trace elements in sea water is unreliable owing to the use of insufficiently tested analytical methods. Furthermore, few of the published analyses are those of waters from below the thermocline which are less affected by biological processes, and which are more representative of the oceans as a whole. Selected data on the average abundances of trace elements in ocean water are presented in Table 4.1. However, it should be realized that considerable variations from these values may be encountered in the sea. Systematic studies of geographical distribution have so far been carried out only for a few trace elements, but interesting patterns have been revealed. It is generally observed that the concentrations of most trace-elements increase with depth (see p. 94). As befits the fact that the Atlantic, north of 10° S. receives 65% of the total dissolved matter supplied to the world's oceans by rivers, its surface and intermediate waters contain greater concentrations of some trace elements than those of the Pacific. However, the deep waters of the Pacific, because of their stability, act as a sink for trace metals.

A. FACTORS AFFECTING THE DISTRIBUTION OF TRACE ELEMENTS IN THE SEA

Owing to the high geochemical reactivity of many trace elements in the sea it is convenient to consider the factors influencing their distribution separately from those affecting the conservative elements. This high reactivity is manifested particularly in the manner in which they are accumulated by authigenic minerals and by marine organisms.

(i) Processes by which trace elements are introduced into the sea

Since the concentrations of most trace elements in the sea are of the order of micrograms or less per litre, the introduction of even quite small amounts of them by geological or human agencies may produce detectable local variations.

Run-off from land is probably the principal pathway by which trace elements reach the sea. The trace element contents of such waters are controlled by the nature of the rocks of the catchment area and may vary greatly. Trace elements are present not only in dissolved form in stream waters, but also occur adsorbed on the finely divided suspended detrital matter. When these particles enter the sea a large proportion of their adsorbed trace metals is displaced by the action of the major cations of the sea water (Kharkar et al., 1968). High concentrations of certain trace elements may be noted many kilometres seawards of the mouths of rivers which have flowed through ore bearing formations, for example, arsenic at twice its normal sea water concentration is found 10 km seawards of the mouth of the River Plym which flows through the Cornish ore-bearing rocks.

Trace elements may enter the sea by leaching of the rock "flour", which enters the Southern Ocean in enormous quantities as the result of glacial attrition in Antarctica. This mechanism has been cited by Schutz and Turekian (1965b) as a possible explanation of the enrichment which they found for Co, Ni and Ag in the waters of moderate depth south of 68° S.

Injection of trace elements may also occur through the action of submarine volcanoes. The relatively high concentrations of cobalt, nickel and barium found in the water in certain regions of the Pacific have been attributed to this source. (Schutz and Turekian, 1965b; Turekian and Johnson, 1966). The enrichment of Fe, Mn, Cu, Cr, Ni and

Pb in the sediments of the equatorial parts of the East Pacific Rise has been attributed (Bostrom and Peterson, 1966) to precipitates formed from trace elements introduced by solutions of magmatic origin. Many trace metals such as Zn, Mn and Cu are enriched in geological waters, and where these issue from the sea floor local trace metal enrichment will occur (e.g. the Red Sea hot brine (p. 83) contains 1·1 mg Mn/l and ca 1 mg Zn/l).

The marine distribution of lead has been considerably modified by human activities. It has been estimated (Tatsumoto and Patterson, 1963) that the rate of introduction of lead into the sea has increased 27-fold since the Pleistocene period. Part of this has been caused by increased leaching of the land caused by the development of agriculture, and later by industrial activity. However, by far the greatest increase has taken place in the last four decades as the result of the use of lead tetraethyl as an additive to petrol. In all, ca $2·5 \times 10^{11}$ g of lead are washed out of the atmosphere into the oceans annually. Owing to the comparatively slow rate of replacement of the mixed surface layer of the oceans, much of the lead added from the atmosphere is present in this layer which at present contains ca 0·2 μgPb/l, in contrast with the 0·03 μg/l present in the deeper waters. This distribution contrasts markedly with that of many other heavy metals, such as Ba, Hg, Co, which are enriched in the deep waters by biological mechanisms (see p. 94).

(ii) Interaction between trace elements and precipitates

The geochemical balances of many of the trace elements in the sea are controlled by adsorption onto inorganic precipitates which are formed in the sea and which are later incorporated into the sediments. The elements concerned are those with relatively short residence times. The nature of the precipitates formed is controlled by the redox potential and pH prevailing. Since pH varies only slightly in the marine environment, it is appropriate to consider uptake of trace elements by solid phases formed in both oxidizing and reducing environments. These processes are best studied by considering their effect on the compositions of the underlying sediments.

(a) *Oxidizing environments.* Manganese and iron at sub-micro molar concentrations are removed from the ocean during the formation of the ferro-manganese nodules and micro-nodules, which are widely dispersed over the deep ocean floor, especially that of the Pacific (see p. 360ff.). During their slow growth at a rate of a few atomic layers per day, these ferromanganese minerals accumulate a range of trace elements from the water by adsorption. When comparison is made between their concentrations in near-continental sediments, it is found that elements such as Cu, Ni, Co, Pb, Zn, Cd, Mn, Mo, W, Th and Tl are strongly concentrated in the nodules, and in the micro-nodules contained in the deep ocean sediments. However, readily hydrolysable elements such as Ti, Ga, Cr, and Zr are not appreciably enriched. It appears probable that this concentration process and perhaps adsorption processes involving clay minerals are important factors controlling the geochemical balance of the former group of elements in the oceans.

There is evidence that under marine conditions hydroxy-apatite ($Ca_5(PO_4)_3OH$) is the only stable form of calcium phosphate. Solubility product calculations suggest that at equilibrium with this solid phase sea water would contain rather more than $10^{-6.5}$ M of phosphorus. This is in reasonable agreement with the concentrations of phosphorus actually found in the deep oceans.

Precipitation of thorium and protactinium with their very short residence times leads to the oceans not being in radioactive equilibrium with respect to the decay series of either uranium–235 or 238. Daughters formed by radioactive decay of thorium isotopes are only slowly returned from the sediments to the bottom waters and this apparently accounts for the enrichment of the deeper waters with radium (p. 97).

(b) *Reducing environments.* When the rate at which oxygen is consumed by oxidation of organic matter present in the water exceeds the rate at which it is supplied by currents, reducing conditions are set up which may result in the formation of hydrogen sulphide. Anoxic conditions are found in regions where the bottom water is dense and stagnant, particularly if the productivity of the overlying surface water is high (see p. 76). Anoxic waters commonly contain relatively high concentrations of iron and manganese (both in the 2+ oxidation state) which have sulphides that are comparatively soluble at pH 6–7.

Metals such as copper and lead, which have sulphides with low solubility products are generally present at extremely low levels, as they are precipitated and thereby become enriched in the sediments. Thus, the sediments of the edges of the anoxic basins of the Baltic Sea consist of black organic-rich clay enriched in Zn, Cu, Ag, Mo and U. The sediment lying at the centres of these basins consists of a solid solution of manganese and calcium carbonates and contains only low concentrations of trace elements, perhaps because the water which reaches the centres of the basins is impoverished in these elements (Manheim, 1961).

Extensive deposits of phosphorite nodules are found off the West African and Southern California coasts (see p. 353). In the regions where these deposits are found, upwelling causes the productivity in the surface layer to be high. When the organic detritus which accumulates in the sediment decomposes under the influence of a restricted supply of oxygen, phosphate is liberated and the conditions of low pH and E_h which are thought to be necessary for phosphate deposition result. The deposits are composed of skeletal apatite together with authigenic francolite $Ca_5(F,PO_4,CO_3)_3$ (Arrhenius, 1963). The latter acts, by both cationic and anionic substitution, as host for a wide range of trace elements, such as Zr, Mn, Pb, V, U and rare earth elements which are taken up during its growth. The presence of zirconium (up to 0·3%) and rare earth elements in francolite considerably reduces its solubility and prevents it being redissolved like skeletal apatite. The mineral contains substantial amounts of uranium, mainly in the 4+ oxidation state, substituting for calcium in the lattice. This adds confirmation to the belief that francolite is formed under strongly reducing conditions.

(iii) Interaction of trace elements with marine organisms

Living cells are able to take up elements from solution against a concentration gradient. This is well demonstrated by marine organisms, many of which contain trace elements at concentrations as high as 10^6 times their sea water concentrations. The concentration of trace elements by marine organisms has been reviewed by Bowen (1966) who has pointed out that:

  (i) Chlorine is definitely rejected;

 (ii) Na, Mg, Br, F and S have concentration factors of the order of 1;

(iii) Other elements (with the exception of the noble gases) are more strongly concentrated in living tissue;

(iv) The orders of affinity of cations for organisms is 4+ and 3+ elements >2+ transition metals >2+ Group IIA metals >1+ Group I metals.

The order of affinity of the plankton and brown algae for the 4+ and 3+ groups and for the 2+ transition metals are somewhat different.

e.g. Plankton Fe >Al >Ti >Cr,Si >Ga

                Zn >Pb >Cu >Mn >Co >Ni >Cd

Brown Algae Fe >La >Cr >Ga >Ti >Al >Si

                Pb >Mn >Zn,Cd >Co >Ni

In contradiction to the suggestion of Goldberg (1957), the order of the concentration factors does not agree well with the order of stability of the chelates formed with ligands (the Irving-Williams order)

$$Mn^{2+} < Fe^{2+} < Co^{2+} < Ni^{2+} < Cu^{2+} > Zn^{2+}$$

(v) In general, the heavier elements in a particular Group of the periodic table are taken up more strongly than the lighter;

(vi) The affinity for anions increases with increasing ionic charge, and in related species, with increasing atomic weight of the central atom, viz.

$$SO_4^{2-} < MoO_4^{2-} < WO_4^{2-} \text{ and } F^-(0\cdot86) > Cl^-(0\cdot062)$$
$$< Br^-(2\cdot8) < I^-(6,200)$$

(where the figures in parentheses denote the average concentration factor).

The behaviour of fluorine appears to be exceptional.

Nitrate and phosphate are very strongly concentrated.

Usually the lower organisms concentrate trace elements more strongly than do higher ones. The power to concentrate a particular element or group of elements is often an attribute of a particular family of organisms, or even of a particular species. Thus, the sponge *Dysidea crawshayi* accumulates titanium whereas *D. etheria* does not. Vanadium is concentrated by certain ascidians (e.g. *Ascidia ceratodes*) which have a vanadium porphyrin complex as a blood pigment. Some individuals of *Molgula manhattensis*, a member of the same family, also concentrate vanadium, but others concentrate only the related element niobium. Another ascidian *Eudiostoma ritteri* contains considerable amounts of titanium. The trace element contents of fish and molluscs vary considerably from one organ to another. In general, the transition and d^{10} metals are concentrated most strongly in the digestive and renal organs, and to a much smaller extent in the flesh, heart, muscles, gonads and shell. For example, the concentration factors for cadmium in the shell, muscle and digestive gland of the scallop *Pecten maximus*, have been found to be 10, 16,000 and 5×10^6 respectively (Mullin and Riley, 1956).

The sites of uptake of trace elements appear to be the mucus surfaces (e.g. those of the digestive glands or gills). These contain macromolecules such as glycoproteins, which have the ability to form fairly strong chelates with transition metals and thus to remove these metals fairly efficiently from sea water. Many organisms, particularly filter feeders, are able to satisfy their trace metal requirements by chelating the adsorbed metals present on marine particulate matter. The ability of organisms to concentrate specific trace elements may be associated with the presence of glycoprotein molecules possessing selective chelating powers. Many of the trace elements concentrated by marine organisms are essential for their growth (e.g. Fe, Mn, Cu, Co, Zn, Mo, B, V) (Bowen, 1966): however, others are probably not essential. It is probable that specific enzymes are present in the organism which break down the chelates of the essential metals and allow them to be assimilated. The non-essential elements are retained in chelated, and thus detoxicated, form until gradually discarded to the water during renewal of the mucus surfaces.

When the organisms die, bacterial attack returns trace elements to the water, perhaps initially in the form of organic complexes. Further decomposition of these complexes liberates ionic or colloidal species of the elements. After death the organisms sink rapidly to the thermocline beyond which they fall only slowly. Biologically concentrated trace elements (such as Hg, Co, Ni, Ag and Ba) will therefore tend to accumulate in the deep water, particularly in regions of high productivity (see Schutz and Turekian, 1965). Entrained air bubbles in the upper layers of the sea gather biologically derived dissolved and suspended organic matter and carry it to the surface as aggregates. When the supporting air bubbles burst, this material sinks, forming the "marine snow" which is an important source of food for marine animals (see p. 209). It has been suggested that this aggregate will adsorb trace metals from the water and carry them to the lower layers of the sea. However, laboratory experiments have shown that the aggregate formed from the *dissolved* organic constituents does not carry zinc, manganese or strontium (Alvin and Burke, 1965).

There is some net loss of trace elements from the sea through biological activity since resistant parts of some organisms reach the sea floor and are incorporated in the sediments. Some bacterial decomposition of these organic materials occurs in the upper, oxygenated, layers of the sediment column resulting in the enrichment of the interstitial waters with trace elements (Brooks *et al.*, 1968). From these waters they may return to the sea or be incorporated in the sediment, probably, by adsorption on ferromanganese phases.

Marine organisms may concentrate trace elements not only in their tissues but also in their skeletal parts, which may consist of hydrous silicon dioxide (opal), calcium carbonate (both calcite and aragonite), iron oxides, calcium fluorophosphate (apatite) or strontium sulphate (Lowenstam, 1963). However, from the small amount of information available it appears that the degree of trace element concentration in the skeletal structures is much less than that in the soft tissues.

Concentration of trace elements may also take place after the death of the organism. Arrhenius *et al.* (1957) have examined the organic and apatite phases of fish debris recovered from the ocean floor. They found that lanthanum, strontium, barium and perhaps thorium and uranium were concentrated in the apatite phase, whereas zinc, tin, lead, titanium, copper, chromium and nickel were enriched in the organic phases. Much lower concentrations of these trace metals were present in bathypelagic fish, and hence concentration must have taken place after the death of the organism. The sediments under zones of high productivity are often rich in barite; Chow and Goldberg (1960) have suggested that this may be precipitated by a "common ion" effect caused by the generation of high concentrations of sulphate ions in the bio-mass during its oxidation. However, it should be pointed out that since barium is known to be strongly concentrated by many organisms, at least some of the barite may be organic in origin.

(iv) Behaviour of trace metals in anoxic environments

Bacterial decomposition of organic matter in ill-ventilated basins (e.g. the Black Sea) leads to the exhaustion of the oxygen and enrichment of the water with both carbon dioxide and hydrogen sulphide. Elements such as copper, having sulphides with low solubility products are precipitated as their sulphides and usually cannot be detected in the sulphide-containing water. Anoxic waters frequently contain relatively high concentrations of dissolved iron and manganese. Since both the pH and E_h of the water are low, both elements are present in the 2+oxidation state and do not precipitate since the solubility products of their sulphides are not exceeded. The abundant manganese and iron which these waters contain may result from decay of organic matter or from reductive attack on the sediments of the periphery of the anoxic basin.

VII. RADIOACTIVE NUCLIDES IN THE SEA

The radioactive nuclides occurring in the sea can be divided into three classes (Burton, 1965): (i) the long-lived primary radio-nuclides which have existed since the elements were formed, and their daughter nuclides; (ii) cosmogenic radionuclides of relatively short half-life which are being formed continuously by bombardment of matter with cosmic rays; (iii) artificial radionuclides produced by human activities, such as fission products from nuclear weapons and waste products from nuclear power stations. With the exception of potassium–40, rubidium–87 and uranium, the concentrations of all the radionuclides are extremely low. It is only in the last few years that improvements in counting techniques have made it possible to study the concentrations of these nuclides in the natural environment in any detail. This has made it practicable to use some of them as tracers to investigate the water circulation system in the ocean, and to study the exchanges occurring between various water masses, and in some instances between the sea and the atmosphere, or sediments. In addition, certain of the longer-lived nuclides can be used for the dating of marine sediments. Marine radiochemistry has been extensively reviewed by several authors (Koczy and Rosholt, 1962; Burton, 1965; Lal and Peters, 1967; Lal and Suess, 1968) and these references should be consulted for detailed information.

A. PRIMARY RADIONUCLIDES

More than 90% of the total radioactivity of sea water arises from ^{40}K (half-life $1 \cdot 27 \times 10^{10}$ years), which constitutes 0·0118% of natural-potassium; this nuclide decays both by β-emission and by K electron capture, yielding two stable nuclides ^{40}Ar and ^{40}Ca respectively. Less than 1% of the activity of sea water is caused by the presence of ^{87}Rb (half-life $4 \cdot 7 \times 10^{10}$ years) which comprises 27·9% of natural rubidium.

Much more important from the standpoint of marine science are the members of the uranium, thorium, and actino-uranium decay series. The sea water and surface sediment average concentrations of the most important of these are shown in Table 4.8.

Uranium series

$$\underset{(99 \cdot 27\% \text{ of U})}{^{238}U} \xrightarrow[\alpha]{4 \cdot 51 \times 10^9 y} {}^{234}Th \xrightarrow[\beta^-]{24 \cdot 1 d} {}^{234}Pa \xrightarrow[\beta^-]{6 \cdot 66 h} \underset{(0 \cdot 0056\% \text{ of U})}{^{234}U} \xrightarrow[\alpha]{2 \cdot 48 \times 10^5 y} {}^{230}Th$$

$$\Big\downarrow {\alpha}\big|{8 \cdot 0 \times 10^4}$$

$$\xleftarrow[\beta^-]{22 \cdot 0 y} {}^{210}Pb \leftarrow \overset{\text{short-lived}}{\underset{\text{daughters}}{}} \leftarrow {}^{218}Po \xleftarrow[\alpha]{3 \cdot 83 d} {}^{222}Rn \xleftarrow[\alpha]{1622 y} {}^{226}Ra$$

Thorium series

$$\underset{(100\% \text{ of Th})}{^{232}Th} \xrightarrow[\alpha]{1 \cdot 39 \times 10^{10} y} {}^{228}Ra \xrightarrow[\beta^-]{6 \cdot 7 y} {}^{228}Ac \xrightarrow[\beta^-]{6 \cdot 13 h} {}^{228}Th \xrightarrow[\alpha]{1 \cdot 91 y}$$

Actino-uranium series

$$\underset{(0 \cdot 72\% \text{ of U})}{^{235}U} \xrightarrow[\alpha]{7 \cdot 1 \times 10^8 y} {}^{231}Th \xrightarrow[\beta^-]{25 \cdot 6 h} {}^{231}Pa \xrightarrow[\alpha]{3 \cdot 43 \times 10^4 y} {}^{227}Ac \xrightarrow[\beta^-]{21 \cdot 8 y} {}^{227}Th \xrightarrow[\alpha]{18 \cdot 4 d}$$

These series contain elements having a wide range of geochemical behaviour and it is therefore not surprising that the radionuclides in the oceans are not in secular equilibrium. In particular, thorium and pro-tactinium are rapidly lost from the sea to the sediments and this causes the ratios of the concentrations of the isotopes of these elements to that of uranium (Table 4.8) to be much lower than the equilibrium values, the greatest disequilibria are found for the ^{230}Th: ^{238}U and ^{231}Pa : ^{235}U ratios which are only 0·05–0·2% of the equilibrium value. This indicates an average residence time for these nuclides of ca 100 years

(Moore and Sackett, 1964; Somayajulu and Goldberg, 1966). The concentration of ^{226}Ra in the sea increases with depth (Fig. 4.1); at all depths considerably more of it is present than could be accounted for by equilibrium with the dissolved ^{230}Th. This distribution and lack of balance is probably attributable to the diffusion of radium from the sediment. However, as with barium (p. 94), the sinking of skeletal

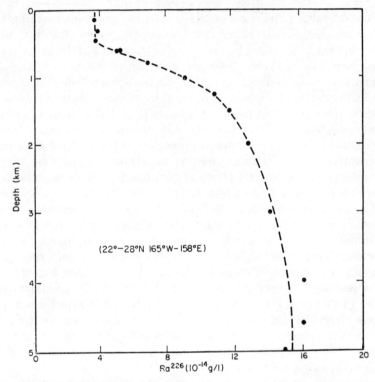

Fig. 4.1. Distribution of radium in the north-east Pacific (Broecker, 1966).

TABLE 4.8

Approximate average concentrations of some radionuclides of the uranium and thorium series in ocean water and surface sediments

	^{238}U	^{235}U	^{230}Th	^{232}Th	^{231}Pa	^{226}Ra	^{210}Po	^{210}Pb
Ocean (g/l) water	$3\cdot2\times10^{-6}$	$2\cdot1\times10^{-8}$	$<3\times10^{-13}$	10^{-9}	$1\cdot0-2\cdot5$ $\times10^{-14}$	1×10^{-13}	$2\cdot2\times10^{-17}$	8×10^{-19}
Surface sediment (g/g)	$1\cdot0\times10^{-6}$	$7\cdot0\times10^{-9}$	2×10^{-10}	5×10^{-6}	$1\cdot0\times10^{-11}$	3×10^{-11}	10^{-15}	3×10^{-17}

material from the surface and its redissolution in deep water may also play a part (Broecker *et al.*, 1967). There is a marked increase in the ^{222}Rn content of the sea water in the immediate neighbourhood of the ocean bed as a result of the decay of ^{226}Ra in the sediment (Broecker *et al.*, 1967).

B. COSMOGENIC RADIONUCLIDES

A considerable proportion of the cosmic ray particles reaching the earth have energies exceeding that binding the nuclei of atoms. Most of their energy is absorbed by the nuclei of the atoms of the gases in the atmosphere (e.g. oxygen, nitrogen and argon) which are thereby fragmented into stable or unstable nuclei lighter than the parent nuclide. Some nuclides are also produced by neutron capture reactions. All these cosmogenic nuclides are gradually distributed to the lower atmosphere, to the oceans and to the land; the fractions in each of these reservoirs depend on the chemical properties of the particular nuclide. Comparatively little cosmic radiation penetrates to the base of the atmosphere, and the rate of production of cosmogenic radionuclides in surface rocks must therefore be small (Lal and Peters, 1967). A significant contribution of certain nuclides, e.g. ^{26}Al, may accrete in extraterrestrial material, such as cosmic dust, which is irradiated in space. The radionuclides continuously produced in the atmosphere by cosmic ray interactions have half-lives ranging from a fraction of a second to more than 10^6 years. So far 20 of them have been detected, but of these only a few are sufficiently long-lived to be of value in oceanographic studies (Table 4.9). Since the specific activities of the nuclides in the various reservoirs are extremely low their determination is difficult. It is usually necessary to concentrate them from a large quantity of sample before measuring the activity with sophisticated counting equipment having a low background count.

Most oceanographic work on cosmogenic nuclides has centred around carbon-14 and tritium, both of which are also produced during the explosion of nuclear weapons. These nuclides have been used as indicators to determine the rates of mixing processes.

Tritium—hydrogen-3—the radioactive isotope of hydrogen is produced by spallation of air molecules and by interaction of secondary neutrons with nitrogen. It also probably accretes from the sun. After formation, most of it is oxidized and falls in rain and snow. The steady state level of the nuclide existing before 1954 in the surface layer of the ocean was ca 1 atom per 10^{18} atoms of hydrogen. By 1961 tritium produced in thermonuclear weapon tests had raised the average level in the northern hemisphere about 8-fold. However, with the halt in nuclear testing the

TABLE 4.9

Data on cosmic-ray-produced nuclides of oceanographic interest (from Lal, 1968 and Lal and Suess, 1968)

Radio-nuclide	Half-life (years)	Total atmospheric production rate (atoms cm⁻² sec⁻¹)	Average specific activity[a] in oceans		Relative proportions of radionuclide in various reservoirs				
			dpm/ton of water	dpm/g of element	Stratosphere	Troposphere	Mixed oceanic layer	Deep oceanic layer	Oceanic sediments
^{7}Be	0·146	$8·1 \times 10^{-2}$	—	$3·3 \times 10^{-4}$	0·60	0·11	0·28	3×10^{-3}	0
^{3}H	12·5	0·25	36	$5·0 \times 10^{-3}$	$6·8 \times 10^{-2}$	4×10^{-3}	$0·50^{b}$	0·43	0
^{39}Ar	270	$5·6 \times 10^{-3}$	$2·9 \times 10^{-3}$	$8·0 \times 10^{-3}$	0·16	0·83	5×10^{-4}	3×10^{-3}	0
^{32}Si	500	$1·6 \times 10^{-4}$	$2·4 \times 10^{-2}$		$1·9 \times 10^{-3}$	$1·1 \times 10^{-4}$	5×10^{-3}	0·96	4×10^{-2}
^{14}C	5,730	2·5	260	10	3×10^{-3}	6×10^{-2c}	$2·3 \times 10^{-2}$	0·91	10^{-2}
^{36}Cl	$3·1 \times 10^{5}$	$1·1 \times 10^{-3}$	0·55	$3·0 \times 10^{-5}$	10^{-6}	6×10^{-8}	2×10^{-2}	0·98	0
^{26}Al	$7·4 \times 10^{5}$	$1·4 \times 10^{-4}$	$1·2 \times 10^{-5}$	$1·2 \times 10^{-3}$	$1·3 \times 10^{-6}$	$7·7 \times 10^{-8}$	2×10^{-5}	10^{-4}	0·999
^{10}Be	$2·5 \times 10^{6}$	$4·5 \times 10^{-2}$	10^{-3}	$1·6 \times 10^{-3}$	$3·7 \times 10^{-7}$	$2·3 \times 10^{-8}$	8×10^{-6}	$1·4 \times 10^{-4}$	0·999

[a] Expressed in terms of disintegrations per minute (dpm).
[b] Includes ^{3}H present in continental hydrosphere.
[c] Includes ^{14}C present in biosphere and humus.

level has subsequently declined somewhat. The comparatively short half-life of tritium (12·7 years) limits its usefulness in oceanography to the investigation of rapid processes, such as those occurring in the mixed layer. Thus, Bainbridge (1963) has deduced from a study of the distribution of this nuclide that the maximum residence time of water in the mixed layer in the North Pacific is 3·5 years. Measurements of the distribution of this nuclide have also been used to investigate vertical movements of water. For example, Bainbridge and O'Brien (1962) have concluded from data on the abundance of tritium in the water column in a region of convergence off New Zealand, that the water there took about 12 years to reach a depth of 2,000 m.

Carbon-14 originates through the capture of neutrons by nitrogen in the upper atmosphere. It soon becomes oxidized to carbon dioxide and eventually participates in the marine carbon dioxide cycle. This nuclide is sufficiently long-lived (half-life 5,730 years) for it to be used for studying the circulation pattern of deep waters. However, absolute "ages" of ocean water masses based on carbon-14 measurements must be accepted with caution since they are subject to many uncertainties (Cooper, 1956). In addition to their value for the dating of water masses, carbon-14 measurements provide a very useful means of studying the deposition rates of deep-sea sediments (see p. 293), although the relatively short half-life restricts the time scale.

By comparison with carbon-14 and tritium other cosmogenic nuclides have so far been of much less value in oceanography partly because of their extremely low concentrations, e.g. the total amount of cosmogenic ^{32}Si on earth is ca 1 kg (Lal, 1963). If technical difficulties can be overcome both ^{32}Si (Lal, 1968) and ^{39}Ar may prove a useful supplement to ^{14}C measurements in the investigation of the oceanic circulation patterns. In a similar way data on ^{7}Be activity may perhaps be used for short term mixing studies (Lal and Suess, 1968).

C. ARTIFICIALLY PRODUCED NUCLIDES

A great variety of radionuclides are produced when nuclear energy is released during the explosion of nuclear weapons or in atomic power stations. The radioactive debris from nuclear explosions has frequently not been contained and much finds its way into the sea either directly (as with submarine explosions), or indirectly via the atmosphere as fall out. This has constituted the main source of the artificial radioactivity present in the sea. However, low activity wastes are discharged into coastal waters at some nuclear power stations (e.g. a few hundred curies a day are discharged into the Irish Sea from the Windscale Station in Cumberland); much of this discharge consists of relatively short-lived nuclides. In addition some dumping of sealed high activity wastes into the deep ocean has occurred.

More than 200 nuclides ranging in atomic number from 30(Zn) to 66(Dy) are produced during nuclear explosions. Many of these are very short-lived and make no contribution to the long term activity of sea water. In the first weeks after the explosion short-lived nuclides such as ^{143}Pr (13·7 days), ^{140}Ba (12·8 days), ^{147}Nd (11·3 days) and ^{131}I(8·05 days) are responsible for most of the fission product activity. However, after a year ^{144}Ce and ^{95}Zr, with their daughters ^{144}Pr and ^{95}Nb account for ca 75% of the total activity. After 20 years, more than 90% of the activity arises from ^{90}Sr (28 years), ^{137}Cs (30 years) and their daughters. Several studies have been made of the distribution in the oceans of these two long-lived and potentially hazardous nuclides. These have shown (see Burton, 1965) that during, and immediately subsequent to the first major period of nuclear testing, the levels of ^{90}Sr and ^{137}Cs were considerably higher in the surface waters of the North West Pacific (80–480 ×10^{-12} Ci of ^{137}Cs/l) than they were in those of the East Pacific and North Atlantic (5–23 ×10^{-12} Ci of ^{137}Cs/l). The rates of vertical penetration of these radionuclides have been the subject of some controversy. According to Broecker et al. (1966) by 1963 they were found only in a few locations at depths below 500–700 m in the East Pacific and North Atlantic. The input of these nuclides at the surface of the ocean, which was essentially free from them previously, provides a useful means of studying the rates of mixing of surface and deeper waters.

In addition to the fission products themselves, the explosion of nuclear weapons leads to the production of other radionuclides. These are formed by the capture of neutrons by materials such as the bomb casings, earth, water and air. The principal of these induced nuclides are ^{3}H, ^{14}C, ^{32}P, ^{35}S, ^{51}Cr, ^{54}Mn, ^{55}Fe, ^{59}Fe, ^{58}Co, ^{60}Co and ^{65}Zn. Their relative proportions will vary considerably according to the nature of the materials irradiated. Tritium and carbon–14 produced in this way have considerably augmented the natural levels of the nuclides present in the troposphere. Peak increases of more than ten-fold and ca 50% respectively occurred in the northern hemisphere. However, following the end of nuclear tests, the amounts have decreased as a result of exchange with the natural reservoirs and radioactive decay.

REFERENCES

Ahrens, L. H. (1965). "Distribution of the Elements in our Planet." McGraw-Hill, New York.

Alvin, S. and Burke, B. (1965). *Deep-Sea Res.* **12**, 789.

Andersen, N. R. and Hume, D. N. (1968). *Anal. Chem. Acta.*, **40**, 207.

Andersen, N. R., Gassaway, J. D. and Maloney, W. E. (1970) *Limnol. Oceanogr.* **15**, 467.

Angino, E. E., Billings, G. K. and Andersen N. (1966). *Chem. Geol.* **1**, 145.

Arrhenius, G. (1963). *In* "The Seas" (M. N. Hill, ed.) Vol. 3. Interscience, New York

Arrhenius, G., Bramlette, M. and Picciotto, E. (1957). *Nature, Lond.* **180**, 85.

Bainbridge, A. E. (1963). *In* "Nuclear Geophysics" (P. M. Hurley, G. Faure and C. Schnetzler, eds), Publication 1075, p. 129. National Academy of Sciences—National Research Council, Washington.

Bainbridge, A. E. and O'Brien, B. J. (1962). *In* "Tritium in the Physical and Biological Sciences", Vol. I. International Atomic Energy Agency, Vienna.

Barger, W. R. and Garrett, W. D. (1970). U.S. Naval Research Laboratory, Report 7079, Washington, D.C.

Barić, A. and Branica, M. (1967). *J. polarog. Soc.* **13**, 4.

Barth, T. W. (1952). "Theoretical Petrology." Wiley, New York.

Berner, R. A. (1965). *Geochim. cosmochim. Acta* **29**, 947.

Berner, R. A. (1964). *Geochim. cosmochim. Acta* **28**, 1497.

Biedermann, G. and Chow, T. J. (1966). *Acta Chem. Scand.* **20**, 1376.

Blanchard, D. C. (1964). *In* "Progress in Oceanography," Vol. 1. (M. Sears, ed.). Pergamon Press, New York.

Borchert, H. (1965). *In* "Chemical Oceanography" (G. Skirrow and J. P. Riley, eds), Vol. 2. Academic Press, London.

Bostrom, K. and Peterson, M. N. A. (1966). *Econ. Geol.* **61**, 1258.

Bowen, H. J. M. (1966). "Trace Elements in Biochemistry." Academic Press, London.

Branica, M., Petek, M., Barić, A. and Jeftić L. (1969). *Rapp. Comm. int. Mer. Medit.* **19**, 929.

Brewer, P. G., Culkin, F. and Riley, J. P. (1965). *Deep-Sea Res.* **12**, 554.

Brewer, P. G., Spencer, D. W. and Wilkniss, P. E. (1970). *Deep-Sea Res.*, **17**, 1.

Broecker, W. S. (1966). *J. geophys. Res.* **71**, 5827.

Broecker, W. S., Bonebakker, E. R. and Rocco, G. G. (1966). *J. geophys. Res.* **71**, 1999.

Broecker, W. S., Li, Y. H. and Cromwell, J. (1967). *Science, N.Y.* **158**, 1307.

Brooks, R. R., Presley, B. J. and Kaplan, I. R. (1968). *Geochim. cosmochim. Acta* **32**, 397.

Burton, J. D. (1965). *In* "Chemical Oceanography" (J. P. Riley and G. Skirrow, eds), Vol. 2. Academic Press, London.

Chau, Y. K. and Riley, J. P. (1965). *Anal. Chim. Acta* **33**, 36.

Chow, T. J. and Goldberg, E. D. (1960). *Geochim. cosmochim. Acta* **20**, 192.

Cooper, L. H. N. (1956). *J. Mar. biol. Ass. U.K.* **35**, 341.

Cox, R. A. and Culkin, F. (1966). *Deep-Sea Res.* **13**, 789.

Craig, H. (1961a). *Science, N.Y.* **133**, 3467.

Craig, H. (1961b). *Science, N.Y.* **133**, 1702.

Craig, H. (1969). *In* "Hot Brines and Recent Heavy Metal Deposits in the Red Sea". Springer Verlag, Berlin.

Craig, H. and Gordon, L. I. (1965). Deuterium and oxygen—18 variations in the ocean and marine atmosphere. Proc. Congr. on stable isotopes in oceanographic studies and paleotemperatures, Spoleto, 26–30 July, 1965, pp. 9–130.

Craig, H. and Gordon, L. I. (1966). *Trans. Am. Geophys. Un.* **47**, 112.

Degens, E. T. and Ross, D. A. (1969). "Hot Brines and Recent Heavy Metal Deposits in the Red Sea." Springer Verlag, Berlin. 600 pp.

Dittmar, W. (1884). "Report on the Scientific Results of the Exploring Voyage of H.M.S. *Challenger*. Physics and Chemistry", Vol. 1. H.M.S.O., London.

Duce, R. A., Winchester, J. W. and Van Nahl, T. W. (1965). *J. Geophys. Res.* **70**, 1775.

Durrum, W. H. and Haffty, J. (1963). *Geochim. cosmochim. Acta* **27**, 1.

Elderfield, M. (1970). *Earth Planet. Sci. Letters*, **9**, 10.

Elgquist, B. (1970). *J. Inorg. Nucl. Chem.* **32**, 937.

Eriksson, E. (1959). *Tellus* **11**, 400.

Fisher, F. H. (1967). *Science, N.Y.* **157**, 823.

Fukai, R. (1969). *J. oceanog. Soc. Japan*, **25**, 47.

Fukai, R. and Shiokawa, F. (1955). *Bull. Chem. Soc. Japan* **38**, 636.

Garrels, R. M. and Christ, C. L. (1965). "Solutions, Minerals and Equilibria." Harper and Row, New York. 450 pp.

Garrels, R. M. and Thompson, M. E. (1962). *Am. J. Sci.* **206**, 57.

Goldberg, E. D. (1957). *Mem. Geol. Soc. Am.* **67**, 345.

Goldberg, E. D. (1965). *In* "Chemical Oceanography" (J. P. Riley and G. Skirrow, eds), Vol. 1. Academic Press, London.

Goldberg, E. D. and Arrhenius, G. O. S. (1958). *Geochim. cosmochim. Acta* **13**, 153.

Goldberg, E. D. and Koide, M. (1962). *Geochim. cosmochim Acta*, **26**, 417.

Goldschmidt, V. M. (1933). *Fortsch. Min. Krist. Petrog* **17**, 112.

Grim, R. W. (1952). "Clay Mineralogy" McGraw-Hill, New York.

Gripenberg, S. (1937). *J. Cons. perm. int. Explor. Mer.* **12**, 293.

Hanor, J. S. (1969). *Geochim. cosmochim. Acta* **33**, 894.

Holland, N. D. (1965). *Proc. Nat. Acad. Sci. U.S.* **53**, 1173.

Horne, M. K. and Adams, J. A. S. (1966). *Geochim. cosmochim. Acta* **30**, 279.

Jacobs, M. B. and Ewing, M. (1969). *Science, N.Y.* **163**, 380.

Kester, D. R. and Pytkowicz, R. M. (1969). *Limnol. Oceanogr.* **14**, 686.

Kharkar, D. P., Turekian, K. K. and Bertine, K. K. (1968). *Geochim. cosmochim. Acta* **32**, 285.

Koczy, F. F. (1958). *Proc. second United Nations int. Conf. peaceful uses Atomic Energy*, **18**, 351.

Koczy, F. F. and Rosholt, J. N. (1962). *In* "Nuclear Radiation in Geophysics" (H. Israel and A. Krebs, eds). Springer Verlag, Berlin.

Krauskopf, K. B. (1956). *Geochim. cosmochim. Acta* **9**, 1.

Kwiecinski, B. (1965). *Deep-Sea Res.* **12**, 797.

Lal, D. (1963). *In* "Earth Science and Meteoritics" (J. Geiss and E. D. Goldberg, eds). North-Holland, Amsterdam.

Lal, D. (1968). *In* "Collected Plenary Talks delivered at 2nd Intern. Oceanogr. Congr. Moscow, 1966." North-Holland, Amsterdam.

Lal, D. and Peters, B. (1967). "Handbuch der Physik", Vol. 46(2), 551.

Lal, D. and Suess, H. E. (1968), *Am. Rev. Nucl. Sci* **18**, 407.

Livingstone, D. A. (1963). Data of Geochemistry 6th Ed. U.S. Geol. Surv. Prof. Paper 440G.

Lloyd, R. M. (1966). *Geochim. cosmochim. Acta* **30**, 801.

Longerich, L. L. and Hood, D. W. (1968) Progress Report Inst. Mar. Sci. University of Alaska, Part 1, p. 54.

Lowenstam, H. A. (1963). *In* "Earth Sciences" (T. W. Donnelly, ed.), p. 137. Rice University Press, Houston, Texas.

Mackenzie, F. T. (1964). *Science, N.Y.* **146**, 517.

Manheim, F. T. (1961). *Geochim. cosmochim. Acta* **25**, 52.

Marcet, A. M. (1819). *Phil. Trans.* **109**, 161.

Miyake, Y. and Tsunogai, S. (1963). *J. Geophys. Res.* **13**, 3989.

Moore, W. S. and Sackett, W. M. (1964). *J. geophys. Res.* **69**, 5401.

Morris, A. W. and Riley, J. P. (1966). *Deep-Sea Res.* **13**, 699.

Morris, J. C. and Stumm, W. (1967). *Adv. Chem. Ser.* **67**, 270.

Mullin, J. B. and Riley, J. P. (1956). *J. mar. Res.* **15**, 103.

Platford, R. F. (1965a). *J. Fish. Res. Bd. Can.* **22**, 885.

Platford, R. F. (1965b). *J. Fish. Res. Bd. Can.* **22**, 113.

Pytkowicz, R. M. and Gates R. (1968). *Science, N.Y.* **61**, 690.

Riley, J. P. (1965). *Deep-Sea Res.* **12**, 219.

Riley, J. P. (1967). *Ann. Rev. Oceanogr. Mar. biol.* **5**, 141.

Riley, J. P. and Tongudai, M. (1967). *Chem. Geol.* **2**, 263.

Rubey, W. W. (1951). *Bull. Geol. Soc. Am.* **62**, 1111.

Schutz, D. F. and Turekian, K. K. (1965a). *J. Geophys. Res.* **70**, 5519.

Schutz, D. F. and Turekian, K. K. (1965b). *Geochim. cosmochim. Acta* **29**, 259.

Siever, R., Beck, K. C. and Berner, R. A. (1965). *J. Geol.* **73**, 39.

Sillén, L. G. (1961). "The Physical Chemistry of Sea Water" (Invited lectures presented at the International Oceanographic Congress in New York, 31 August–15 September 1959), pp. 549–581.

Sillén, L. G. (1963), *Svensk. Kemisk. Tidskr.* **75**, 161.

Sillén, L. G. (1964). "Stability Constants," *Chem. Soc. Spec. Publ.* **17·**

Sillén, L. G. (1965) *Arkiv. Kemi.*, **24**, 431.

Sillén, L. G. (1967). *Science, N.Y.* **156**, 1189.

Skopintsev, B. A., Gubin, F. A., Vorob'eva, R. V. and Vershinina, O. A. (1958). *C.R. Acad. Sci. U.R.S.S.* **119**, 121.

Somayajulu, B. L. K. and Goldberg, E. D. (1966). *Earth Sci. Planet. Letters.* **1**, 102.

Stefansson, U. (1966). *J. mar. Res.* **24**, 241.

Sugawara, K. (1965). *Oceanogr. Mar. Biol. Ann. Rev.* **3**, 59.

Tatsumoto, M. and Patterson, C. C. (1963). *In* "Earth Science and Meteoritics" (J. Geiss and E. D. Goldberg, eds). North Holland Publishing Co., Amsterdam.

Thompson, M. E. (1966). *Science, N.Y.* **1953**, 866.

Traganza, E. D. and Szabo, B. J. (1967). *Limnol. Oceanog.* **12**, 281.

Turekian, K. K. (1969). *In* "Handbook of Geochemistry" (K. H. Wedepohl, ed.), Vol. 1. Springer Verlag. Berlin.

Turekian, K. K. and Johnson, D. G. (1966). *Geochim. cosmochim. Acta* **30**, 1153.

van't Hoff, J. H. (1912). "Untersuchungen über die Bildungsverhältnisse der ozeanischen Salzablegerungen, insbesonders des Stassfurter Salzlagers." Leipzig.

Wedepohl, K. H. (1966). *Naturwiss* **53**, 352.

Whitfield, M. (1969). *Limnol. Oceanog.* **14**, 547.

Williams, P. M. (1969). *Limnol. Oceanog.* **14**, 156.

Williams, P. M. and Solorzano, L. (1967). Unpublished manuscript U.C.S.D.-34P–108–52.

Williams, P. M. and Strack, P. M. (1966). *Limnol. Oceanog.* **11**, 401.

Yoshio, H. and Nobuko, O. (1968). *J. geophys. Res.* **73**, 1239.

Chapter 5

The Dissolved Gases in Sea Water.
Part I. Gases Other than Carbon Dioxide

I. ATMOSPHERIC GASES

The gases present in the atmosphere may be divided into the non-variable components (Table 5.1), of which the most abundant are oxygen, nitrogen and argon, and the variable components such as water vapour and the gases produced at least in part by human activities (e.g. NO_2, CO and NH_3). Interchange of these gases takes place between the atmosphere and the surface waters of the sea with the result that these waters are generally at, or near, equilibrium with respect to them. Distribution of the gases throughout the oceans as a whole is brought about by advection and eddy diffusion during oceanic circulation. Some of the gases behave in a conservative manner, but the concentrations of others such as oxygen and carbon dioxide are influenced by biological or other processes.

TABLE 5.1

Proportions of non-variable gases in the atmosphere (Glueckauf, 1951)

Gas Partial pressure (atm)	N_2	O_2	CO_2	Ar	Ne	He	Kr
	0·7808	0·2095	0·000330*	0·00934	$1·82 \times 10^{-5}$	$5·24 \times 10^{-6}$	$1·14 \times 10^{-6}$
Gas Partial pressure (atm)	Xe	H_2	CH_4	N_2O			
	$8·7 \times 10^{-8}$	5×10^{-7}	2×10^{-6}	5×10^{-7}			

* Subject to slight variations (see p. 141).

II. SOLUBILITIES OF GASES

The solubility of a pure gas in a liquid is usually expressed in terms of the Bunsen coefficient (α). This is the volume of the gas at S.T.P. which can be dissolved by unit volume of the liquid, at the given temperature under a gas pressure of 760 Torr (1 atm). The solubility can be measured by equilibrating a known volume of the degassed liquid with the gas, and either measuring the volume of gas absorbed or determining the amount of gas present in the solution.

Solubilities of gases in sea water are functions of temperature, salinity and pressure. Solubility falls with increasing temperature, in reasonably close agreement with the relationship $\log \alpha \propto 1/T$. The presence of the dissolved ions in sea water decreases the solubility of gases in fair accordance with the empirical Setchenow (1875) equation which states that at constant temperature, $\log (\alpha_{sw}/\alpha_{pw}) = -Ks$ where α_{sw} and α_{pw} are the Bunsen coefficients of the gas in sea water (of salinity s) and pure water respectively, and K is a constant for the particular gas at the specified temperature. The effect of pressure is described by Henry's Law, which states that the solubility of the gas is directly proportional to its partial pressure in the gas phase (p). At equilibrium, the latter is equal to its partial pressure in the liquid phase (P). Thus, if the liquid is equilibrated with a mixture of gases (e.g. air) and there is no interaction between the individual gases, the concentration (c) of a particular gas of partial pressure p will be given by the expression.

$$c = 1000 \cdot \alpha \cdot p \text{ ml/l} \tag{5.1}$$

Solubilities of atmospheric gases in sea water of given salinity and temperature (designated O_2', N_2', etc.) are generally defined in terms of the volume (in ml) of gas, reduced to S.T.P., which will be dissolved by 1 l of the water from an atmosphere saturated with water vapour at the temperature in question, under a total pressure of 760 Torr (1 Torr = 1 mm of mercury). If the water is saturated with respect to air at a total pressure different from 760 Torr and if the air is incompletely saturated with water vapour, then the observed solubility of, for example, oxygen (O_2'') is related to the solubility (O_2') from a saturated atmosphere having a total pressure of 760 Torr by the expression

$$O_2'' = \frac{O_2' \cdot [p - p_s(h/100)]}{760 - p_s} \tag{5.2}$$

where p is the barometric pressure (in Torr), p_s is the vapour pressure of the sea water at the given temperature* and h is the percentage humidity of the air. It is probable that $h = 100\%$ in the layer immediately above the sea surface. Other units are also used for expressing dissolved gas concentrations (e.g. mg/l), and the use of mg-atoms/l has been recommended recently for international adoption.

---

* The vapour pressure of sea water (p_s) can be calculated from that of pure water (p_0) using the equation $p_s = p_0(1 - 0.000969 \text{ Cl } {}^0/_{00})$. For most practical purposes the differences between p_0 and p_s are negligible.

For comparative purposes oceanographers often express findings in terms of the percentage saturation of the sample with the gas concerned

$$\% \text{ saturation} = 100 \text{ G/G}' \qquad (5.3)$$

where G is the observed concentration of the gas and G' is its solubility in water of the appropriate *in situ* temperature and salinity. The solubility of oxygen has been reinvestigated recently by Carpenter (1966) and by Murray and Riley (1969) whose results are in substantial

TABLE 5.2

Solubility of oxygen (ml/l) from an atmosphere of 20·94% O_2 and 100% relative humidity (Murray and Riley, 1969)

Temp. (°C)	Chlorinity (°/₀₀)										
	0	2	4	6	8	10	12	14	16	18	20
1	9·94	9·68	9·45	9·23	9·01	8·79	8·58	8·37	8·17	7·97	7·78
2	9·66	9·42	9·20	8·99	8·77	8·56	8·36	8·16	7·96	7·77	7·58
3	9·39	9·17	8·96	8·76	8·55	8·35	8·15	7·95	7·76	7·57	7·39
4	9·14	8·93	8·73	8·53	8·33	8·13	7·94	7·75	7·56	7·39	7·22
5	8·90	8·69	8·50	8·31	8·12	7·93	7·74	7·56	7·38	7·21	7·05
6	8·68	8·48	8·28	8·10	7·91	7·73	7·55	7·37	7·20	7·04	6·88
7	8·47	8·27	8·08	7·90	7·72	7·55	7·37	7·20	7·03	6·87	6·72
8	8·27	8·07	7·88	7·71	7·54	7·37	7·20	7·03	6·87	6·71	6·56
9	8·07	7·88	7·69	7·52	7·36	7·19	7·03	6·87	6·71	6·56	6·42
10	7·88	7·70	7·52	7·35	7·19	7·03	6·87	6·71	6·56	6·42	6·28
11	7·71	7·52	7·35	7·18	7·03	6·87	6·72	6·57	6·42	6·28	6·14
12	7·54	7·35	7·18	7·02	6·87	6·72	6·57	6·43	6·28	6·14	6·01
13	7·37	7·19	7·03	6·87	6·72	6·57	6·43	6·29	6·15	6·01	5·88
14	7·21	7·03	6·88	6·72	6·57	6·43	6·29	6·15	6·02	5·89	5·76
15	7·05	6·88	6·73	6·58	6·44	6·30	6·16	6·03	5·89	5·77	5·65
16	6·90	6·74	6·59	6·45	6·30	6·17	6·03	5·91	5·78	5·66	5·54
17	6·75	6·60	6·46	6·32	6·18	6·04	5·91	5·79	5·66	5·54	5·43
18	6·61	6·47	6·33	6·19	6·05	5·92	5·80	5·68	5·56	5·44	5·53
19	6·48	6·34	6·20	6·06	5·93	5·80	5·69	5·57	5·45	5·34	5·23
20	6·36	6·21	6·07	5·94	5·81	5·69	5·58	5·47	5·35	5·24	5·14
21	6·23	6·09	5·96	5·83	5·70	5·58	5·47	5·37	5·25	5·15	5·05
22	6·11	5·97	5·85	5·72	5·60	5·48	5·37	5·26	5·16	5·06	4·96
23	6·00	5·86	5·74	5·61	5·49	5·38	5·27	5·17	5·07	4·97	4·88
24	5·89	5·76	5·63	5·51	5·39	5·28	5·18	5·08	4·98	4·88	4·79
25	5·77	5·65	5·53	5·41	5·29	5·19	5·09	4·99	4·89	4·80	4·71
26	5·67	5·54	5·43	5·32	5·20	5·10	5·00	4·91	4·81	4·72	4·63
27	5·57	5·44	5·33	5·23	5·11	5·01	4·91	4·82	4·73	4·64	4·55
28	5·47	5·34	5·24	5·14	5·02	4·93	4·83	4·74	4·65	4·56	4·47
29	5·37	5·25	5·15	5·05	4·94	4·85	4·76	4·66	4·57	4·48	4·39
30	5·28	5·16	5·06	4·97	4·86	4·77	4·68	4·59	4·50	4·41	4·32

agreement (Table 5.2). For nitrogen and argon the solubility data of Murray *et al.* (1969) and Murray and Riley (1970) should be consulted respectively.

The biologically induced changes in oxygen concentration which have taken place after the water has ceased to be in direct exchange with the atmosphere are usually expressed in terms of "apparent oxygen utilization" (A.O.U.) (Redfield, 1942) where A.O.U. $= O_2' - O_2$, where O_2 is the oxygen concentration of the submerged water.

III. DETERMINATION OF DISSOLVED GASES IN SEA WATER

Because of its biological importance and the ease with which it can be determined, oxygen has been the most frequently studied of all the dissolved gases. The analysis can be carried out even on shipboard by means of the Winkler (1888) method, or one of its modifications. The sample, contained in a glass stoppered bottle, is treated with a strong solution of manganous sulphate and a concentrated reagent containing sodium iodide and sodium hydroxide. The bottle is then carefully stoppered so as to exclude bubbles of air and shaken. The dissolved oxygen reacts with the precipitated manganous hydroxide, and under these highly alkaline conditions manganese is oxidized to a higher oxidation state. The precipitate is allowed to settle and dissolved by the addition of acid. Iodine liberated by the reaction of the oxidized manganese with iodide ion is titrated with standard thiosulphate, using starch as an indicator.

$$Mn^{2+} + 2OH^- \rightarrow Mn(OH)_2$$

$$Mn(OH)_2 + O \rightarrow MnO(OH)_2$$

$$MnO(OH)_2 + 4H^+ + 3I^- \rightarrow Mn^{2+} + I_3^- + 3H_2O$$

$$I_3^- + 2S_2O_3^{2-} \rightarrow 3I^- + S_4O_6^{2-}$$

For practical details of the Winkler method the monograph by Strickland and Parsons (1968) and the paper by Carpenter (1965) should be consulted. The method gives reliable results with ocean waters, but interference from oxidizing agents (e.g. NO_2^-) or reducing agents (e.g. Fe^{2+}, S^{2-} or organic matter) may cause inaccuracies with polluted estuarine waters unless modified procedures are used.

The development of physico-chemical methods of analysis now permits dissolved gases other than oxygen to be determined on a routine basis with a precision of better than $\pm 1\%$. The foremost of these techniques is gas chromatography which can be used for the determination of the major dissolved gases (O_2, N_2, Ar) (Swinnerton, *et al.*, 1962 a,b; Craig *et al.*, 1967). In these procedures, the gases are stripped from the water by a current of helium or by evacuation using a Toepler pump. The gases are adsorbed in a column containing molecular sieve, preferably cooled to $-78°C$, and then eluted successively by means of a current of helium. A catharometer coupled to an electronic integrator is used to determine the eluted gases. Gas-solid chromatography has also been used for the determination in sea water of certain minor dissolved gases which occur at levels of ca 1×10^{-4} ml/l such as carbon monoxide and hydrocarbons (Swinnerton et al., 1968). Mass spectrometry, because of its specificity and very high sensitivity, has been used in studies of the distribution of the noble gases in the sea (Bieri et al., 1966; Craig et al., 1967).

Electrochemical procedures are available which allow the biologically important dissolved gases to be determined both in the laboratory (Carritt and Kanwisher, 1959) and in situ down to depths of more than 1,000 m (Beckman Instruments, 1969). Oxygen detectors are normally polarographic in principle. In a typical instrument the polarographic cell containing a gel of potassium chloride as electrolyte is isolated from the sea water by means of a thin PTFE membrane which is impermeable to electrolytes and pollutants in the water, but which permits oxygen to diffuse through it rapidly. The cell contains two electrodes—(i) a gold cathode at which oxygen is reduced according to the reaction $O_2+2H_2O+4e^- \rightleftharpoons 4OH^-$; (ii) a silver anode at which the reaction $4Ag+4Cl^- \rightleftharpoons 4AgCl+4e^-$ occurs. A potential of ca 1 V is applied between the electrodes. The diffusion current, which is proportional to the partial pressure of oxygen, is measured with a recording instrument. The use of a thermistor enables automatic compensation to be made for the high temperature coefficient of the cell.

The in situ determination of carbon dioxide is carried out in a cell having a membrane permeable to the dissolved gas. The cell contains a pH-sensitive glass electrode and calomel reference cell. The carbon dioxide which diffuses from the sea water into the electrolyte in the cell causes a pH change because of the reaction $H_2O+CO_2 \rightleftharpoons HCO_3^- +H^+$. The change in E.M.F. of the cell is proportional to $\log P_{CO_2}$ in the sea water.

IV. EXCHANGE OF ATMOSPHERIC GASES ACROSS THE SEA SURFACE

The interchange of a gas between the atmosphere and water is a dynamic process, and at saturation, when the partial pressure of the gas is the same in both media, molecules enter and leave each phase at the same rate. When, as is more usual, the partial pressure in one medium is higher there will be a net flow of gas from that phase to the other. The rates of transfer of a gas from the atmosphere to the sea and vice versa are proportional to its partial pressures in the atmosphere

(p_g) and sea (P_g) respectively. The net rate of accumulation of gas per unit area of sea surface will be given by

$$dQ/dt = k_a \cdot p_g - k_s \cdot P_g$$

where k_a and k_s are the appropriate velocity constants. At equilibrium, $dQ/dt = 0$, and therefore $k_a \cdot p_g = k_s \cdot P_g$.

It seems likely that the rate of interchange between the gaseous and liquid phases is controlled by the rate of molecular diffusion of the gas through two boundary layers; one in the gaseous phase, and the other in the liquid. Since rates of diffusion in the gas phase are several thousand times greater than those in the liquid phase, the latter boundary layer is the more important. This layer is thought to be stable and non-turbulent, and has a thickness which decreases with increasing turbulence of the water; it varies inversely with the wind speed, being ca 0·1 mm thick at an air speed of 1·5 m/sec. The diffusion of gas through this film is analogous to gaseous diffusion through a semi-permeable membrane, for which

$$\frac{dQ}{dt} = \frac{A \cdot D \cdot (C_1 - C_2)}{dz}$$

where A is the surface area, D is the molecular diffusion coefficient, dz is the thickness of the film and $C_1 (= p_g)$ and $C_2 (= P_g)$ are the concentrations of the gas in the two phases separated by the membrane. Hoover and Ibert (1966) have shown that the dissolution of carbon dioxide in acidified sea water closely fits this model. Studies of the diffusion of oxygen and nitrogen into sea water by Wyman et al. (1952) indicate that in the sea the average thickness of the layer is 35 μm. Under highly turbulent conditions the system deviates increasingly from this simple model, and the exchange rate is greatly increased, either because the surface film disappears, or because it is frequently renewed by eddy diffusion.

The fluxes of the various atmospheric gases will vary considerably for a constant partial pressure difference since their solubilities differ widely (e.g. the fluxes for N_2, O_2 and CO_2 would be in the ratio 1 : 2 : 70). The volumes of gas entering and leaving the sea surface seasonally are very considerable. Thus, Redfield (1948) has estimated that on average 30×10^4 ml of oxygen/m² enters the surface of the Gulf of Maine during the autumn and winter. A roughly corresponding volume leaves the water during the spring and summer, about 2/5 because of photosynthetic activity and the remainder because of the lower solubility of the gas in warmer water. Keeling (1965) has found that the annual exchange rate for carbon dioxide in the Pacific (30° N.–65° S.) averages 18 moles/cm²/atm.

The flux of dissolved gas between the boundary film and the upper layer of the sea is brought about by eddy diffusion (see Chapter 3). It is known that, in the upper metre of the sea, appreciable vertical concentration gradients can exist under non-turbulent conditions for oxygen, and presumably also for other gases; gradients in the horizontal direction are generally much less pronounced. Thus, since the flux depends on the concentration gradient, it will be much greater in the vertical than in the horizontal direction. However, under the conditions usually prevailing in the sea, turbulence produced by wind ensures that the layer above the seasonal thermocline is in thermodynamic equilibrium with the sea surface.

The dissolved gases are transported from this upper layer to the deeper parts of the oceans mainly by advection; since the coefficients of molecular diffusion of gases are very low (ca 2×10^{-5} cm^2 sec^{-1}), transport by molecular diffusion is negligible. In advection, cold gas-saturated water sinks at high latitudes and is carried at depth to lower latitudes. Eddy diffusion along surfaces of equal density may also play a subsidiary role in the transport of gas in the deep waters. Broecker *et al.* (1967) give coefficients of vertical eddy diffusion of 1·5–50 cm^2/sec for the bottom water of the South Atlantic. Eddy diffusion is also involved in the release of gas to the atmosphere when the water subsequently reaches the surface again.

V. FACTORS AFFECTING THE CONCENTRATIONS OF GASES IN SEA WATER

For purposes of discussion it is helpful to consider the factors influencing the concentrations of dissolved gases in sea water under two headings: those which affect the surface water and those which occur in the body of the water.

A. SURFACE FACTORS

As we have seen above, the dissolved gas concentrations are usually close to their saturation values in the surface waters of the sea (but see also p. 114). The gross factors controlling their concentrations at any location are therefore those which control their solubilities. The principal of these are the prevailing temperature and salinity of the water. Barometric pressure and atmospheric humidity* also influence the solubility since the solubility is proportional to the partial pressure of the gas in the atmosphere (Henry's Law) (see equation 5.2). The seasonal

---

* It is probable that the effect of atmospheric humidity is negligible since the relative humidity of the air close to the sea surface is usually close to 100%.

variations in atmospheric pressure at high latitudes, where most of the deep waters of the oceans are formed by sinking, may exceed 22 Torr; this corresponds with a variation in dissolved gas concentration of 2·9% (Carritt, 1954). The observed variations in the argon content of the intermediate and deep waters of the South Pacific are probably to be explained by pressure variations at the time of saturation (Craig *et al.*, 1967).

The closeness with which surface waters approach saturation with gases is probably controlled by the interplay of several factors such as surface conditions, the kinetics of dissolution, surface heating and the extent to which the surface is covered by organic or inorganic films. Of these, the first is probably the most important. It might be thought that equilibrium would be most nearly attained under still conditions. However, the surface will not be able to attain equilibrium if the rate of transport of dissolved gas between the surface film and the under- or over- saturated subsurface water exceeds the rate of diffusion of gas through the surface film. Turbulence increases the latter (the rate of exchange is approximately proportional to the square of the wind velocity, above a critical speed of 3 cm/sec) and this assists in the ventilation of the mixed layer. Increasing turbulence leads to the production of bubbles in the surface layer of the water; these with their high surface/volume ratios and high internal pressures speed the attainment of equilibrium. During violent storms, bubbles are carried down to considerable depths, and since the water surrounding them will be undersaturated with respect to the prevailing hydrostatic pressure they will tend to dissolve. The resulting water if brought to the surface will then be supersaturated.

B. INTERNAL FACTORS

When surface waters are advected to the depths, they retain the salinity, (potential) temperature and dissolved inert gas concentration which they possessed at the surface. These properties can then be used as tracers for an individual water mass until mixing causes it to lose its identity. If mixing takes place between two masses of sea water which have been equilibrated with the atmosphere at different temperatures, the resultant water will be supersaturated when raised to the surface because of the non-linear relationship between temperature and solubility. In the extreme case of mixing equal volumes of water ($Cl = 20^0/_{00}$) equilibrated at $0°C$ and $30°C$ the mixture would contain $11·06$ ml N_2/l compared with the saturation value of $10·27$ ml/l at $15°C$. The effects of such mixing have been noted for argon and krypton by Bieri *et al.* (1964) who observed that marked supersaturation (i.e.

4–6%) of these gases occurred in, or adjacent to, the thermocline in both the Atlantic and Pacific Oceans. In the same way, heating of the subsurface water by, for example, submarine vulcanism, although it would not alter the actual concentration of gases, would cause the percentage saturations (based on the new temperature) to be different.

The discussion has so far concerned only processes which change the saturation of gases in subsurface waters without altering their concentrations. However, there are a number of physical and biological processes which add or extract gas from the water; of these the latter are by far the most important.

(i) *Physical processes*

Radioactive decay leads to the introduction of helium, radon and argon into the sea. Helium–4 is produced by the decay of uranium and its daughters and mainly enters the sea from the interior of the earth through the sediments (Bieri et al., 1964). The deep waters of the ocean are therefore enriched with the element (Bieri et al., 1966, 1968). The deep waters of the South Pacific are considerably more enriched with helium than are those of the North Atlantic as befits their much longer residence time (1,000 years compared with 200–300 years in the Atlantic). These waters eventually reach the surface when the excess helium is lost, first to the atmosphere, and then to space at the same rate as it is produced (Revelle and Suess, 1963). Clarke et al. (1969) have recently observed that the ^{3}He/^{4}He ratio in deep sea water is appreciably higher than that in water from the mixed layer. They have attributed this enrichment to the leakage of primordial ^{3}He through the ocean floor. However, Fairhall (1969) has pointed out that a significant proportion, if not all, of this excess ^{3}He may be produced through radioactive decay of tritium (^{3}H). Radon is formed by decay of the radium present not only in the water (p. 98), but also in the sediments, and for this reason, easily measured excesses of this gas are present in the water within a few metres of the bottom (Broecker, et al., 1967). The decay of ^{40}K in the sea gives rise to ^{40}Ar*, but the short period effect is negligible. Unusual conditions may give rise to localized changes in dissolved gas concentrations. In some regions submarine vulcanism may inject volcanic gases (e.g. N_2 and CO_2) into the sea, in which they dissolve under the prevailing high hydrostatic pressure. Loss of oxygen and accretion of carbon dioxide and nitrogen occur in the layers lying immediately over euxinic waters (see p. 114–5).

* Revelle and Suess (1963) estimate that in 1 l of sea water 50 atoms of argon accumulate per minute.

(ii) Biological processes

(a) Oxygen and carbon dioxide. The competing processes of photosynthesis and respiration are the main causes of *in situ* changes in the concentrations of dissolved oxygen and carbon dioxide in the sea. In the upper layers of the sea, where there is sufficient light, photosynthesis by phytoplankton may predominate and lead to the removal of carbon dioxide and to the liberation of oxygen, in rough accordance with the equation

$$xCO_2 + xH_2O \rightleftharpoons (CH_2O)_x + xO_2$$

where $(CH_2O)_x$ represents carbohydrate (see p. 230). Under optimum conditions, this will lead to supersaturation of the surface layers with oxygen, and saturation values of over 120% have been recorded (Richards, 1957).

Below the compensation depth (p. 237) the reverse process predominates, and dissolved oxygen is consumed by the respiration of plants, animals and bacteria. The ultimate factor limiting the consumption of oxygen in the sea is the supply of organic matter. In the open oceans oxygen is supplied so rapidly by advection and diffusion that it does not become completely exhausted, although it is reduced to very low levels (ca 0·1 ml/l) in the intermediate and deep waters in some regions of the Pacific. Without such ventilation the deep waters would soon become stripped of their oxygen and therefore be unable to support normal forms of life. In waters where the circulation is very sluggish, e.g. the subsurface waters of the Black Sea and the Cariaco Trench, the rate of supply of organic matter exceeds that of oxygen, and anoxic conditions are set up under which hydrogen sulphide is liberated (see p. 115). Such waters are unusually rich in carbon dioxide.

(b) Nitrogen. Little is known about the processes of nitrogen fixation and denitrification in the sea. Dugdale *et al.* (1961) have demonstrated that the blue-green alga *Trichodesmium thiebautii* is able to utilize the dissolved gas to supply its nitrogen requirements when other inorganic forms of the element have become exhausted. This organism grows abundantly in tropical and subtropical waters, e.g. the Indian Ocean, at times when nutrient elements are at low levels because of phytoplankton growth. However, it has only recently been established that fixation by this class of algae is significant in maintaining the nutrient balance in these waters (Dugdale and Goering, 1964), although its role in limnology has been known for some time (see also p. 156).

The reverse process, denitrification, which appears to take place only in anoxic, or almost anoxic waters, may occur either by reduction of nitrate, or by arrested oxidation of organic nitrogen or ammonia. At present, evidence suggests that nitrate reduction predominates (see also p. 163). Thus, in a study using ^{15}N–labelled nitrate, Goering (1968) has shown that both molecular nitrogen (5–10 μg N/1/hr) and nitrite are formed simultaneously by bacterial reduction of nitrate in water containing less than 0·2 ml O_2/l taken from below the thermocline in the east tropical Pacific. About 0·1–0·8 atoms of nitrogen are produced per nitrate ion reduced. A direct indication that denitrification is occurring in the anoxic water of the Cariaco Trench was obtained by Richards and Benson (1961), who showed by an examination of the N_2 : Ar and [14]N : [15]N ratios that the dissolved gas is enriched in biogenic nitrogen. Fiadeiro and Strickland (1968) from a consideration of plots of phosphate and nitrate concentrations against apparent oxygen utilization (see p. 108) have demonstrated that denitrification is, in fact, occurring by this mechanism in the south-flowing, almost anoxic water near the Peruvian coast.

(c) *Hydrogen sulphide*. Formation of hydrogen sulphide in the sea takes place only in basins which are so poorly ventilated that their waters have become completely anoxic; for example, the Black Sea, the Cariaco Trench (Fig. 5.1) and certain Norwegian Fjords. Such waters cannot support normal forms of life, and the only living organisms which they contain are specialized forms of bacteria, such as *Desulphovibrio desulphuricum* (see Jajic, 1969). These utilize the oxygen from sulphate ions instead of dissolved oxygen for their metabolic processes. This leads to the liberation of hydrogen sulphide and to the setting up of conditions of negative redox potential. Thus, hydrogen sulphide concentrations as high as 7·3 ml/l and redox potentials as low as −129 mV are encountered at depth in the Black Sea. Hydrogen sulphide is oxidized rapidly by dissolved oxygen by a purely chemical mechanism. The euxinic layer is therefore separated from the upper well-ventilated water by an intermediate zone of unstable redox potential, in which dissolved oxygen diffusing from above reacts with the hydrogen sulphide. The kinetics of the oxidation have been found to be complex by Cline and Richards (1969) who have shown that thiosulphate and sulphate are the major products, with sulphite occurring as an unstable, but long-lived intermediate,

$$2HS^- + 2O_2 \rightarrow S_2O_3^{2-} + H_2O$$

$$HS^- + 2O_2 \rightarrow SO_4^{2-} + H^+$$

$$2HS^- + 3O_2 \rightarrow 2SO_3^{2-} + 2H^+$$

When oxygen was the limiting reactant, the maximum concentrations of thiosulphate-sulphur and sulphite-sulphur reached 30–35% and 10–15% respectively of the sulphide oxidized. The oxidation is catalysed by iron and the yield of thiosulphate is thereby greatly increased.

(d) *Methane.* Methane is fairly uniformly distributed in the atmosphere at a concentration in the range 1·2–1·5 ppm (Junge, 1963). Although other mechanisms may also be involved in its production, it

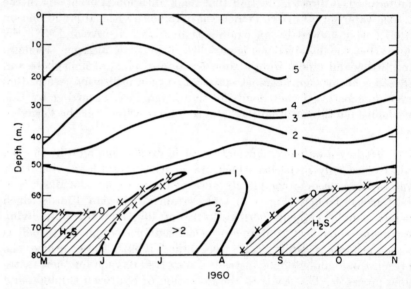

Fig. 5.1. Dissolved oxygen and hydrogen sulphide in the Gulf of Cariaco, May–November, 1960, after Gade (1961) and Richards (1965). Oxygen isopleths are in ml/l and are indicated by continuous lines. Hydrogen sulphide isopleths are represented by — × — × — × .

is probable that much of it originates during the anoxic or almost anoxic decay of organic matter, e.g. in the soil and in marshes. For this reason, its concentration in river waters (Swinnerton *et al.*, 1969) and euxinic marine basins may be relatively high. In contrast, in the surface water of the open ocean, it is almost at equilibrium with the atmosphere. The geochemical balance of methane is probably maintained by photochemical oxidation in the upper atmosphere (Junge, 1963).

(e) *Carbon Monoxide.* Because most of the carbon monoxide in the atmosphere originates from the burning of fuels, its distribution is not uniform. Values of over 1 ppm may be found in urban areas, whereas over the North Atlantic concentrations as low as 0·01 ppm have been recorded (Junge, 1963; Robbins *et al.*, 1968). The concentration of the gas in the surface water of the western Atlantic has been studied by

Swinnerton *et al.* (1969), who observed that the carbon monoxide content of the water is considerably in excess of that which would be present if it were at equilibrium with the atmosphere above it. They therefore concluded that the excess gas must be produced in the water either by photo-chemical oxidation of dissolved organic matter, or by a biological mechanism (certain algae and siphonophores are known to produce carbon monoxide (Loewus and Delwicke, 1963; Pickwell *et al.*, 1964)).

VI. DISTRIBUTION OF OXYGEN IN THE SEA

More is known about the distribution of oxygen in the sea than other gases because of the ease with which it can be determined. Only the

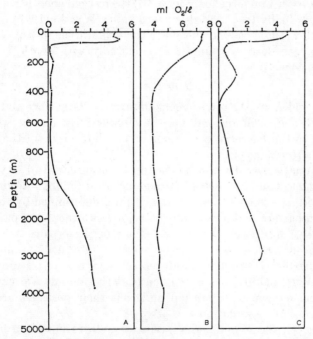

Fig. 5.2. Some representative profiles for the concentration of dissolved oxygen in the ocean. (A) eastern tropical Pacific (Albatross Station 74; 11° 39′ N., 114° 5′ W.) (B) Antarctic Convergence (Discovery Station 1054; 30° 08′ S., 35° 9′ W.) (C) eastern tropical Pacific, showing two minima (Albatross Station 65; 0° 21′ N., 103° 42′ W.)

main features of its distribution can be presented here, and the reader is referred for further details to the review by Richards (1957).

The observed vertical and horizontal distributions of the gas in the oceans result from the interplay between biochemical processes and those by which the gas enters and is transported in the water. The vertical distribution pattern, which is similar for most oceans, shows the following principal features:

(1) A wind-mixed surface layer, extending down to the thermocline has a uniform oxygen content, and is essentially close to equilibrium with the sea surface. Under conditions producing a stable water column (e.g. in temperate regions such as the English Channel in spring) a subsurface maximum may be set up, at a depth coinciding with that of maximum productivity (see p. 239).*

(2) Beneath the surface layer, the oxygen content decreases with increase in depth as a result of oxidation of organic matter. In some regions, e.g. the east tropical North Pacific Ocean (Fig. 5.2A) there may be a rapid decrease to almost anoxic conditions, whereas in others (e.g. in the Antarctic Convergence, Fig. 5.2B) the oxygen concentration may fall only gradually to $\frac{1}{2}-\frac{1}{3}$ of its surface value. Wyrtki (1962) using the data of Riley (1951) has shown that in the Atlantic the rate of consumption of oxygen decreases almost exponentially with depth, according to the relationship

$$R = R_0 e^{-\alpha z}.$$

Where R_0 and R are the rates of consumption at the surface and depth respectively; R_0 will depend on the productivity and α will vary according to the bacterial conditions and to how resistant the organic material is to oxidation.

(3) Minima caused by the biochemical consumption of oxygen are present in most oceans, except in regions of upwelling. According to Wyrtki (*loc. cit.*) the depth of the minimum is determined by the water circulation pattern and it normally coincides with the upper part of the layer of least advection. In many regions of the oceans two oxygen minima occur; these are separated by an intermediate maximum (Fig. 5.2C) which is probably produced by advection of water richer in oxygen. Wyrtki (1962) has concluded that the minima are caused by biochemical oxygen consumption and that their positions are determined by oceanic circulation.

(4) Beneath the minima, the oxygen concentration often increases gradually as a result of the circulation of water which has sunk from the surface at high latitudes. The consumption of oxygen in these deep waters is small since little organic matter reaches them from above, and most of that which does is resistant to oxidation.

The horizontal distribution throughout the oceans also reflects the pattern of oceanic circulation. Thus, the near surface water masses or

* In summer throughout the entire North Pacific there is a subsurface maximum which results from the stripping of oxygen from the water in the layer above it because of surface heating (Pytkowicz, 1964).

he eastern sides of the oceans in subtropical regions are almost devoid
f oxygen because an upward movement of deoxygenated water from
elow predominates over the sluggish horizontal circulation in this
ayer.

REFERENCES

Benson, B. B. and Parker, P. D. M. (1961). *Deep Sea Res.* **7**, 237.
Beckman Instruments Inc. (1969) Technical Memorandum, Fullerton, Calif.
Bieri, R. H., Koide, M. and Goldberg, E. D. (1964). *Science, N.Y.* **146**, 1035.
Bieri, R. H., Koide, M. and Goldberg, E. D. (1966). *J. geophys. Res.* **71**, 5243.
Bieri, R. H., Koide, M. and Goldberg, E. D. (1968). *Earth Planet.Sci. Letters*, **4**, 329.
Broecker, W. S., Li, Y. H. and Cromwell, J. (1967), *Science, N.Y.* **158**, 1307.
Carpenter, J. H. (1965). *Limnol. Oceanog.* **10**, 141.
Carpenter, J. H. (1966). *Limnol. Oceanog.* **11**, 264.
Carritt, D. E. (1954). *Deep Sea Res.* **2**, 59.
Carritt, D. E. and Kanwisher, J. W. (1959). *Anal. Chem.* **31**, 5.
Clarke, W. B., Beg, M. A. and Craig, H. (1969). *Earth Planet. Sci. Letters*, **6**, 213.
Cline, J. D. and Richards, F. A. (1969). *Environ. Sci. Tech.* **3**, 838.
Craig, H., Weiss, R. F. and Clarke, W. B. (1967). *J. geophys. Res.* **72**, 6165.
Dugdale, R. C. and Goering, J. J. (1964). *Limnol. Oceanog.* **9**, 507.
Dugdale, R. C., Menzel, D. W. and Ryther, J. M. (1961). *Deep Sea Res.* **7**, 297.
Fairhall, A. W. (1969). *Earth Planet. Sci. Letters*, **7**, 249.
Fiadeiro, M. and Strickland, J. D. H. (1968). *J. mar. Res.* **26**, 187.
Fade, H. G. (1961). *Bol. Inst. Oceanogr. Univ. Orient. Cumana, Venez.* **1**, 21.
Glueckauf, E. (1951). *In* "Compendium of Meteorology" (T. F. Malone, ed.), pp. 3–10. American Meteorological Society, Boston.
Goering, J. J. (1968). *Deep Sea Res.* **15**, 157.
Hoover, T. E. and Ibert, E. R. (1966). Texas A & M Univ. Tech. Rept. 66–10T.
Vajic, J. E. (1969). "Microbial Biogeochemistry." Academic Press, New York.
Junge, C. E. (1963) "Air Chemistry and Radioactivity " Academic Press, New York.
Keeling, C. D. (1965). *J. geophys. Res.* **70**, 6099.
Loewus, M. W. and Delwicke, C. C. (1963). *Plant Physiol.* **38**, 371.
Murray, C. N. and Riley, J. P. (1969). *Deep Sea Res.*, **16**, 311.
Murray, C. N. and Riley, J. P. (1970). *Deep Sea Res.* **17**, 203.
Murray, C. N., Riley, J. P. and Wilson, T. R. S. (1969). *Deep Sea Res.* **16**, 297.
Pickwell, G. V., Barham, E. G. and Wilton, J. W. (1964). *Science, N.Y.* **144**, 860.
Pytkowicz, R. M. (1964). *Deep Sea Res.* **11**, 381.
Redfield, A. C. (1942). *Pap. phys. Oceanog.* **7**, 1.
Redfield, A. C. (1948). *J. mar. Res.* **7**, 347.
Revelle, R. R. and Suess, H. E. (1963). *In* "The Sea, Ideas and Observations on Progress in the Study of the Seas" (M. N. Hill, ed.), Vol. I. Interscience, New York and London.
Richards, F. A. (1957). *In* "Treatise on Marine Ecology and Palaeoecology" (J. W. Hedgpeth, ed.). Vol 1. Memoir 67, Geological Society of America, New York.
Richards, F. A. (1965). *In* "Chemical Oceanography" (J. P. Riley and G. Skirrow, eds), Vol. 1. Academic Press, London.

Richards, F. A. and Benson, B. B. (1961). *Deep Sea Res.* **7**, 254.

Riley, G. A. (1951). *Bull. Bingham oceanogr. Coll.* **13**, 1.

Robbins, R. C., Borg, K. M. and Robinson, E. (1968). *J. Air Pollut. Contr. Ass* **18**, 106.

Setchenow, J. (1875). *Mem. Acad. Imp. Sci. St. Petersb.* 7 ser. No. 6.

Strickland, J. D. H. and Parsons, T. R. (1968). A practical handbook of sea wate analysis *Fish. Res. Bd Can. Bull.* **167**, 311 pp.

Swinnerton, J. W., Linnenbom, V. J. and Cheek, C. H. (1962a). *Anal. Chem* **34**, 483.

Swinnerton, J. W., Linnenbom, V. J. and Cheek, C. H. (1962b). *Anal. Chem.* **34** 1509.

Swinnerton, J. W., Linnenbom, V. J. and Cheek, C. H. (1968). *Limnol. Oceanog* **13**, 193.

Swinnerton, J. W., Linnenbom, V. J. and Cheek, C. H. (1969). *Environ. Sci. Tech* **3**, 836.

Winkler, L. W. (1888). *Ber. dtsch. chem. Ges.* **21**, 2843.

Wyman, J., Scholander, P. F., Edwards, G. A. and Irving, L. (1952). *J. mar. Re* **11**, 47.

Wyrtki, K. (1962). *Deep Sea Res.* **9**, 11.

Chapter 6

The Dissolved Gases in Sea Water.
Part 2. Carbon Dioxide

I. INTRODUCTION

The discussion of dissolved gases in the sea in Chapter 5 was concerned mainly with those which do not react with water. However, carbon dioxide, which from the biological standpoint is one of the most important gases, participates in the following reactions when it comes into contact with aqueous solutions:

$$CO_2(g) + H_2O \rightleftharpoons CO_2(s) + H_2O$$

$$CO_2(s) + H_2O \rightleftharpoons H_2CO_3$$

$$H_2CO_3 \rightleftharpoons HCO_3^- + H^+$$

$$HCO_3^- \rightleftharpoons CO_3^{2-} + H^+$$

Coexisting with these equilibria is that for the ionization of water $(2H_2O \rightleftharpoons H_3O^+ + OH^-)$. It will be shown later that account must be taken of the ionization of boric acid $(H_3BO_3 + H_2O \rightleftharpoons B(OH)_4^- + H^+)$[*] when attempts are made to estimate the carbonate alkalinity of sea water. The picture is complicated by the presence of ion pairs (such as $NaCO_3^-$) formed between bicarbonate and carbonate ions and the major cations of sea water (p. 72). Further complications arise in organic-rich waters, because of the presence of hydrogen sulphide and other weak acids, and because of the existence of carbamino-carboxylic acids formed by reaction of carbon dioxide with amino acids

$$(RNHCH_2COOH + H_2CO_3 \rightleftharpoons RN(COOH)CH_2COOH + H_2O).$$

Each of these reactions has associated with it an equilibrium constant which is expressed in terms of the thermodynamic activities of the

---

[*] Information from raman spectroscopy suggests that this is the ionized form of boric acid present in aqueous solution not $H_2BO_3^-$.

appropriate species (p. 70). The concentrations of the reactants will be influenced by changes in salinity not only because activity coefficients are functions of ionic strength (see p. 71) but also because of complexing of the reactants (p. 72). Since equilibrium constants are functions of temperature and pressure, the *in situ* equilibrium concentrations of the reactants in the sea will be affected by these parameters.

Apart from its intrinsic value, a detailed knowledge of the carbonic acid equilibrium system is of considerable practical importance. In principle it enables the concentrations of the various species, such as HCO_3^-, CO_3^{2-}, H_2CO_3 and total CO_2 to be calculated from the results of two relatively simple measurements—those of pH and alkalinity (see below Sections IIa3 and IIa5). In the account which follows (Section IIB), the theoretical treatment has been considerably simplified and idealized in order to draw attention to the main principles. The agreement between measured parameters and those predicted by this simple theory is therefore often only approximate. The complicating features which must be taken into consideration in a more rigorous treatment are outlined in Section IIB(ii). However, no fully comprehensive mathematical solution has yet been produced which takes all these factors into account. Although much further work is needed before the theoretical aspects of the carbon dioxide system are fully understood, a good deal is known about the distribution of the gas in the oceans (see Section III).

II. Theory of Carbon Dioxide Equilibria in Sea Water Systems

A. SOME RELEVANT CONCEPTS

Before dealing with the carbon dioxide system itself it is convenient to discuss a number of relevant concepts.

(i) Ionization of water

The ionization of water, which is a very weak electrolyte, takes place according to the equation*

$$H_2O \rightleftharpoons H^+ + OH^-$$

for this reaction the equilibrium constant is

$$K_{H_2O} = \frac{a_H \cdot a_{OH}}{a_{H_2O}}^\dagger = c_H \cdot c_{OH} \cdot \frac{\gamma_H \cdot \gamma_{OH}}{a_{H_2O}} = k_w \cdot \frac{\gamma_H \cdot \gamma_{OH}}{a_{H_2O}} \tag{1}$$

* Hydrogen ions are present in aqueous media in solvated forms, e.g. H_3O^+, $H_5O_2^+$ etc. For simplicity, the symbols H^+, a_H and c_H are used in the text, but should be assumed to refer to the solvated ions.

† For simplicity the charges on the various species in equation (1) and following equations have been omitted.

where a, c and γ are the thermodynamic activities, concentrations and activity coefficients respectively of the various species, and k_w is the (concentration) ionic product. In pure water a_{H_2O}, γ_H and γ_{OH} are all unity, and the thermodynamic ionic product $k_{w(TD)} = a_H \cdot a_{OH}$, which is then identical with k_w, has the value of 10^{-14} at 22°C. The neutral point, at this temperature is where $c_H = c_{OH} = 10^{-7}$. k_w is a function of both temperature and pressure, decreasing to 0.12×10^{-14} at 0°C and increasing by ca 2·5 fold under a pressure of 1,000 atm. k_w is also influenced by the presence of electrolytes which produces changes in the activity coefficients of the ions and of water. Buch (1938) has estimated that k_w in ocean water is about 1·75 times its value in pure water. This means that the neutral point in this medium is at pH 7·33 at 0°C and at pH 6·98 at 16°C.

(ii) Activity of water

The dissolved ions in sea water depress the activity of the water. The latter can be assessed from the ratio of the vapour pressures (V.P.) of sea water and pure water according to the relationship

$$a_{H_2O} = \frac{\text{V.P.}_{\text{sea water}}}{\text{V.P.}_{\text{pure water}}} = 1 - 0.000969\ \text{Cl}^0/_{00}. \qquad (2)$$

Thus, in a sea water of chlorinity $20^0/_{00}$ $a_{H_2O} = 0.981$.

(iii) The pH concept

(a) General. On a thermodynamic basis pH is defined according to the relationship

$$\text{pH} = -\log_{10} a_H$$

where a_H is the activity of the hydrogen ion, although the meaningfulness of single ion activities or ionic activity coefficients is a matter of some doubt (see p. 71). Unfortunately, fundamental difficulties are encountered when attempts are made to measure a_H. If an E.M.F. cell of the type

$$\text{H}_2(\text{Pt})\ |\ \text{H}^+\text{X}^-(\text{aq})\ \vdots\ \text{KCl(aq)}\ |\ \text{Hg}_2\text{Cl}_2\ |\ \text{Hg}$$
$$\qquad\qquad\text{solution X}\qquad\text{salt bridge}$$

is set up, in which the hydrogen electrode (at 1 atm pressure) is contained in the solution X, the a_H of which is to be determined, and the latter is separated from the calomel reference electrode by a salt bridge (SB), then, according to the Nernst equation,

$$E = E° - \frac{RT}{F} \ln a_{H(X)} \cdot a_{Cl(SB)} - E_l \qquad (3)$$

Where E is the measured E.M.F. of the cell and $E°$ is the standard potential. E_l, which is the potential associated with the liquid junction in the cell, cannot be measured, and the individual ionic activity of chloride in the salt bridge ($a_{Cl(SB)}$) cannot be determined. For these reasons it is not possible to make estimates of a_H without making assumptions not based on thermodynamics. Provided that E_l is assumed to remain constant, it is possible to compare the a_H values, and thus the pH values, of two solutions by making E.M.F. measurements on them in a cell of the type represented above. If a value is *assumed* for the pH of one of these solutions, an operational pH scale, such as that recommended by the U.S. National Bureau of Standards, can be set up. This scale is based on the supposition that 0·05 M potassium hydrogen phthalate buffer solution has a pH value (pH(S)) of 4·00 at 15°C. Then, if E_s and E_x are the E.M.F. values for the standard buffer and unknown solution of pH(X) respectively

$$pH(X) - pH(S) = \frac{E_x - E_s}{2·303 \ RT/F} \qquad (4)$$

Up to this point the liquid junction potential (E_l) has been assumed to be constant. This is probably a reasonable approximation where the standard buffer and the unknown solution have similar pH values and ionic strengths. However, when the ionic strength of the medium is high, as it is with sea water, the difference in E_l will produce a significant and indeterminable error in the measured pH value. For this reason Smith and Hood (1964) have suggested that an alternative operational scale should be set up for oceanographic purposes using buffers made up in sea water. It should be appreciated that although it is unwise to make theoretical interpretations from measured pH values in sea water medium, highly reproducible values can be obtained which are very useful for comparative purposes.

(b) *Effect of temperature and pressure on pH.* The dissociation constants of carbonic acid are functions, not only of salinity, but also of temperature and pressure (see pp. 132 and 137). Although changes in the physical parameters of waters out of contact with the atmosphere will not influence the total concentration by weight of carbon dioxide present, they will affect both the pH of the water and the distribution of carbon dioxide among its various species. At a constant total carbon dioxide concentration pH falls with increasing temperature. The change in pH can be calculated from the equation

$$pH_{t_2} = pH_{t_1} + x(t_2 - t_1)$$

where pH_{t_1} and pH_{t_2} are the pH values of the water at temperatures t_1 and t_2°C respectively. The value of the constant x has been determined by C. N. Murray (private communication) and has been found to be 0·0111±0·0010 pH units/°C, for all pH values in the range 7·5-

8·4, for salinities between 10 and 40⁰/₀₀, and over the temperature range 1–30°C. This agrees well with the value $0·0118\pm0·0006$ calculated by Gieskes (1969) from Lyman's (1956) dissociation constants for carbonic acid, and with his experimental value of $0·0112\pm0·0015$ for the pH range 7·2–8·1.

The pH of sea water falls with increasing hydrostatic pressure. Table 6.1 (p. 131) shows the value that must be subtracted from the pH measured at 1 atm to obtain the pH value at pressures up to 1,000 atm (Culberson and Pytkowicz, 1968). Application of the above temperature and pressure compensations enables the *in situ* pH to be readily calculated.

(c) *Measurement of pH*. In the past, indicator methods were used for the determination of the pH of sea water. They have now been completely superseded by more convenient and precise potentiometric procedures using the glass electrode. The latter is a thin bulb of conductive glass containing a buffer in which a silver : silver chloride electrode is immersed. It is used in conjunction with a calomel reference electrode and high impedance valve voltmeter (pH meter). The pH response of the glass electrode usually does not correspond with that of the hydrogen electrode, and it is necessary to standardize it at frequent intervals using buffers having pH values lying on each side of that of the unknown solution. Glass electrode systems capable of measuring pH *in situ* down to depths of several hundreds of metres are available (see e.g. Disteche and Dubuisson, 1960).

(iv) *Buffering action*

Solutions of weak acids or bases with their salts exhibit a marked pH buffering action. This means that the addition of small amounts of acid or alkali will produce much smaller changes in pH than they would if added to pure water. If we consider a solution of a weak acid (HA), dissociating according to the equation $HA \rightleftharpoons H^+ + A^-$, together with its totally ionized salt then, if the thermodynamic dissociation constant of the acid (K_{HA}) is small, the pH of the solution will be given fairly closely by

$$pH = pK_{HA} + \log\frac{c_A}{c_{CA}} + \log\frac{\gamma_A}{\gamma_{HA}} \tag{5}$$

where c_A and c_{HA} are the concentrations of A^- and HA respectively and γ_A and γ_{HA} are the relevant activity coefficients. When a small addition of acid (Δc_H) is made to the system, only the ratio c_A/c_H in the expression will change and will now become $(c_A - \Delta c_H)/(c_{HA} + \Delta c_H))$. As a result, the change in pH will be slight provided that c_A and c_{HA} are relatively large compared with Δc_H. Consideration of the final term in the equation indicates that if the ionic strength of a buffered solution is altered the pH will change since γ_{HA} and particularly γ_A are functions of this variable.

The ability of a buffer to resist pH changes on addition of strong acid or alkali is expressed quantitatively in terms of the buffer capacity $\beta = dc_a/d\mathrm{pH}$, where dc_a represents an infinitessimally small addition of strong acid. When a solution of a weak monobasic acid is titrated with a strong base the maximum buffer capacity occurs at the half titration point at a pH value equal to pK_{HA} where $c_A = c_{HA}$. The buffer capacity curve of a weak dibasic acid (H_2A), such as carbonic acid, shows an additional maximum corresponding to the point where $c_{HA} = c_{H_2A}$.

In the open ocean the pH of sea water rarely falls outside the limits 7·8–8·2, although more extreme values may be encountered in restricted locations such as rock pools, or where industrial contamination occurs. On a geological time scale it has been proposed that the pH of the sea is regulated by equilibria involving suspended clay minerals (Sillén, 1963), e.g.

$$KAl_3Si_3O_{10}(OH)_2(s) + 1 \cdot 5\ H_2O + H^+ \rightleftharpoons 1 \cdot 5\ Al_2Si_2O_5(OH)_4(s) + K^+$$
potassium mica   kaolinite

However, more immediate control is exerted by the buffering action of the carbonic acid system and, to a lesser extent, by the boric acid-borate equilibrium (no other weak acids occur at sufficiently high concentrations to exert a significant effect).

Some pH variations do occur in the ocean despite the buffering action of the carbonic acid system. These differences are almost entirely brought about by changes in the concentration of the dissolved carbon dioxide (see below p. 144–7).

(v) Alkalinity and carbonate alkalinity

According to the Brønsted-Lowry theory those ions which are able to accept protons are to be regarded as bases. The anions of weak acids and the hydroxyl ion are to be placed in this category since they can react with water (which can be considered in this context to be an acid as it donates a proton) yielding an OH^- ion and the conjugate acid, for example

$$HCO_3^- + H_2O \rightleftharpoons H_2CO_3 + OH^- \tag{6}$$
conjugate acid

In ordinary sea waters the only anions of weak acids present at significant concentrations are the bicarbonate, carbonate and borate ions, and it is to the presence of these that sea water owes its alkalinity.* Quantitatively the alkalinity is the sum of the equilibrium concentrations of these anions expressed in g ions/l (e.g. c_{HCO_3} etc.).

$$\text{alkalinity (equivalents/l.)} = c_{HCO_3} + 2c_{CO_3} + c_{B(OH)_4} + (c_{HO} - c_H) \tag{7}$$

---

* In anoxic waters the HS^- and NH_4^+ ions also contribute to the alkalinity and additional terms must be added to equation 7 to correct for their presence. In this context NH_4^+ is not a proton acceptor, but associates with OH^- ions.

The final term in equation (7) is necessitated by the fact that the concentrations of the various anions determined by titration are greater than their equilibrium concentrations by $(c_{OH} - c_H)$ because of reactions of the type shown in equation (6). In the pH range 5·5–8·5—covering that of normal sea waters—this term can be neglected since it is very small compared with the other terms in this equation.

The alkalinity of sea water can be determined directly by potentiometric titration of the sample with standard acid (Edmond, 1970). The equivalence point is best determined from the potentiometric data using the graphical Gran plot method (Dyrssen, 1965) as it is difficult to determine accurately by conventional means. A simpler and much faster method (Anderson and Robinson, 1946) is to treat a 100 ml aliquot of the sample with an accurately measured excess (25 ml) of 0·01 N hydrochloric acid and to measure the pH of the mixture, preferably after removing carbon dioxide with a current of air (Culberson et al., 1970). This pH is a measure of the excess of acid remaining after neutralization of the alkalinity, and a knowledge of this enables the alkalinity itself to be calculated.

The term *specific alkalinity* has been introduced to facilitate the comparison of the alkalinities of various bodies of water; where

$$\text{Specific alkalinity} = \frac{\text{alkalinity} \times 10^3}{\text{chlorinity} \,^0/_{00}} \tag{8}$$

Since it is a "mixed" unit Dyrssen and Sillén (1967) have suggested that it should be redefined entirely in terms of weight units as it would then be independent of pressure. According to Koczy (1956) different water masses frequently have characteristic specific alkalinities, these usually lie in the range 0·119–0·130 (weighted mean 0·126); higher values being found in the Pacific and lower ones in the Atlantic. Such variations appear to be associated in a complex fashion with differences in the calcium contents of the waters. Thus, low specific alkalinities are found in high productivity surface waters, in which calcium utilizing organisms proliferate (e.g. equatorial waters), or in waters in which chemical precipitation of calcium carbonate is occurring (e.g. Bahama Banks). High values are found in deep waters, in which the calcareous tests of dead plankton raining from the surface dissolve rapidly because of the prevailing high carbon dioxide concentration.

That part of the alkalinity which is associated with the carbonic acid system is termed the carbonate alkalinity $(c_{HCO_3} + 2c_{CO_3})$. This can be calculated from the alkalinity by deduction of the borate ion contribution. Boron occurs in sea water only as free boric acid and borate ion (reports of the existence of complexes of boric acid with carbohydrate in sea water appear to be ill-founded). There is a linear relationship in sea water between chlorinity and the total boron concentration ($c_{\Sigma B} = c_{H_3BO_3} + c_{B(OH)_4}$), such that

$$\frac{c_{\Sigma B}(\text{moles/l})}{\text{Cl}^0/_{00}} = 2 \cdot 1 \times 10^{-5} \qquad (9)$$

Since boric acid is a weak acid it ionizes appreciably only at pH values above 7; the contribution of the borate ion to total alkalinity is therefore significant only at pH values above ca 8. The concentration of borate ion can be evaluated by considering the equilibrium

$$H_2O + H_3BO_3 \rightleftharpoons H^+ + B(OH)_4^-$$

for which the apparent dissociation constant is

$$K_B^1 = \frac{a_H \cdot c_{B(OH)_4}}{c_{H_3BO_3}} \qquad (10)$$

the concentration of borate ions in sea water is therefore

$$c_{B(OH)_4} = \frac{K_B^1 \cdot c_{\Sigma B}}{(K_B^1 + a_H)} \qquad (11)$$

The carbonate alkalinity (CA) is thus given by

$$CA = c_{HCO_3} + 2c_{CO_3}$$
$$= \text{alkalinity} - \frac{K_B^1 \cdot c_{\Sigma B}}{(K_B^1 + a_H)} - (c_{OH} - c_H) \qquad (12)$$

For normal sea waters the final term in equation (12) is very small and may be neglected. Values of the apparent dissociation constant of boric acid as a function of chlorinity and temperature are given in Table 6.4. (p. 134).

(vi) Total carbon dioxide and P_{CO_2}

Only two parameters of the carbon dioxide system in sea water can be measured directly—total carbon dioxide (ΣCO_2) and the partial pressure of carbon dioxide (P_{CO_2})—all the others (such as c_{HCO_3} and c_{CO_3}) can only be calculated.

ΣCO_2, which is the sum of the concentrations of all the species of dissolved carbon dioxide (viz. $\Sigma CO_2 = c_{CO_2} + c_{H_2CO_3} + c_{HCO_3} + c_{CO_3}$), is usually estimated by acidifying the sample and determining the amount of carbon dioxide liberated by I. R. spectrometry or by gas chromatography. Alternatively, ΣCO_2 can be determined by potentiometric titration of the sample (Edmond, 1970).

If a sample of sea water is shaken with a bubble of air until equilibrium is attained, the partial pressure of carbon dioxide in the gas phase

(p_{CO_2}) is equal to that in the water P_{CO_2} (see p. 106). Analysis of the gas phase for carbon dioxide therefore enables P_{CO_2} to be determined (and this is related by Henry's Law, to the concentration of *free* dissolved carbon dioxide in the sample). This analysis can be most conveniently performed by making use of the strong absorption of infra red radiation by carbon dioxide. Instruments with a precision of ca $\pm 0\cdot3$ ppm are available for the continuous monitoring of the P_{CO_2} of sea water. In these, a small volume of air is equilibrated with the flowing sample and after drying is circulated through a gas analyser in which the I.R. absorption is measured differentially against a standard sample of air. P_{CO_2} can also be measured *in situ* by an electrometric method see p. 109.

B. EQUILIBRIA IN THE SEA WATER–CARBON DIOXIDE SYSTEM

When carbon dioxide gas is allowed to equilibrate with water a series of equilibria is set up. The first of these can be represented as

$$CO_2(gas) \rightleftharpoons CO_2(dissolved) \tag{13}$$

In dissolving in the water, the gas diffuses into the thin boundary layer lying just below the surface (p. 110). In this layer the predominant reaction at pH 8 of the dissolved molecular carbon dioxide is direct hydration

$$CO_2 + H_2O \rightleftharpoons H_2CO_3 \tag{14}$$

This process, and the reverse reaction (dehydration), which both have first order kinetics, are slow. Thus, the half-life of carbon dioxide in water is of the order of minutes. However, catalytic agents, particularly the enzyme carbonic anhydrase, can greatly increase the rates of both reactions. The carbonic acid produced in this way undergoes almost instantaneous ionization.

$$H_2CO_3 \rightleftharpoons H^+ + HCO_3^- \tag{15}$$
$$HCO_3^- \rightleftharpoons H^+ + CO_3^{2-} \tag{16}$$

At pH values above 8, as the hydroxyl ion concentration in the water becomes significant, bicarbonate ion is formed directly by reaction of dissolved carbon dioxide with the hydroxyl ion.

$$CO_2 + OH^- \rightleftharpoons HCO_3^-$$

This reaction, which is rapid in comparison with that for the hydration and dehydration reactions, predominates at pH values above 10.

In aqueous solutions, the balance between the various components of the carbon dioxide equilibria is controlled by the pH. Addition of carbon dioxide causes the pH to fall as the result of the formation of hydrogen ions by the ionization of carbonic acid. Figure 6.1 shows how the relative proportions of bicarbonate, and carbonate ions and dissolved carbon dioxide vary with change of pH in pure water and sea water. It will be seen that in pure water at pH values below 6, almost all the carbon dioxide is present as the dissolved gas. In this context the latter is the dissolved molecular carbon dioxide plus unionized carbonic acid (this amounts to only ca 1/600 of the dissolved gas). As the pH of the water is raised the proportion of bicarbonate increases to a maximum of ca 98% in pure water at a pH value of 8·5 at 0°C. Further increase in pH causes the proportion of bicarbonate to fall as carbonate ions are produced by equilibrium (16). With saline waters the curves for all three components are displaced towards lower pH values.

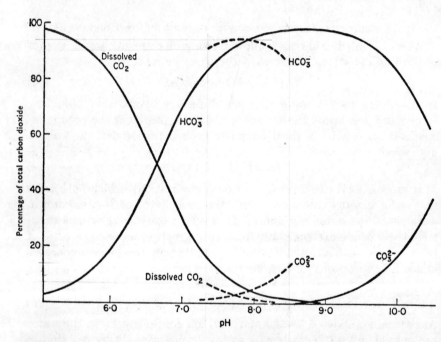

Fig. 6.1. Percentages of dissolved carbon dioxide, bicarbonate ion and carbonate ion as a function of pH at 0°C (data plotted from tables by Saruhashi, 1955). Continuous line, distilled-water; broken line sea water (Cl = 20⁰/₀₀) from Skirrow (1965).

(i) Dissolved carbon dioxide

When carbon dioxide dissolves in pure water the system obeys Henry's Law, i.e.

$$c_{CO_2} = p_{CO_2} \cdot \alpha_0$$

where c_{CO_2} is the concentration of the dissolved gas in moles/l, p_{CO_2} is the applied pressure of the gas in atmospheres and α_0 is the solubility co-efficient in moles/l/atm at the given temperature. Under the slightly alkaline conditions prevailing in sea water, P_{CO_2} is proportional only to the concentration of dissolved molecular gas and not to the total carbon dioxide concentration as ca 99% of the carbon dioxide will be present as bicarbonate and carbonate ions. Measurement of the P_{CO_2} of sea water (see p. 128) thus provides a convenient way of determining the concentration of the dissolved gas.

(ii) Dissociation of carbonic acid

When considering the ionization of carbonic acid in sea water it is helpful to combine equations 14 and 15, since the equilibrium constant for reaction (14) in sea water is not known accurately. The equilibrium for the first ionization is therefore

$$CO_2(\text{dissolved}) + H_2O(\text{liquid}) \rightleftharpoons H^+(\text{aq}) + HCO_3^-(\text{aq}) \qquad (17)$$

TABLE 6.1

Calculated values of $(pH_1 - pH_p)$ — at $34.8°/_{oo}$ salinity
(Culberson and Pytkowicz, 1968)

Temp. (°C)	Depth (m)	pH at atmospheric pressure				
		7·6	7·8	8·0	8·2	8·4
0	2,500	0·112	0·107	0·103	0·100	0·098
	5,000	0·222	0·213	0·205	0·200	0·196
	7,500	0·330	0·318	0·308	0·300	0·294
	10,000	0·437	0·422	0·409	0·399	0·391
5	2,500	0·107	0·102	0·098	0·096	0·094
	5,000	0·212	0·203	0·197	0·192	0·189
	7,500	0·316	0·304	0·294	0·288	0·283
	10,000	0·417	0·402	0·391	0·383	0·376
10	2,500	0·102	0·098	0·094	0·092	0·091
	5,000	0·203	0·195	0·189	0·185	0·182
	7,500	0·302	0·291	0·283	0·277	0·272
	10,000	0·401	0·387	0·376	0·369	0·362

As the thermodynamic activities (p. 70) of some of the species are not known, it is of practical value to substitute the molar concentration of these in the expression for the first and second dissociation constants.

The *apparent* dissociation constants (K_1^1 and K_2^1 respectively) are thus given by

$$K_1^1 = \frac{a_H \cdot c_{HCO_3}}{a_{H_2O} \cdot a_{CO_2}} \tag{18}$$

$$K_2^1 = \frac{a_H \cdot c_{CO_3}}{c_{HCO_3}} \tag{19}$$

The Finnish chemist Buch, after a series of painstaking investigations of the carbon dioxide system extending over more than 30 years, has published tables giving values of K_1^1 and K_2^1 as a function of salinity and temperature (Buch, 1951). More recently, Lyman (1956) has re-examined the dissociation constant of carbonic acid and his treatment will be

TABLE 6.2

First apparent dissociation constant of carbonic acid* in sea water
(expressed as pK$_{L1}'$) (Lyman, 1956)

Cl(⁰/₀₀)	t(°C)							
	0°	5°	10°	15°	20°	25°	30°	35°
0	6·58	6·52	6·47	6·42	6·38	6·35	6·33	6·31
1	6·47	6·42	6·37	6·33	6·29	6·26	6·24	6·23
4	6·36	6·32	6·28	6·24	6·21	6·18	6·16	6·15
9	6·27	6·23	6·19	6·15	6·13	6·10	6·08	6·07
16	6·18	6·14	6·11	6·07	6·05	6·03	6·01	5·99
17	6·17	6·13	6·10	6·06	6·04	6·02	6·00	5·98
18	6·16	6·12	6·09	6·06	6·03	6·01	5·99	5·97
19	6·15	6·11	6·08	6·05	6·02	6·00	5·98	5·97
20	6·14	6·10	6·07	6·04	6·01	5·99	5·97	5·96
21	6·13	6·09	6·06	6·03	6·00	5·98	5·96	5·95
25	6·09	6·05	6·02	6·00	5·97	5·95	5·93	5·92
36	6·00	5·97	5·94	5·92	5·89	5·87	5·86	5·84
49	5·92	5·88	5·86	5·84	5·82	5·80	5·78	5·77
64	5·84	5·80	5·78	5·76	5·74	5·72	5·71	5·70

* Values based on the N.B.S. pH scale.

used below. In this, pH measurements are made in terms of the U.S. National Bureau of Standards scale (p. 124).* The second dissociation constant is defined according to equation 19 but the first dissociation constant K_{L1}^1 is related to K_1^1 by

$$K_{L1}^1 = K_1^1 (p_s/p_0) \cdot (\alpha_0/\alpha_s) \tag{20}$$

* It should be pointed out that although the measured pH values of sea water have no absolute significance (p. 124), it is valid to use them in marine carbonate equilibrium calculations since the pH measurements used in determining the apparent dissociation constants of carbonic and boric acids were themselves measured by the same technique.

where the ratio of the vapour pressure of sea water (p_s) to that of pure water (p_0) is the activity of sea water (a_{H_2O}) (see p. 123), and the ratio of the solubility coefficients of carbon dioxide in pure water (α_0) and in sea water (α_s) is the activity coefficient of carbon dioxide in sea water (γ_{CO_2}) (see Table 6.5 on p. 136).

Thus,
$$K_{L1}^1 = \frac{a_H \cdot c_{HCO_3}}{c_{CO_2}} \qquad (21)$$

TABLE 6.3

Second apparent dissociation constant of carbonic acid* in sea water (expressed as pK'_2) *(Lyman, 1956)*

Cl(⁰/₀₀)				t(°C)				
	0°	5°	10°	15°	20°	25°	30°	35°
0	10·62	10·55	10·49	10·43	10·38	10·33	10·29	10·25
1	10·06	9·99	9·93	9·87	9·81	9·76	9·71	9·66
4	9·78	9·72	9·67	9·61	9·54	9·49	9·43	9·38
9	9·64	9·58	9·52	9·46	9·40	9·34	9·27	9·21
16	9·46	9·40	9·35	9·29	9·23	9·17	9·10	9·02
17	9·44	9·38	9·32	9·27	9·21	9·15	9·08	9·00
18	9·42	9·36	9·30	9·25	9·19	9·12	9·06	8·98
19	9·40	9·34	9·28	9·23	9·17	9·10	9·02	8·95
20	9·38	9·32	9·26	9·21	9·15	9·08	9·01	8·92
21	9·36	9·30	9·25	9·19	9·13	9·06	8·98	8·89
25	9·29	9·23	9·17	9·11	9·05	8·98	8·91	8·82
36	9·12	9·06	8·99	8·93	8·86	8·79	8·72	8·63
49	8·95	8·89	8·82	8·75	8·68	8·61	8·53	8·43
64	8·77	8·71	8·64	8·57	8·50	8·42	8·34	8·23

* Values based on the N.B.S. pH scale.

If a weak dibasic acid is titrated with an alkali, the pH-titre curve shows two end points, the pH values of which are numerically equal to the $pK(= \log 1/K)$ values of the acid. The presence of boric acid in sea water complicates the application of this technique to the determination of the apparent dissociation constants of carbonic acid in this medium. To overcome this difficulty Lyman (1956) carried out three successive titrations on each water sample, using a glass electrode to follow pH changes. The first titration was carried out with hydrochloric acid to an endpoint slightly beyond the stoichiometric point, at which all carbon dioxide was present as the dissolved gas. Carbon dioxide was removed by boiling, and the solution was then titrated to pH 10 with carbonate-free sodium hydroxide. The resulting solution was treated with sufficient hydrochloric acid to neutralize the sodium hydroxide and titrated to pH 10 with sodium hydroxide after the addition of mannitol which reacts with the boric acid to form mannitoboric acid—a strong acid. The second and third titrations enabled the concentration and apparent dissociation constants of boric acid to be determined. The first titration showed two end points, the first corresponding to the conversion of carbonate to bicarbonate and the second to that of

TABLE 6.4

First apparent dissociation constant of boric acid* in sea water (expressed as pK'_B*) (Lyman, 1956) (see p. 128)*

Cl(⁰/₀₀)	t(°C)							
	0°	5°	10°	15°	20°	25°	30°	35°
0	9·50	9·44	9·38	9·33	9·28	9·24	9·20	9·16
1	9·40	9·34	9·28	9·23	9·18	9·14	9·10	9·06
4	9·28	9·22	9·16	9·11	9·06	9·02	8·98	8·94
9	9·14	9·08	9·03	8·98	8·93	8·88	8·85	8·82
16	9·00	8·95	8·89	8·84	8·80	8·76	8·72	8·69
17	8·98	8·93	8·88	8·83	8·78	8·74	8·70	8·67
18	8·96	8·91	8·86	8·81	8·76	8·72	8·69	8·66
19	8·95	8·90	8·85	8·80	8·75	8·71	8·67	8·64
20	8·94	8·88	8·83	8·78	8·74	8·69	8·65	8·63
21	8·92	8·87	8·82	8·77	8·72	8·68	8·64	8·61
25	8·85	8·80	8·75	8·70	8·66	8·62	8·59	8·56
36	8·71	8·66	8·61	8·57	8·53	8·49	8·46	8·43
49	8·56	8·52	8·47	8·43	8·39⁵	8·36	8·33	8·30
64	8·41	8·37	8·33	8·30	8·26⁵	8·23	8·20	8·17

* Values based on the N.B.S. pH scale.

bicarbonate to carbonic acid. Some borate ions were converted to boric acid during this titration and correction had to be made for this before the first and second apparent dissociation constants of carbonic acid could be evaluated. The values found by Lyman for K^1_{L1}, K^1_2 and K^1_B at various temperatures and salinities are given in Tables 6.2–6.4. These constants can be used in conjunction with two or more of the following parameters: pH, carbonate alkalinity (CA), ΣCO_2 and P_{CO_2} to calculate the equilibrium concentrations of the participants in the carbonic acid system (Park, 1969). In the classical treatment by Buch (1930), pH and CA are used for this purpose:

(i) Bicarbonate ion concentration

$$c_{HCO_3} = \frac{CA}{1+2K_2^1/a_H} \text{ moles/l} \tag{22}$$

(ii) Carbonate ion concentration

$$c_{CO_3} = \frac{CA \cdot K_2^1}{[a_H + 2K_2^1]} \text{ moles/l} \tag{23}$$

(iii) Dissolved carbon dioxide concentration

$$c_{CO_2} = \frac{CA \cdot a_H}{K_{L1}^1[1+2K_2^1/a_H]} \text{ moles/l} \tag{24}$$

(iv) Partial pressure of carbon dioxide in the water

$$P_{CO_2} = \frac{CA \cdot a_H}{K_{L1}^1 \cdot \alpha_s[1+2K_2^1/a_H]} \text{ atm} \tag{25}$$

where α_s is the solubility of carbon dioxide in sea water, expressed in terms of moles/l/atm (see Table 6.5).

(v) Total carbon dioxide (ΣCO_2)

$$\Sigma CO_2 = c_{CO_2} + c_{HCO_3} + c_{CO_3}.$$

Combining equations 22, 23 and 24

$$\Sigma CO_2 = CA \cdot \left[\frac{1+K_2^1/a_H+a_H/K_{L1}^1}{1+2K_2^1/a_H}\right] \text{ moles/l.} \tag{26}$$

The only parameters of the carbon dioxide system which can be determined directly in sea water are ΣCO_2 and P_{CO_2}, of which the latter is probably the easier to determine (see p. 128). Unfortunately, it has been found that the observed P_{CO_2} values are often different from those calculated by application of equation 25. Thus, Hood (1963) has found that the observed values are up to 80% greater than the calculated ones in Batanabo Bay where carbonate precipitation is occurring. The calculated values of both alkalinity and P_{CO_2} are highly sensitive to pH, and some of this lack of accordance may be caused by failure to measure the pH with sufficient precision. However, it is likely that oversimplification of the theoretical treatment of the equilibrium is the principal cause of the disagreement. For this reason it is preferable to measure P_{CO_2} directly when accurate values of it are required. Park (1969) has suggested that, when the analytical precisions of the various measurements are taken into account, a combination of pH and P_{CO_2} may provide more reliable data for the oceanic CO_2

TABLE 6.5

Solubility of carbon dioxide in pure water and in sea water in moles/l/atmosphere (Murray and Riley, 1971)

Temp. °C / Cl‰	0	2	4	6	8	10	12	14	16	18	20	22	24	26
0	771	713	661	614	572	535	499	467	440	413	389	367	347	328×10^{-4}
15	670	620	576	536	500	468	437	409	385	364	344	325	308	291
16	664	615	571	531	496	464	433	406	382	360	341	322	305	289
17	658	609	565	526	491	460	429	402	378	357	338	319	303	287
18	652	603	559	521	486	455	425	398	374	354	335	316	300	284
19	645	597	554	515	481	450	421	394	371	350	332	313	297	281
20	638	591	548	510	476	445	416	390	367	347	328	310	294	278

system than the classical treatment based on pH and carbonate alkalinity given above.

$$c_{HCO_3} = \frac{K_{L1}^1 \alpha_s P_{CO_2}}{a_H} \tag{27}$$

$$c_{CO_3} = \frac{K_{L1}^1 \cdot K_2^1 \alpha_s P_{CO_2}}{a_H^2} \tag{28}$$

$$c_{CO_2} = \alpha_s P_{CO_2} \tag{29}$$

(iii) Complicating factors in the carbon dioxide system

The above discussion presents a much simplified picture of the carbon dioxide system in sea water. In a comprehensive treatment it would be necessary to take into account the existence of species of carbon dioxide other than CO_2, H_2CO_3, HCO_3^- and CO_3^{2-}. The principal of these are the ion pairs which the carbonate and bicarbonate ions form with sodium, calcium and magnesium (e.g. $MgHCO_3^+$, $MgCO_3^\circ$, $NaCO_3^-$). Garrels and Thompson (1962) have given grounds for thinking that more than 90% of the carbonate and 30% of the bicarbonate in sea water are present as ion pairs (see p. 72).* There is also some evidence of complex formation between carbon dioxide and amino acids, leading to the formation of carbamino carboxylic acids (see p. 121).

(iv) Effect of pressure on dissociation of carbonic acid

Up to this point the dissociation of carbonic acid at a pressure of 1 atm has been considered. When a body of water is removed from the surface of the sea to great depths, there will, of course, be conservation of ΣCO_2, if allowance is made for the compression of the water. However, since the dissociation constants of boric and carbonic acids vary with pressure the relative concentrations of the species arising from these two acids will change with depth and this will alter the pH of the water. However, since alkalinity is a conservative property and there will be no change in alkalinity and carbonate alkalinity other than that produced by the change in volume resulting from compression.

The effect of pressure on the dissociation constants of the two acids in sea water has been investigated by Culberson and Pytkowicz (1968). Their results are shown in Table 6.6, expressed in terms of pressure coefficients $(K_i^1)_d/(K_i^1)_1$] where $(K_i^1)_1$ and $(K_i^1)_d$ are the apparent dissociation constants at the surface and at the depth of d metres respectively).

By combination of the data in Table 6.6 with those given in Tables 6.2–6.4 the apparent dissociation constants at any depth may be evaluated. The corrected value of K_B^1 may be substituted in equation

---

* The existence of these ion-pairs does not detract from the correctness of equations 22–29, since the concentration terms in the equations refer to total concentrations,

$$\text{e.g. } c_{CO_3} = c_{CO_3^{2-} \text{ free}} + c_{NaCO_3^-} + c_{CaCO_3^\circ} + c_{MgCO_3^\circ}$$

12 to calculate the *in situ* value of the carbonate alkalinity. The latter may then be substituted together with the *in situ* values of K_{L1}^1, K_2^1 and a_H (p. 125) in equations 22, 23, 24 and 26 to estimate the equilibrium values of c_{HCO_3}, c_{CO_3}, c_{CO_2} and ΣCO_2 at the chosen depth. An example will help to illustrate the magnitude of the changes produced by pressure (Culberson and Pytkowicz, 1968). A water sample of salinity $34 \cdot 69^0/_{00}$ taken at a depth of 7,210 m (*in situ* temperature $1 \cdot 86 \,^\circ C$) was found to have a pH of $7 \cdot 78$ at $25 \,^\circ C$ and an alkalinity of $2 \cdot 60$ meq/kg. Compensating for the effect of temperature and pressure, the *in situ* pH

TABLE 6.6

Pressure coefficients of apparent dissociation constants of carbonic and boric acids in sea water ($S = 34 \cdot 2{-}35 \cdot 2^0/_{00}$), after Culberson and Pytkowicz (1968)

Depth (m)	$(K_{L1}^1)_d/(K_{L1}^1)_1$			$(K_2^1)_d/(K_2^1)_1$			$(K_B^1)_d/K_B^1)_1$		
	0 °C	5 °C	10 °C	0 °C	5 °C	10 °C	0 °C	5 °C	10 °C
1,000	1·12	1·11	1·11	1·07	1·07	1·07	1·14	1·13	1·13
2,000	1·25	1·24	1·23	1·15	1·15	1·15	1·30	1·28	1·27
4,000	1·55	1·53	1·50	1·34	1·33	1·32	1·67	1·64	1·61
6,000	1·92	1·88	1·84	1·55	1·53	1·51	2·14	2·09	2·03
8,000	2·37	2·30	2·23	1·79	1·76	1·73	2·73	2·64	2·55
10,000	2·91	2·80	2·70	2·07	2·03	1·99	3·45	3·31	3·18

would be $7 \cdot 73$. The effect of the pressure causes c_{CO_2} and c_{CO_3} to decrease by ca 15% and c_{HCO_3} to increase by ca $0 \cdot 6\%$ with respect to their values at the surface. For further information on the effect of pressure on carbon dioxide equilibria in the marine environment the reader should consult Pytkowicz (1968).

(v) The dissolution of calcium carbonate in the sea

Calcium carbonate occurs in the marine environment in two crystalline forms, calcite and aragonite, which have different solubilities. In considering the solubility of these forms of calcium carbonate in sea water it is helpful to make use of their apparent solubility products. (K_{SP}^1) where

$$K_{SP}^1 = c_{Ca} \cdot c_{CO_3}$$

K_{SP}^1 is a function of temperature, pressure and salinity, typical values for K_{SP}^1 at $25 \,^\circ C$ and $S = 36^0/_{00}$ for calcite and aragonite at a pressure of 1 atmosphere are $0 \cdot 6 \times 10^{-6}$ and $1 \cdot 1 \times 10^{-6}$ respectively.[*] Consideration of these values suggests that most surface and near-surface waters of normal salinity and pH are apparently supersaturated by as much as $100{-}300\%$ with respect to both minerals. However, *inorganic* precipita-

[*] Hawley and Pytkowicz (1969) have recently determined the solubility product of calcium carbonate at $2 \,^\circ C$ under high pressure.

tion of calcium carbonate from sea water only occurs in a few restricted areas where there are abundant nuclei for seeding the crystals. One of these localities is the Bahama Banks where aragonite drewite precipitates from the warm highly saline waters.

The degree of saturation decreases with increasing depth owing to the increased content of dissolved carbon dioxide, and the decreased temperature, pH and carbonate ion concentration as well as the effect of pressure and temperature on the various equilibrium constants. Figure 6.2 shows depth profiles of the carbonate saturation with respect to calcite and aragonite at stations in the Atlantic and Pacific (Li *et al.*, 1969). It indicates that there is a cross-over from supersaturation to undersaturation at a depth of 500–3,000 m for calcite, at ca 300 m for aragonite in the Pacific, at 4,000–5,000 m for calcite and at 1,000–2,500 m for aragonite in the Atlantic (see also Hawley and Pytkowicz, 1969). These differences may be caused by the considerably greater

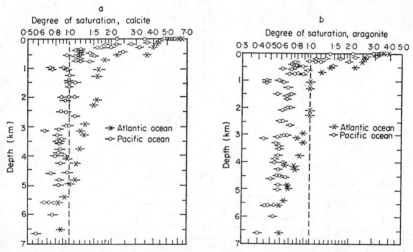

Fig. 6.2. Degree of saturation of calcite (Fig. 6.2a) and aragonite (Fig. 6.2b) as a function of depth in the Atlantic and Pacific Oceans. The figures for the degree of saturation are the ratio obtained by dividing the product of the total molar concentrations of calcium and carbonate ions by the solubility products of calcite and aragonite respectively at the temperature and pressure corresponding to the temperature and depth of the sample. Values above unity represent supersaturation, those below correspond with undersaturation (Li *et al.*, 1969).

concentration of total carbon dioxide in the deep waters of the Pacific. The tests of calcareous organisms falling from the surface layers will dissolve as they settle through the undersaturated parts of the water column. For this reason, the relationship between carbonate saturation and depth exerts an important influence on the pattern of carbonate sedimentation in the oceans (see p. 374–5).

III. Distribution of Carbon Dioxide in the Atmosphere and Sea

For living creatures the sea and the atmosphere represent the most important reservoirs of carbon dioxide. The amount of carbon in these two reservoirs is only a very small fraction of the earth's total stock of carbon compounds (Fig. 6.3). Although much of this stock must have passed through the atmosphere in the geological past, it now exerts very little influence on the atmospheric or marine carbon dioxide

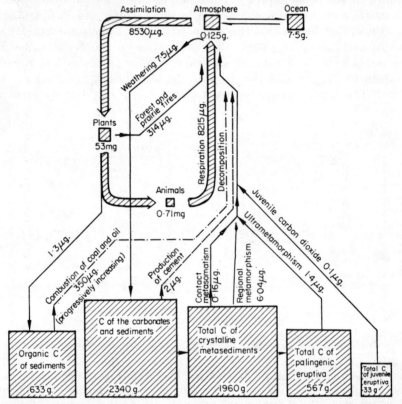

Fig. 6.3. Geochemical carbon cycle in Nature (after Dietrich, 1957) showing relative weights of carbon in the various reservoirs and the transport routes.

reservoirs. In the present discussion it is only these two reservoirs and the exchange between them which needs to be considered. However, it should be borne in mind that there is a slow loss of carbon dioxide from the oceans as a result of the sedimentation of calcareous oozes (p. 372) and organic-carbon, and that this is offset to some extent by that introduced to the sea by weathering of carbonate minerals.

A. DISTRIBUTION OF CARBON DIOXIDE IN THE ATMOSPHERE

The concentration of carbon dioxide in the atmosphere, which averages ca 320 ppm, varies with time as well as geographically. These variations result from plant growth, industrial activity, the burning of fossil fuels and exchange with the sea.

Over the land surfaces the carbon dioxide concentration of the lower 200 m of the atmosphere exhibits both diurnal and seasonal variations. These are caused by changes in the relative rates of photosynthesis and respiration in the plant cover; as a consequence, the maximum concentration occurs during the night and the minimum concentration is found at about midday. The range of variation varies with the type of

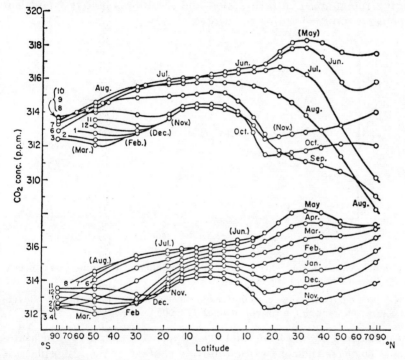

Fig. 6.4. Smoothed curves of the atmospheric carbon dioxide concentration as a function of latitude for each calendar month (Bolin and Keeling, 1963).

vegetation, and is least over barren land and greatest in rain forests, in which the daily range may be as large as 100 ppm (Keeling, 1958, 1961). As would be expected, the effect decreases as the height of sampling increases, but it can still be detected at heights of 1,500 m.

The diurnal and seasonal variation in the carbon dioxide concentration of the atmosphere are very much smaller over the sea than over the land. A detailed picture of the seasonal and latitudinal variations of the concentration of atmospheric carbon dioxide is only available for the Pacific Ocean, where Bolin and Keeling (1963) have carried out a series of measurements by infra red gas analysis extending over 5 years. The smoothed average monthly concentrations at various latitudes are shown diagrammatically in Fig. 6.4. Bolin and Keeling (1963) suggest that their data show that at the equator there is a net release of carbon dioxide from the ocean surface; this is compensated by absorption at high latitudes. Combustion of fossil fuels accounts for the high concentrations found in the middle latitudes of the northern hemisphere. The land vegetation north of 45° N. removes ca $1·5 \times 10^{10}$ tons of CO_2 during the summer growth period and an approximately equivalent amount is returned during the winter.

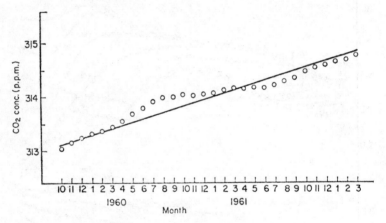

Fig. 6.5. Annual mean concentration of carbon dioxide at Mauna Loa Observatory, Hawaii (Bolin and Keeling, 1963).

Key: o o o o o o average carbon dioxide concentration;———— carbon dioxide concentration assuming a rate of increase of 0·006 ppm per month.

Comparison of data from the last century with more recent measurements suggests that the carbon dioxide content of the atmosphere is gradually increasing. This belief appears to be confirmed by short-term studies carried out by means of an infra red gas analyser (see Fig. 6.5 and Bolin and Keeling, 1963). The increase appears to have two main causes. The first arises from changes in the biosphere which have resulted from deforestation and land clearance. These are thought by Hutchinson (1954) to have increased the carbon dioxide content of the atmosphere by ca 4% over the last 100 years. The second is associated with the burning of coal and oil which supplies carbon dioxide to the atmosphere at a rate which has increased in an approximately exponential fashion since the beginning of the nineteenth century. Revelle and Suess (1957) have estimated that by 1970 the burning of fossil fuels will have produced ca $1\cdot3 \times 10^{11}$ tons of carbon dioxide, an amount equal to ca 20% of that present in the atmosphere in 1800.

Further confirmation that the carbon dioxide content of the atmosphere is increasing has been obtained from the examination of the ^{14}C activity of the annual rings of trees. It has been found (see, e.g. Fergusson, 1958) that between 1850 and 1954, the specific activity has steadily declined as a result of the dilution of the atmospheric cosmic ray produced $^{14}CO_2$ with "old" carbon dioxide devoid of ^{14}C (Suess effect). Testing of nuclear weapons since 1954 has markedly increased the $^{14}CO_2$ specific activity of the tree rings laid down since that time.

Because equilibration with the oceans as a whole is a slow process most of the added carbon dioxide is still present in the atmosphere. Bolin and Eriksson (1959) have postulated that complete equilibrium would only be attained after more than 10^5 years, after which time 90% of it would be present in the ocean (see below p. 149).

B. DISTRIBUTION OF CARBON DIOXIDE IN THE SEA

(i) *Surface waters*

Equilibrium is rarely attained between carbon dioxide in the surface water of the sea and in the atmosphere above it (i.e. $P_{CO_2} \neq p_{CO_2}$). Biological processes, such as photosynthesis and respiration, water transport, and surface heating and cooling ensure that equilibrium, if attained, is of only short duration. P_{CO_2} is very sensitive to small changes in ΣCO_2, and Kanwisher (1960) has shown experimentally that a 10% change in P_{CO_2} is produced by a variation in ΣCO_2 of ca $0\cdot6\%$ (see also Fig. 6.6). The magnitude of the change in P_{CO_2} and pH produced by addition or removal of a given amount of carbon dioxide depends on the alkalinity of the water. P_{CO_2} increases with increase in temperature. With constant ΣCO_2, as would prevail during short-term temperature changes, P_{CO_2} increases by ca 10 ppm/°C in the range $3°$–$30°C$, the value depending on the pH (Takahashi, 1961).

Seasonal and diurnal variations of P_{CO_2} can frequently be observed in surface waters of the oceans (Takahashi, 1961; Cooper, 1933). During the spring, photosynthesis exceeds respiration, and this causes P_{CO_2} to fall. However, the rise in water temperature tends to increase P_{CO_2} and this partially compensates for the fall produced by biological action; as a result P_{CO_2} decreases to a minimum in June. In the winter, respiration and decay dominate the biological cycle, and the effect of these exceeds that produced by the fall in water temperature with the result that P_{CO_2}

attains its maximum value in about January. Except in restricted bodies of water, the diurnal variations are much smaller than the seasonal ones, and are restricted to the upper few metres of the water column. As the maximum values are attained during the day these variations are probably produced by temperature changes.

The variations in P_{CO_2}, ΣCO_2 and pH are most extreme in isolated bodies of water, such as rock pools, in which the relative concentration of living organisms is high. This effect is often accentuated by the stratification which may develop in the summer. Thus, Orr (1947) found

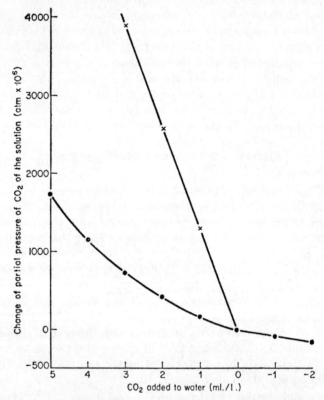

Fig. 6.6. Change of partial pressure of carbon dioxide as a function of change of total carbon dioxide in distilled water ($—\times—\times—$) and sea water of salinity $31^0/_{00}$ ($—\cdot—\cdot—$) from Kanwisher (1963a, b).

that production of carbon dioxide by photosynthesis increased the pH of the water at the surface of a marine pool to 9.9, whereas in the oxygen deficient water at a depth of 3 m the water was at pH 7·4. Seasonal and diurnal variations are also particularly strongly developed in restricted areas, such as bays where it is frequently observed that the pH may range over 0·7 units from its maximum value during daylight hours to its minimum at night (Fig. 6.7).

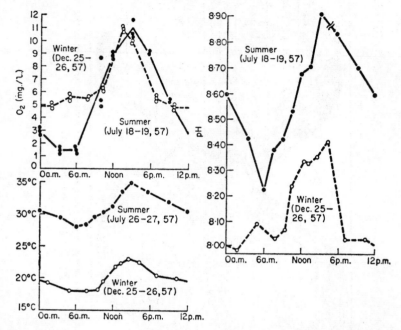

Fig. 6.7. Diurnal variation of oxygen, pH and temperature in Redfish Bay, Texas. Continuous lines for summer data. Broken lines for winter data (Park *et al.*, 1958).

The geographical distribution of P_{CO_2} in the surface waters of the oceans has been reviewed by Keeling (1968), whose conclusions are summarized diagrammatically in Fig. 6.8. P_{CO_2} is low relative to p_{CO_2} in the waters of the northern Pacific and in those of high latitude in the North and South Atlantic. This produces a net transport of carbon dioxide from the atmosphere to the sea. In warmer regions, the partial pressure is higher in the sea than in the air above it and the net transport is in the opposite direction. High values of P_{CO_2} are found in the surface waters in regions where CO_2-rich intermediate waters upwell (e.g. off the Peruvian coast and along the equator).

ii) Deep waters

When waters sink below the thermocline their carbon dioxide contents can be altered only by processes involving advection, eddy diffusion or biological regeneration. The most important of these are

7

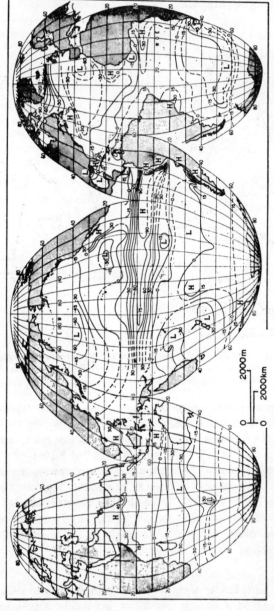

Fig. 6.8. The distribution of P_{CO_2} in oceanic surface water expressed as the difference in ppm from the value at equilibrium with atmospheric carbon dioxide. H = high; L = low (from Keeling, 1968).

the regenerative processes which occur as a result of the oxidation of organic material falling from the surface layers. Most of this oxidation occurs in the upper few hundred metres of the water column, and causes ΣCO_2 to rise sharply below the thermocline and to increase to a maximum at a depth of 500–1,000 m. The depth of the ΣCO_2 maximum (and

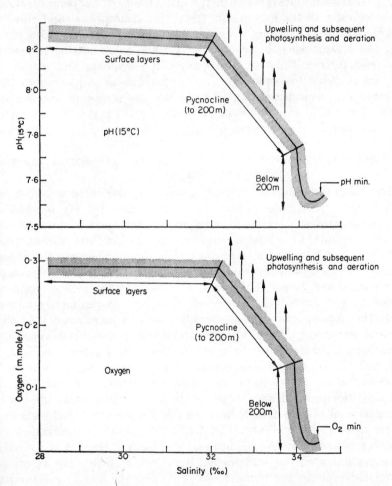

Fig. 6.9. Simplified pH-oxygen-salinity relationships off Newport, Oregon Park, 1968).

therefore that of the pH minimum) coincides with that of the oxygen minimum. The shapes of the pH-depth and O_2-depth profiles are similar, but vary from one location to another according to the water circulation pattern (see, e.g. Park, 1968 and Fig. 6.9).

Keeling (1968) has pointed out the general similarity between the distributions of P_{CO_2} and PO_4^{3-}—P in the oceans. This reflects the roughly constant C : P ratio in marine organisms. When dead organisms fall from the surface layers, decaying as they do so, they liberate CO_2—C and PO_4^{3-}—P to the water in the same ratio. There are two factors influencing P_{CO_2} in the surface layers which do not affect the PO_4^{3-}—P concentration —the transfer of the gas to and from the atmosphere, and change in temperature. Upwelling water is rich in phosphate and has a high P_{CO_2} value, and when it reaches the surface, solar heating increases P_{CO_2} still further. Only small amounts of phosphate and dissolved CO_2 are removed from the water by photosynthesis and to relieve the elevated P_{CO_2} considerable amounts of the gas escape to the atmosphere. Thus, high P_{CO_2} and high concentrations of PO_4^{3-}—P occur over the same areas because of the upwelling of deep water.

C. CIRCULATION OF CARBON DIOXIDE BETWEEN THE ATMOSPHERE AND THE MARINE ENVIRONMENT

The natural production of radioactive nuclides by interaction of cosmic rays with gases in the upper atmosphere (p. 98) provides a convenient source of tracers for studying the transport of water in the oceans. Carbon–14 has been the most widely used of these tracers. This isotope, which is formed by neutron activation of nitrogen in the upper atmosphere, decays by beta emission, and has a half life of ca 5,600 years. The radioactive atoms thus produced react rapidly with oxygen or ozone to give carbon dioxide which soon becomes uniformly mixed with the atmospheric carbon dioxide and is incorporated into the natural reservoirs. When the ^{14}C activities of these reservoirs are considered allowance must be made for the slight fractionation of the ^{14}C and ^{12}C isotopes which occurs during the exchange processes. Broecker *et al.* (1960) have used data on atmospheric and oceanic ^{14}C activities to build up a steady state model representing the inter-circulation of carbon dioxide between the atmosphere and the various water masses of the oceans (Fig. 6.10). This model is constituted by nine reservoirs, including the atmosphere. Each of these is in a steady state with respect to carbon–14, the transfer into each reservoir being balanced by the transfer out of it, plus the loss by radioactive decay. Figures are presented for the average residence times of carbon dioxide in the nine reservoirs; these range from 7 years in the atmosphere to ca 700 years in the deep waters. The model postulates that negligible mixing takes place between the waters above the thermocline and those below it. Both the surface and deep waters of the Pacific and Indian Oceans (combined in the model) exchange with the vertically mixed Antarctic Ocean. There is a flux of water from the latter, via the South and North Atlantic surface waters, to the vertically mixed Arctic waters, from which some of the water returns to the south via the Atlantic Deep Water.

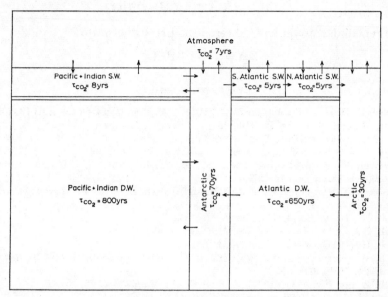

Fig. 6.10. Model of the steady-state cycle in the atmosphere and oceans. The arrows indicated possible modes of transfer between reservoirs. The τ_{CO_2} values represent the mean residence times of carbon dioxide in the various reservoirs (after Broecker *et al.*, 1960). SW = surface water; DW = deep water.

Although this model is based on a number of approximations (see, e.g. Broecker, 1962) it does provide a useful indication of the circulatory pattern of carbon dioxide. Such two layer models are helpful when considering the fate of the carbon dioxide introduced into the atmosphere by the burning of fossil fuels. As a first approximation it may be assumed that only water lying above the thermocline (average depth 75 m) is at equilibrium with the atmosphere. This water represents ca 2% of the total volume of the ocean and contains about twice as much carbon dioxide as the atmosphere. Bolin and Eriksson (1959) have shown that a 12·5% change in P_{CO_2} produces, at equilibrium, a ca 1% change in ΣCO_2 in the water. Thus, less than 10% of any increase in atmospheric CO_2 will be taken up by the mixed layer at any time. Renewal of this layer takes 10–15 years, and it therefore acts as a bottleneck restricting the equilibration of the oceans as a whole. At complete equilibration, which would take $>10^5$ years, ca 92% of the atmospheric increase would have entered the sea. Of this, 87% would have reacted with the calcium carbonate on the ocean floor, causing an increase in the carbonate alkalinity of the water (Bolin and Eriksson, 1959)

$$CaCO_3(s) + H_2CO_3 \rightleftharpoons Ca^{2+} + 2HCO_3^-$$

the remainder would have lowered the pH of the water

$$H_2CO_3 \rightleftharpoons H^+ + HCO_3^-$$

REFERENCES

Anderson, D. H. and Robinson, R. J. (1946). *Ind. Eng. Chem. (Anal. Ed.)* **18**, 767.

Bolin B. and Eriksson E. (1959). *In* "Rossby Memorial Volume", p. 130. Rockefeller Institute Press, New York.

Bolin, B. and Keeling, C. D. (1963). *J. geophys. Res.* **68**, 3899.

Broecker, W. S. (1962). *In* "The Sea" (M. N. Hill. ed.), Vol. II, p. 88. Interscience Publishers Inc., New York.

Broecker, W. S., Gerard, R., Ewing, M. and Heezen, B. C. (1960). *J. geophys. Res.* **65**, 2903.

Buch, K. (1930). *Rapp. Cons. Expl. Mer.* **67**.

Buch, K. (1938). *Acta. Acad. åbo. Math. Phys.* **11**, No. 5.

Buch, K. (1951). *Meeresforsch.* No. 151.

Buch, K., Harvey, H. W., Wattenberg, H. and Gripenberg, S. (1932). *Rapp. Cons. Expl. Mer.* **79**, 1.

Buch, K. and Nynäs O. (1939). *Acta Acad. åbo Math. Phys.* **12**, No. 3.

Cooper, L. H. N. (1933). *J. mar. biol. Ass. U.K.* **18**, 677.

Culberson, C. and Pytkowicz, R. M. (1968). *Limnol. Oceanog.* **13**, 403.

Culberson, C., Pytkowicz, R. M. and Hawley, J. E. (1970). *J. mar. Res.* **28**, 15.

Dietrich, G. (1957). "General Oceanography." 588 pp. Interscience Publishers Inc., New York.

Disteche, A. and Dubuisson, M. (1960). *Bull. Inst. Oceanog. Monaco*, **57**, No. 1174.

Dyrssen, D. (1965). *Acta. chim. Scand.* **19**, 1265.

Dyrssen, D. and Sillén, L. G. (1967). *Tellus* **19**, 113.

Edmond, J. M. (1970). *Deep Sea Res.* **17**, 737.

Fergusson, G. J. (1958). *Proc. R. Soc.* A **243**, 561.

Garrels, R. M. and Thompson, M. E. (1962). *Am. J. Sci.* **260**, 57.

Garrels, R. M., Thompson, M. E. and Siever, R. (1961). *Am. J. Sci.* **259**, 24.

Gieskes, J. M. (1969). *Limnol. Oceanog.* **14**, 679.

Hawley, J. and Pytkowicz, R. M. (1969). *Geochim. cosmochim. Acta*, **33**, 1557.

Hood, D. W. (1963). Great Lakes Research Division, University of Michigan, Publ. **10**, 91.

Hutchinson, G. E. (1954). *In* "The Earth as a Planet" (G. Kuiper, ed.). Chicago.

Kanwisher, J. (1960). *Tellus*, **12**, 209.

Kanwisher, J. (1963a). *J. geophys. Res.* **68**, 3921.

Kanwisher, J. (1963b). *Deep Sea Res.* **10**, 195.

Keeling, C. D. (1958). *Geochim. cosmochim. Acta*, **13**, 322.

Keeling, C. D. (1961). *Geochim. cosmochim. Acta*, **24**, 277.

Keeling, C. D. (1968). *J. geophys. Res.* **73**, 4543.
Koczy, F. F. (1956). *Deep Sea Res.* **2**, 279.
Li, Y-H., Takahashi, T. and Broecker, W. S. (1969). *J. geophys. Res.* **74**, 5507.
Lyman, J. (1956). Ph.D. Thesis, University of California.
Murray, C. N. and Riley, J. P. (1971). *Deep Sea Res.* In the press.
Orr, A. P. (1947). *Proc. R. Soc. Edinb.* B **63**, 3.
Park, K. (1968). *Deep Sea Res.* **15**, 171.
Park, K. (1969). *Limnol. Oceanog.* **14**, 179.
Park, K., Hood, D. W., and Odum, H. T. (1958). *Inst. Marine Sci.* **5**, 47.
Pytkowicz, R. M. (1968). *Oceanog. Mar. Biol. Ann. Rev.* **6**, 83.
Reid, J. L. (1962). *Limnol. Oceanog.* **7**, 287.
Revelle, R. and Suess, H. E. (1957). *Tellus* **9**, 18.
Saruhashi, K. (1955). *Pap. Met. Geophys., Tokyo.* **6**, 38.
Sillén, L. G. (1963). *Svensk. Kem. Tidskr.* **75**, 161.
Skirrow, G. (1965). *In* "Chemical Oceanography" (J. P. Riley and G. Skirrow, eds), Vol. 1. Academic Press, London.
Smith W. H. and Hood, D. W. (1964). *In* "Recent Researches in the Fields of Hydrosphere, Atmosphere and Nuclear Geochemistry", p. 185. Maruzen, Tokyo.
Takahashi, T. (1961). *J. geophys. Res.* **66**, 477.

Chapter 7

Micronutrient Elements

I. Introduction

In common with land plants, marine phytoplankton require certain trace elements for their healthy growth. The most important of these *micronutrients* are nitrogen and phosphorus, which may be taken up by them from the water to such an extent that their further growth is inhibited. Those types of organisms which have siliceous frustules (e.g. diatoms) also require a supply of silicon, and the blooming of these species may reduce the silicon content of the water appreciably

Certain other elements, such as iron, manganese, copper, zinc, cobalt and molybdenum are essential to the growth of marine plants as they occur in their enzyme systems (see p. 246). It is unlikely that phytoplankton growth is ever limited by the *total* concentration of any of these trace metals, but in some waters an essential element may be present in a form in which it is not assimilable by the organism. Thus, there is some evidence that certain ocean waters may be deficient in *available* iron and manganese. (Harvey, 1955; Tranter and Newell, 1963). The very low requirements for the other trace metals makes it improbable that they ever limit marine photosynthesis. The recent introduction of rapid and sensitive atomic absorption methods for the determination of low concentrations of trace elements in sea water (see, e.g. Riley and Taylor, 1968) will undoubtedly add impetus to the study of the biological roles of these elements in the sea. In addition to inorganic micronutrients, many species of phytoplankton find minute amounts of certain organic compounds, such as vitamins, necessary for their growth (see Chapter 8). The remainder of this chapter will be devoted to a description of the marine chemistry of the major micronutrient elements, nitrogen, phosphorus and silicon.

II. Nitrogen

In addition to dissolved molecular nitrogen (see Chapter 5) the sea contains low, but extremely important, concentrations of inorganic and organic nitrogen compounds, the total weight of which (ca $2\cdot2 \times 10^9$ tons) is about a tenth of that of the dissolved gas. The principal inorganic forms of nitrogen are nitrate ion (1–500 μg NO_3^-—N/l), nitrite ion (usual range <0·1–50 μg NO_2^-—N/l) and ammonia* (usual range <1–50 μg NH_3—N/l). Small concentrations of other inorganic nitrogen compounds have also been shown to be present, e.g. nitrous oxide (Craig and Gordon, 1963), and some short-lived species such as hydroxylamine and the hyponitrite ion may also occur, although they have not yet been detected. The sea also contains low concentrations of dissolved and particulate organic nitrogen compounds associated with marine organisms and the products of their metabolism and decay. The occurrence of these compounds and their biochemical significance will be discussed in Chapters 8 and 9.

A. ANALYTICAL CHEMISTRY

Before considering the inter-relationships between the various forms of nitrogen in the sea, it will be of value to mention briefly their analytical chemistry. In practice, only nitrate, nitrite and perhaps ammonia are determined on a routine basis. Determinations may be carried out on board ship since colorimetric methods are used. The tedium of these analyses has been greatly reduced by the development of automatic methods of photometric analysis employing, for example, the Technicon Autoanalyzer. Practical details lie outside the scope of this work, and for these the reader is referred to the monograph on sea water analysis by Strickland and Parsons (1968).

1. Nitrite is determined by treating the water sample with a solution of sulphanilamide; the resultant diazonium ion is coupled with N-(1-naphthyl)-ethylenediamine to give an intensely pink azo dye, the absorbance of which is measured at 543 nm with a spectrophotometer.

$$\underset{\text{sulphanilamide}}{NH_2 . C_6H_4 . SO_2NH_2} + NO_2^- + 2H^+ \rightarrow \overset{+}{N} \equiv N . C_6H_4 . SO_2NH_2 + H_2O \ \underset{\text{diazonium ion}}{}$$

$$\underset{\text{naphthyl–ethylenediamine}}{NH_2CH_2CH_2NH . C_{10}H_7} + \overset{+}{N} \equiv N . C_6H_4SC_2NH_2 \rightarrow$$

$$\underset{\text{pink azo dye}}{NH_2CH_2CH_2NH . C_{10}H_6 . N = N . C_6H_4 . SO_2NH_2} + H^+$$

2. Nitrate is usually determined by reducing it to nitrite which is determined as described above. The reduction is carried out by treating the sample with ammonium chloride, or EDTA and passing it through a glass column packed with amalgamated or copper-coated cadmium filings.

---

* It should be understood that when ammonia is referred to in the text it implies the sum of ammonia+the ammonium ion with which it is in equilibrium, the relative proportions of each depending upon pH.

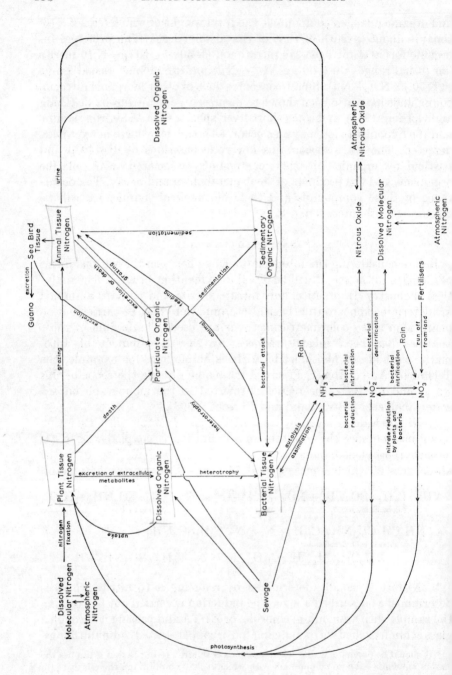

3. Ammonia. No completely satisfactory method as yet exists for the routine determination of ammonia at the extremely low concentrations at which it occurs in oxygenated sea waters. At present two methods are in vogue for the determination, in both of which certain dissolved organic nitrogen compounds may cause interference. In the first of these ammonia is oxidized to nitrite using an alkaline hypochlorite solution. The excess hypochlorite is then reduced with arsenite and nitrite is determined as described above. In the second procedure, which is somewhat less sensitive, the sample in an alkaline citrate medium is treated with sodium hypochlorite and phenol in the presence of catalytic amounts of sodium nitroprusside. A blue indophenol dye is produced, and its intensity is measured photometrically.

B. THE NITROGEN CYCLE IN THE SEA

The concentrations of the various organic and inorganic nitrogen species in the sea are mainly controlled by biological factors. However, physical effects such as the sinking of dead organisms and upwelling (p. 43), tend to bring about a redistribution of these species in the water column. In any body of water the instantaneous balance of the nitrogen compounds represents a dynamic equilibrium between these various processes. At the present time only qualitative information is available about the various processes involved in the marine chemistry of nitrogen. These are most conveniently discussed in the framework of the nitrogen cycle shown schematically in Fig. 7.1 (cf. Vaccaro and Ryther, 1960). Bacteria, plants and animals play complementary parts in the various transformations. Bacteria probably dominate the regenerative processes, in which organic nitrogen compounds are converted into inorganic nitrogen species and eventually to nitrate (see p. 159). Phytoplankton normally synthesize their proteins from nitrate, nitrite and ammonia, but, bacteria usually only use these forms of nitrogen when organic nitrogen is not available. However, the only contribution of living animals to the nitrogen cycle is the excretion into the water of ammonia and to a lesser extent its precursors such as urea, amino acids, trimethylamine oxide and peptides.

It should be emphasized that the nitrogen cycle in the sea is not a closed system. Deposition of organic nitrogen compounds in sediments annually removes ca 9×10^6 tons of nitrogen from the sea. Fixed nitrogen enters the sea in river water and rain water* at a much greater rate (ca 8×10^7 tons/yr) and an unknown amount is fixed by the activities of

---

* According to Hutchinson (1954) most of the nitrate in rainwater results not from the fixation of nitrogen during electrical storms, but from the atmospheric oxidation of ammonia formed by the decomposition of organic matter.

p = partial pressure

blue-green algae *vide infra*. Thermodynamical considerations suggest that nitrate should be formed in the sea at the expense of dissolved nitrogen, and that if equilibrium were to be attained, p_{N_2} would be only $1/10,000$ of its present value (see p. 78). The slowness of this reaction cannot be the sole explanation. There is probably a considerable imbalance in the nitrogen cycle, since p_{N_2} has probably been reasonably constant for millions of years. Sillén (personal communication) has postulated that the balance must be restored by some, as yet unidentified, biological denitrification process which converts fixed nitrogen to the molecular form.

The more important of the various stages of the nitrogen cycle can be considered most usefully under three main headings.

(i) Nitrogen fixation (see also p. 114)

The nodular bacteria, such as *Rhizobium* and *Clostridium* of land plants are not alone in their ability to fix atmospheric nitrogen, and it is now recognized that this function is also shared by certain fresh-water bacteria, moulds and yeasts. Counterparts of some of these have been found in the sea, for example, species of *Clostridium* have been found in sediments, but doubt has been cast on the existence of truly marine species of *Azotobacter* (Wood, 1967). However, since the fixation process is endothermic, and requires a plentiful supply of organic material as an energy source (Nutman, 1959), it is unlikely that these organisms, although abundant in the sea, fix significant amounts of nitrogen. In contrast, certain blue green algae, notably *Trichodesmium* spp. have been shown to fix nitrogen on a large scale in tropical and subtropical waters (Dugdale and Goering, 1964, 1967; Goering *et al.*, 1966), even though the concentrations of organic matter are only low. Fogg and Than-Tun (1958) have postulated that these organisms obtained the energy which they require for nitrogen fixation from solar radiation, the molecular nitrogen acting instead of carbon as an acceptor for hydrogen liberated photochemically from the water. As a consequence, nitrogen fixation does not occur in darkness (Dugdale *et al.*, 1961). Fixation is inhibited if an alternative source of inorganic nitrogen (e.g. nitrate or ammonia) is available (Ramamurthy and Krishnamurthy, 1968), which the organisms use in preference to molecular nitrogen (Goering *et al.*, 1966).

The importance of nitrogen–fixing phytoplankton in the sea is emphasized by the work of Dugdale and Goering (1967) who found that nitrogen fixation could support the growth of *Trichodesmium* sp. at rates which are similar to those of species which rely on nitrate. Nitrogen fixation may also indirectly benefit other organisms, since some species of blue-green algae (e.g. *Calothrix scopulorum*) liberate extracellular nitrogen to the water as they grow. This has been shown to be utilized by a variety of algae, fungi and bacteria (Jones and Stewart, 1969a and b).

(ii) Assimilation of fixed nitrogen

Some form of nitrogen is required by phytoplankton for the synthesis of their cellular amino acids etc. They satisfy most of their needs by utilizing the ammonia, nitrite and nitrate present in the water. Uptake is confined virtually to the euphotic layer of the sea since it is a consequence of photosynthesis. Although all three sources of nitrogen can be absorbed by most species of phytoplankton, ammonia is usually used preferentially (Harvey, 1955). Thus, Vaccaro (1963) has found that even though ammonia is the result of short-term regeneration, it is probably more important than nitrate as a micronutrient in the coastal water off New England, and it is likely that this is the case in many other areas.

The kinetics of the uptake of nitrate and ammonia by natural populations of phytoplankton in the sea have been investigated by MacIsaac and Dugdale (1969) using nitrogen–15 tracer techniques. They observed that there is a hyperbolic relationship between the concentration of nitrate or ammonia and its rate of uptake. Mathematically this is in agreement with the Michaelis-Menten expression used in the study of enzyme kinetics. When these forms of nitrogen become severely depleted in the water (i.e. total N < 10 μg/l), nitrogen-deficient organisms are produced before cell division finally ceases. These deficient cells are able to take up ammonia and nitrate, but not nitrite, in the dark, converting them to organic compounds including chlorophyll (Harvey, 1953); again ammonia is absorbed preferentially (Syrett, 1954).

A few classes of phytoplankton, e.g. phytoflagellates (Schreiber, 1927), appear to be able to satisfy their nitrogen requirements by utilization of dissolved organic nitrogen compounds, such as amino acids, and some others, e.g. diatoms, can do so after attached bacteria have deaminated these compounds (Harvey, 1940). Since most species of phytoplankton use nitrate and ammonia preferentially (Guillard, 1963), it seems unlikely that they will derive significant amounts of nitrogen from the very low concentrations of organic nitrogen compounds present in the open sea (< 20 μgN/l). However, in sewage-polluted waters an appreciable proportion of their nitrogen requirements may be satisfied by utilization of urea and uric acid. Grant *et al.* (1967) have shown that when *Cylindrotheca closterium* was grown on relatively high concentrations of nitrate,

ammonia and urea, the order of preference was generally ammonia >
urea >nitrate. However, if the urea concentration was greater than that
of ammonia both were used simultaneously, but, urea was taken up less
effectively at low concentrations. The ability to use urea appears to
be mainly restricted to coastal and estuarine species of phytoplankton,
although a few open sea species (e.g. *Skeletonema*) are also able to utilize
it (Guillard, 1963).

Before it can be incorporated into amino acids by algae for protein
synthesis, the assimilated nitrate must be first converted to ammonia.
A hydrogen donor, and perhaps a source of high energy phosphate (e.g.
adenosine triphosphate) is required to bring about this endothermic
reduction process. Although these may be produced by respiration
(nitrate reduction can occur in the dark), they are more usually produced by photosynthetic mechanisms. Thus, nitrate reduction proceeds
much faster in the presence of light, and more oxygen is produced by
illuminated *Chlorella* in the presence of nitrate than in its absence. The
reduction of nitrate to ammonia takes place in four stages involving
the consecutive production of nitrite, hyponitrite and hydroxylamine:

$$NO_3^- + 2H^+ + 2e^- \rightarrow NO_2^- + H_2O$$

$$2NO_2^- + 4H^+ + 4e^- \rightarrow N_2O_2^{2-} + 2H_2O$$

$$N_2O_2^{2-} + 6H^+ + 4e^- \rightarrow 2NH_2OH$$

$$NH_2OH + 2H^+ + 2e^- \rightarrow NH_3 + H_2O$$

It is known that in the first step the reduction of nitrate by a molyb-
denum-containing nitrate reductase is catalysed by reduced coenzyme
II which is produced by the chloroplasts. Not much is known about the
other steps in the reduction, but it is possible that direct products
of the photosynthetic process may also be involved in them.

Ammonia produced by nitrate reduction or assimilated directly
from the water, is converted into glutamic acid by reaction with
α-ketoglutaric acid in the presence of reduced adenine nicotinamide
dinucleotide phosphate (NADPH)

$$HOOC . CO . (CH_2)_2 . \underset{\text{ketoglutaric acid}}{COOH} + NH_3 + 2NADPH \rightarrow$$

$$HOOC . CH(NH_2)CH_2CH_2 . \underset{\text{glutamic acid}}{COOH} + 2NADP + H_2O$$

The twenty or so other amino acids required as building blocks for algal
proteins are formed from glutamic acid by transamination. Thus the
reaction of glutamic acid with pyruvic acid gives rise to alanine with
the reformation of α-ketoglutaric acid.

$$CH_3CO\ COOH + HOOC\ .\ CH(NH_2)CH_2CH_2\ .\ COOH \rightarrow$$
pyruvic acid

$$CH_3CH(NH_2)COOH + HOOC\ .\ CO(CH_2)_2\ .\ COOH$$
alanine

Proteins are then produced by linking together of the various amino acids by complicated reaction sequences involving RNA and DNA and energy provided by adenosine triphosphate. Amino acids and their derivatives probably also serve as building units for many of the other organic nitrogen compounds which the algae synthesize. For further information on the bio-synthesis of such compounds the reader is referred to standard textbooks of biochemistry. The proteins of the higher members of the marine food chain—zooplankton and fish—are built up from those of the phytoplankton by proteolysis and recombination of the amino acids.

(iii) Regeneration of nitrate

The regenerative processes by which the organic nitrogen compounds are reconverted via ammonia to nitrate ion are believed to be mainly bacterial. It is appropriate therefore to consider briefly the metabolic requirements and fate of marine bacteria. Many different classes of marine bacteria are known, some of which are specialized forms capable of bringing about particular reactions, e.g. the oxidation of ammonia to nitrite. Bacteria may occur both free floating and attached to organic and inorganic suspended matter. Aerobic species use suspended and dissolved organic matter as food, and satisfy their energy requirements through oxidation of organic matter to carbon dioxide, using dissolved oxygen obtained from the water. Species which grow in anoxic waters use alternative sources of oxygen (e.g. NO_3^- or SO_4^{2-}). Bacterial respiratory processes are much faster than those of phytoplankton or marine animals. If the organic matter on which the bacteria feed contains more nitrogen and phosphorus than they require, the excess is liberated as ammonia and phosphate ion. However, in the absence of an adequate supply of these elements, some species are able to utilize dissolved inorganic species of the elements. The bacteria will continue to flourish and multiply provided that adequate food is available and that conditions are suitable. When food becomes exhausted, or non-viable conditions develop, the cells will die and undergo rapid autolysis, liberating ammonia and phosphate to the surrounding water.

(a) Decomposition of organic nitrogen compounds to yield ammonia. The soluble and particulate nitrogen compounds of dead organisms and those excreted by plants and animals are rapidly broken down to ammonia. This decomposition is brought about by the various species of proteolytic bacteria which occur at all depths in the sea. The decomposition is quite efficient, and at no time do any substantial concentrations of free or combined dissolved amino acids build up. However, some particularly stable organic nitrogen compounds resist attack, and eventually sink to the bottom where they are incorporated in the sediments as marine humus.

An extensive series of experiments on the decomposition of phyto-plankton in sea water was carried out by von Brand and Rakestraw (1937–42). Living diatoms were added to sea water which was then stored in the dark with aeration. The water was analysed initially and at intervals over a period of several months. It was found (Fig. 7.2)

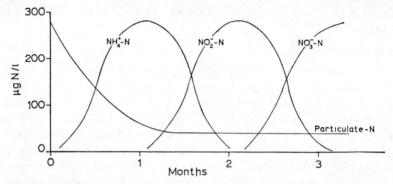

Fig. 7.2. Consecutive production of ammonium-, nitrite- and nitrate-nitrogen in sea water containing diatoms during storage in the dark (after Harvey, 1955).

that the particulate nitrogen concentration fell by 80% in about one month, the exact time depending on the temperature. The remaining 20% was resistant to bacterial attack and remained at a constant level throughout the rest of the experiment. Simultaneously with the decay of the particulate nitrogen compounds ammonia was set free in an amount equivalent to that of the nitrogen present in the decomposed organic matter. Subsequently, the ammonia became oxidized first to nitrite and then to nitrate, except under anaerobic conditions when the decomposition stopped at ammonia. This stepwise decay sequence suggests that it is necessary for specialized bacteria to proliferate before any stage can commence. However, at the end of the experiment sufficient bacteria are present to cause any diatoms added subsequently to decay rapidly and produce ammonia, nitrite and nitrate simul-taneously. Stepwise degradation of organic material is rarely encount-ered outside the laboratory; in the upper layers of the sea, all three reactions usually occur concurrently, since the necessary bacteria are normally already present in sufficient numbers (see, e.g. Fig. 7.3).

(b) *Nitrification.* Ammonia formed by the bacterial decay of marine organisms or excreted by marine animals is, as we have seen above, subsequently oxidized to nitrite and then nitrate. These oxidation processes, which are termed nitrification, may be represented by the following equations (Cooper, 1937)

$$NH_4^+ + OH^- + 1{\cdot}5O_2 \rightleftharpoons H^+ + NO_2^- + 2H_2O \quad \varDelta G^\circ = -59{\cdot}4 \text{ kcal}$$

and

$$NO_2^- + 0{\cdot}5O_2 \rightleftharpoons NO_3^- \quad \varDelta G^\circ = -18{\cdot}0 \text{ kcal}$$

Both reactions are thus exothermic and only require activation to make them proceed. Evidence that nitrification is proceeding in the deep

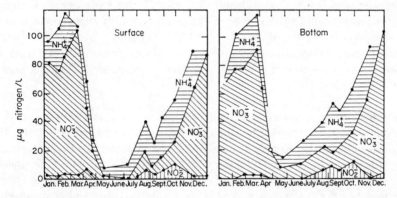

Fig. 7.3. Seasonal variation of ammonium-, nitrite- and nitrate-nitrogen in the upper layers and near the bottom at a station in the English Channel (after Cooper, 1933).

waters of the oceans, and that nitrate is its final product, has been obtained from a consideration of the relationship between nitrate, phosphate and the oxygen deficit in these waters (Wattenberg, 1929; Riley, 1951).

Nitrification in the sea appears to occur largely by a bacterial mechanism. It was thought until recently that bacteria capable of carrying out nitrification occurred only in inshore waters and on sediments. Murray and Watson (1963) provided the first indication that such organisms have a more general distribution when they isolated *Nitrocystis oceanus* from ocean waters. Subsequently, nitrifying bacteria have been found by Carlucci and Strickland (1968) in waters taken from many depths in most of the oceans. Kinetic measurements with a range of ammonia and nitrite-oxidizing bacteria showed that the rate of oxidation was usually too slow to account for the estimated rate of disappearance of ammonia in the sea. These authors therefore concluded that nitrification must also occur by other mechanisms; it is not known at present what these other processes are. Kreps (1934) has postulated that the oxidation of ammonia is induced by the catalytic action of organic matter produced probably by bacteria. Although photochemical oxidation by solar ultra-violet radiation probably does not cause significant oxidation of ammonia (Hamilton, 1964), other processes, such as accelerated oxidation at the air-water interface (Cooper, 1937; Vaccaro, 1963) may be operative in the upper layers of the sea.

Bacterial oxidation of ammonia occurs more rapidly in waters having a low oxygen content even when the concentration of organic substrate is low. However, the significance of these processes in deoxygenated waters is uncertain. Carlucci and McNally (1969) have invoked nitrification to explain the existence of the secondary deep nitrite maximum in the oxygen depleted waters off Peru (Thomas, 1966). However, Fiadeiro and Strickland (1968) consider that reduction of nitrate is a more likely mechanism for its formation (see below). Perhaps both processes occur.

(c) *Reduction of nitrate to nitrite.* Bacteria which are capable of reducing nitrate to nitrite are abundant in the sea and in sediments. It was thought until recently that a much richer organic substrate than is available in the sea was necessary for their growth. However, Carlucci (cited by Fiadeiro and Strickland, 1968) has isolated a bacterium from the waters off Peru which was able to reduce nitrate in almost anoxic seawater containing very low concentrations of organic matter. Carlucci and Strickland (1968) have demonstrated by means of plots of nitrate and phosphate concentration *versus* apparent oxygen utilization (see p. 108) that nitrate reduction occurs in the oxygen-depleted water (<0.1 ml O_2/l) off Peru. Nitrite is produced as an intermediate in the reduction to an extent of 10–20% of the amount of nitrate reduced, and appears as a secondary nitrite maximum at a depth of 200–400 m. The remainder of the nitrogen reduced is probably converted to molecular nitrogen or to nitrous oxide (see p. 163). Considering that this water contains less than 0.5 mg/l of organic material, of which probably less than 20% is suitable for use, it is remarkable that the reduction occurs at all.

Nitrate reduction is not only brought about by bacteria. Many species of marine phytoplankton when growing in the presence of excess nitrate excrete significant amounts of extracellular nitrite. This phenomenon is most pronounced at low light intensities. Vaccaro and Ryther (1960) have suggested that this mechanism may offer an alternative explanation to bacterial nitrification for the nitrite maximum found in the early winter in temperate surface waters, and for that observed near the compensation depth in tropical waters.

(d) *Denitrification.* Denitrification may be considered as the biological reduction of nitrate or nitrite to nitrogen or nitrous oxide. Micro-organisms, such as some *Pseudomonas* species, which are able to reduce nitrate to molecular nitrogen have been known for a long time to exist in sea water and marine sediments. These bacteria when they grow in anoxic, or almost anoxic, water use nitrate ions as alternative electron acceptors instead of oxygen in the oxidation of organic matter. Richards and Benson (1961) have used measurements of the N_2/Ar ratio in the dissolved gas to demonstrate the presence of excess nitrogen, presumably formed by denitrification, in the anoxic sulphide-containing waters of the Cariaco Trench. Carlucci and Strickland (1968) have provided indirect evidence (*vide supra*) that liberation of nitrogen can also take place in oxygen-depleted waters (<0.15 ml O_2/l). Experiments using ^{15}N labelled nitrate have led to the view that denitrification is a significant process in sea water provided that the concentration of oxygen does not exceed 0.10 ml/l (Goering and Dugdale, 1966b).

Nitrous oxide is known to be produced by bacterial denitrification under some conditions (e.g. in soils). Goering and Dugdale (1966a) could find no evidence of its production under anoxic conditions in an Antarctic lake. However, the presence of unusually high concentrations of this gas (0.27 ppmv) in surface and intermediate waters of the tropical and temperate zones of the central Pacific (Craig and Gordon, 1963) suggests that it is being produced in the sea.

C. THE SEASONAL VARIATION AND DISTRIBUTION OF COMBINED INORGANIC NITROGEN IN THE SEA

(i) *Near shore waters*

Seasonal variations in the concentrations of nitrate, nitrite and ammonia occur in the surface layer of the sea as a result of biological activity. These changes are most pronounced in the shallow waters near to the continents in mid and high latitudes (e.g. those of the English Channel (Cooper, 1933 and Fig. 7.3, p. 161) and off the eastern coast of the United States (Rakestraw, 1936)). In spring, in these waters the increased intensity and duration of light causes a proliferation of phytoplankton. This leads to a rapid removal of dissolved inorganic nitrogen species from the euphotic zone. A high proportion of the phytoplankton is consumed by zooplankton and fish. Nitrogen is returned to the water in their excreta, either in the rapidly assimilable forms present in the urine (e.g. ammonia and urea), or in the faecal pellets which must be decomposed bacterially before the nitrogen is again available. During the spring, vertical mixing contributes to the replenishment by bringing up nitrate-rich water from below the euphotic zone. However, in early summer, solar heating causes the development, at a depth of 20–40 m, of a thermocline which inhibits vertical mixing. As a consequence, the rapidly multiplying phytoplankton soon strip the water above the thermocline of its available inorganic nitrogen. The dominant form of nitrogen which is present at this time is usually ammonia which is excreted into the water by zooplankton feeding on the algae; this is rapidly reassimilated by the algae. In many localities nutrient depletion may be sufficiently severe to limit further growth of phytoplankton (see Dugdale, 1967). The excretion of nitrogen by zooplankton has been found (Martin, 1968) to be at a maximum when phytoplankton are scarce and *vice versa*. This probably arises because of utilization, as an energy source, of proteins laid down in time of food abundance. As the organisms die, or are consumed and excreted by zooplankton, bacterial regeneration of their nitrogen sets in. During this process the three stages in the conversion of organic nitrogen compounds to nitrate ion (p. 160) usually proceed simultaneously. The steady regeneration of nitrate is often interrupted by the occurrence of a further algal bloom in the late summer. During the earlier part of the regeneration the nitrite concentration increases progressively to a maximum in the autumn and then falls to low values. Nitrification is usually complete by January when surface cooling and storms have broken down the thermocline, and allowed the nitrate to become distributed homogeneously down the water column.

Very different conditions prevail in regions where upwelling occurs, bringing nutrient-rich water to the surface from an intermediate depth (see p. 43). The waters in these areas are highly fertile and support an abundant growth of phytoplankton. As long as upwelling continues, which it often does for periods of several months at a time, nutrients are never a growth-limiting factor.

(ii) Ocean waters

(a) Nitrate. As in inshore waters, changes in nutrient concentrations in the oceans produced by the growth of phytoplankton are restricted to the euphotic surface layers. However, the regenerative processes occur throughout the upper water column. Dead organisms and other organic detritus are attacked bacterially as they fall from the surface layer. As the particles decay they fall increasingly slowly because of their decreasing size and because of the higher density of the deeper waters.

Oxidation of these particles removes oxygen from the water, and carbon dioxide and nitrate ion which are the end-products of the oxidation of their organic compounds, therefore accumulate in the deeper waters.

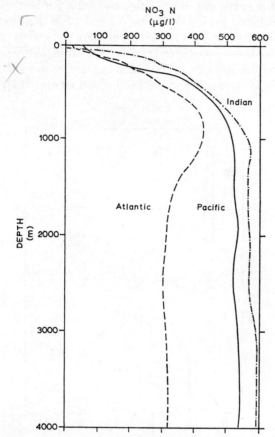

Fig. 7.4. Typical vertical profiles for nitrate concentration in the Atlantic, Pacific and Indian Oceans (Sverdrup *et al.*, 1942).

In the Atlantic there is a well defined nitrate maximum, with nitrate concentrations reaching 400 μg NO_3^-—N/l, at depths just below the oxygen minimum, ranging from 350–800 m depending on location (Fig. 7.4). The general levels of nitrate concentration in the sub-surface waters of the Pacific and Indian Oceans are considerably higher than those in the Atlantic and there is no clearly defined maximum because of differences in the deep water circulation pattern (Fig. 7.4). In all three oceans the nitrate concentration remains reasonably constant from intermediate depths to the bottom.

Although the nitrate and phosphate concentrations in the lower parts of the water column are fairly constant, considerable regional variations are found in the surface layers. These are the main factors controlling the fertility of a particular area. Thus, the most fertile regions of the tropical Atlantic and Pacific lie within a band a few degrees north and south of the equator, where upwelling enriches the surface water with nutrients. In contrast, the upper layers of the tropical oceans between 10° and 40° latitude are almost devoid of nutrients, and only support a low and relatively constant level of production (see p. 264).

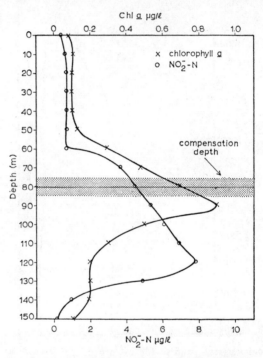

Fig. 7.5. Vertical profile of nitrite and chlorophyll a in relation to the compensation depth in the Sargasso Sea (Vaccaro and Ryther, 1960).

(b) *Nitrite.* Away from land, the nitrite concentration in the oceans increases, from barely detectable levels in the euphotic zone, to a maximum at a depth lying somewhat below the compensation point at ca 75–125 m (Fig. 7.5). This maximum probably arises from the intermediate formation of nitrite in the nitrification process (see p. 162). Beneath this maximum, the nitrite concentration falls rapidly to undetectable levels. In certain parts of the Pacific Ocean a second and greater nitrite maximum occurs in the almost anoxic waters (< 0.15 ml

O_2/l) of intermediate depth (see Fig. 7.6 and Brandhorst, 1959). This is thought to result from bacterial reduction of nitrate (see p. 162).

III. PHOSPHORUS

A. FORMS OF PHOSPHORUS PRESENT IN THE SEA

Phosphorus occurs in sea water in a variety of dissolved and particulate forms.

(i) Dissolved phosphorus

Inorganic phosphate exists in the sea practically entirely in the form of orthophosphate ions. Kester and Pytkowicz (1967) have shown that

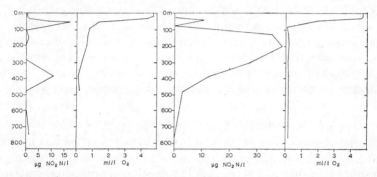

Fig. 7.6. Vertical profiles of dissolved oxygen and nitrite-nitrogen concentrations at two stations in the east tropical Pacific. Station S10 8° 42′ N., 86° 01′ W.; station S24 19° 30′ N., 105° 52′ W. (from Brandhorst, 1959).

in a sea water of average salinity (pH 8·0) at 20°C, 87% of the phosphate occurs as HPO_4^{2-}, 12% as PO_4^{3-} and 1% as $H_2PO_4^-$, and that 99·6% of the PO_4^{3-} and 44% of the HPO_4^{2-} is in the form of ion pairs (see p. 71), presumably with calcium and magnesium. Polyphosphate ions have not been detected in ocean water, but are known to occur in estuarine and coastal waters as a result of pollution with detergents (Solorzano and Strickland, 1968).

Organic phosphorus compounds constitute a significant, but variable, proportion of the dissolved phosphorus present in the upper layers of the ocean. They have not yet been identified, but it is probable that they are mostly decomposition and excretion products of marine organisms. If this is so, one might expect to find sugar phosphates,* phospholipids and phosphonucleotides and their decomposition products. In addition to these phosphate esters containing O—P linkages,

* Strickland and Solorzano (1966) were unable to detect simple sugar phosphates in offshore waters.

aminophosphonic acids, which contain the much more stable C—P bonds (bond energy 65 kcal/mole), may comprise a significant proportion of the dissolved organic phosphorus present in the sea (Kittredge et al., 1969). These compounds have been detected in a variety of marine invertebrates such as coelenterates and molluscs (Quin, 1965; Kittredge et al., 1967) and in a number of species of phytoplankton (Kittredge et al., 1969).

(ii) Particulate phosphorus

Little is known about the nature of the particulate phosphorus compounds occurring in the sea. It has been postulated that flocs of ferric phosphate may be present (Cooper, 1948), and this is supported by solubility product data which suggest that sea water may be supersaturated with this compound. Phosphate may also be present adsorbed onto particulate matter. Particulate organic phosphorus is that associated with living or dead organisms. It may therefore contain the whole range of organic phosphorus compounds involved in the biochemistry of marine organisms and their degradation products.

B. DETERMINATION OF PHOSPHORUS IN SEA WATER

The determination of phosphate is now usually carried out by treatment of an aliquot of the sample with an acidic molybdate reagent containing ascorbic acid and a small proportion of potassium antimonyl tartrate (Murphy and Riley, 1962). Phosphate yields a blue-purple complex, the absorbance of which is measured at 885 nm with a spectrophotometer. The reaction probably takes place with the intermediate formation of phosphomolybdic acid, which is then reduced to a heteropoly acid containing phosphorus, molybdenum and antimony in the ratio 1 : 12 : 1 by atoms. Polyphosphates do not react with the reagent, but can be determined with it after hydrolysis in acid medium at 100°C. (Solorzano and Strickland, 1968). Before total phosphorus can be determined with the reagent it is necessary to break down organic phosphorus compounds to phosphate. This decomposition can be most readily achieved by treating the sample with a small quantity of hydrogen peroxide, and irradiating it for a few hours with high intensity ultra-violet radiation (Armstrong and Tibbitts, 1968). Organically bound phosphorus is determined from the difference between total phosphorus and phosphate-phosphorus concentrations.

C. THE PHOSPHORUS CYCLE

The distribution of the various forms of phosphorus in the sea is broadly controlled by biological and physical agencies that are similar to those which influence the marine chemistry of nitrogen. It is most readily considered in terms of the phosphorus cycle (Fig. 7.7). This is very much simpler than the nitrogen cycle since phosphorus occurs in the marine environment only in the 5+ oxidation state (if the rare occurrence in invertebrates of organic compounds of trivalent phosphorus having C—P linkages is ignored). The cycle is not a completely closed one. When dead organisms sink to the sea floor, much of their phosphorus will be subsequently regenerated to the water. However, a proportion of it will eventually undergo diagenetic changes and be converted into phosphate minerals, such as apatite ($Ca_5(PO_4)_3(OH,F)$) (p. 354.) This loss of phosphorus from the sea is balanced by phosphate produced by rock-weathering which enters the sea in river water.

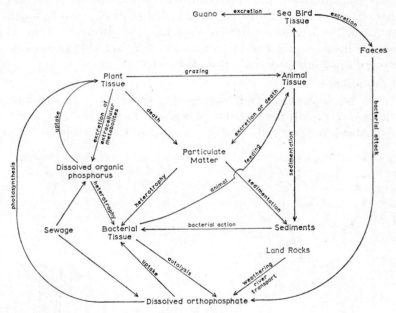

Fig. 7.7. The phosphorus cycle.

(i) Uptake of phosphorus

Phosphorus compounds, such as adenosine triphosphate and nucleotide coenzymes, play a key role in photosynthesis and other processes in plants. Phytoplankton normally satisfy their requirements of the element by direct assimilation of ortho-phosphate. Absorption and conversion to organic phosphorus compounds proceeds even in the dark. At phosphate concentrations above ca 10 μgP/l the rate of growth of many species of phytoplankton is independent of the phosphate concentration. However, as the concentration decreases below this critical level, cell division becomes increasingly inhibited and phosphorus-deficient cells are produced (Ketchum, 1939); eventually photosynthesis ceases (see also p. 244). If these deficient cells are supplied with phosphate they will readily assimilate it.

It is doubtful whether phosphorus deficiency is ever a growth-limiting factor in the sea, since nitrate is usually exhausted before phosphate falls to a critical level. Some species of phytoplankton in both their normal and phosphorus-deficient states are able to utilize dissolved organic phosphates (e.g. glycerophosphate and nucleotides) (Chu, 1946; Provasoli and McLaughlin, 1961). However, it is uncertain if this uptake mechanism is significant in the sea, or whether organic phosphorus compounds are first broken down bacterially before being assimilated. Bacteria normally satisfy their rather large phosphorus requirements from the organic detritus on which they live. However, if this food source is insufficiently rich in phosphorus they are able to assimilate dissolved inorganic phosphate from the water.

(ii) Regeneration

When phytoplankton and bacteria die, the organic phosphorus in their tissue is rapidly converted to phosphate through the agency of the phosphatases in their cells. In the sea most of the phytoplankton are

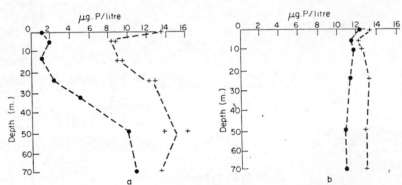

Fig. 7.8. Vertical profiles of phosphate ($\bullet$) and total phosphorus ($+$) in the English Channel. (a) August, (b) January (from Armstrong and Harvey, 1950).

consumed by animals, mainly zooplankton, which obtain their phosphorus requirements in this way. Part of the phytoplankton is assimilated and phosphate may appear in the urine. The unassimilated material is lost in the faecal pellets which contain appreciable amounts of organic phosphates in addition to orthophosphate (see, e.g., Butler *et al.*, 1969). Hydrolysis of the organic phosphates proceeds rapidly in the faecal pellets through the action of phosphorylases which also are present. Inorganic phosphate is soon leached from the faecal pellets and passes into the sea water together with any undecomposed organic phosphorus compounds. The latter which may account for as much as 75% of the phosphorus excretion of zooplankton (Hargrave and Geen, 1968) are rapidly broken down to orthophosphate bacterially or enzymatically. The excretion of phosphorus by zooplankton is minimal when phytoplankton are abundant and *vice versa* (Martin, 1968). It is thought that the low rate of phosphorus excretion when food is abundant arises because phospholipids are being stored and/or used in egg production. When food is scarce these lipids are used as an energy source and phosphorus is excreted.

D. DISTRIBUTION AND SEASONAL VARIATION OF PHOSPHORUS IN THE SEA

(i) Coastal waters

In shallow coastal waters of temperate latitudes, as typified by the English Channel, seasonal variations occur in both the phosphate and dissolved organic phosphorus concentrations. During the winter most of the phosphorus is present as orthophosphate (Fig. 7.8). However, this decreases rapidly in March when it is utilized by phytoplankton as they proliferate. Zooplankton and fish grazing on the phytoplankton return phosphorus to the water in their excretions as both phosphate and organic phosphorus compounds. The latter becomes the predominant form of dissolved phosphorus in May-June, when phosphate may have decreased to levels as low as 2 μgP/l in the euphotic zone. After the bloom, regeneration of phosphate from phytoplankton, detritus and the dissolved organic phosphorus compounds sets in rapidly (Fig. 7.9).

In many tropical regions, the seasonal variations in phosphate concentration of coastal waters are very much smaller than those in temperate regions. However, in those areas subject to monsoon conditions considerable changes do occur. Thus, Seshappa and Jayaramaran (1956) have observed that phosphate and plankton are abundant along the Indian coast during the south-west monsoon. Subsequently, both fall to low levels as grazing by zooplankton sets in. The bottom deposits become enriched with organic detritus, from which phosphorus is returned to the water either through the agency of filter feeders, or by bacterial action.

Very high nutrient levels may build up in estuaries and land-locked bodies of water as a result of the discharge of sewage and of effluents containing detergents rich in polyphosphates. These may be further augmented by nitrate and phosphate introduced from the run-off water from farm-land to which excessive amounts of fertilizer have been applied. Such conditions frequently lead to very rapid proliferation of phytoplankton, which when they die and decay soon strip the water of its dissolved oxygen. It seems probable that phosphorus is the main cause of such eutrophication, since even in the absence of combined inorganic nitrogen, nitrogen-fixing algae (p. 114) will continue to flourish, provided that sufficient phosphate is available.

(*iii*) *Ocean waters*

The distribution of phosphate in the open oceans closely parallels that of nitrate as it is regulated by the same agencies (see p. 164).

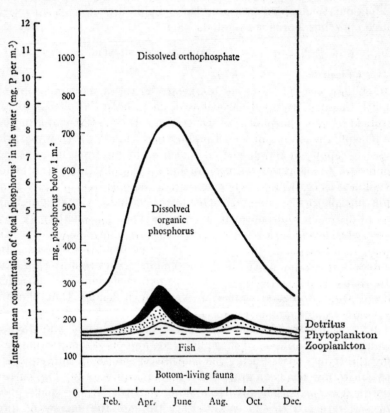

Fig. 7.9. Distribution of phosphorus in a water column 70 m deep in the English Channel (Harvey, 1955).

The concentration in the surface layer depends on the amount of exchange with the deeper water. It is high in regions such as the eastern boundaries of the oceans, where upwelling replenishes the surface layers. Over the rest of the ocean its level is usually low ($< 10\ \mu gP/l$), particularly in regions of convergence. There is a noteworthy correlation between the phosphate concentration of the upper layers of the oceans and their zooplankton content; the latter is an indirect measure of the phytoplankton available for grazing and hence of productivity (Reid, 1962). Phosphate increases rapidly with depth in the neighbourhood of the permanent thermocline (Fig. 7.10), as a result of the oxidation of detritus falling from the surface, and perhaps also because of the activities of zooplankton. The phosphate maximum usually lies close to, or just below, the oxygen minimum and the carbon dioxide maximum. The linear relationship which Postma (1964) has observed between the concentrations of phosphate and ΣCO_2, below the thermocline in the Pacific, confirms that carbon and phosphorus are simultaneously regenerated. At depths below the phosphate maximum, vertical eddy diffusion causes the phosphate concentration to be more or less uniform.

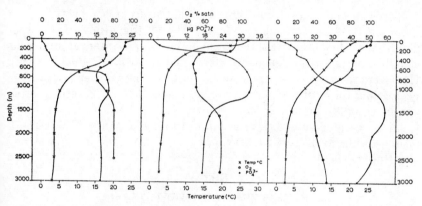

Fig. 7.10. Vertical distribution of phosphate, oxygen (as percent saturation) and temperature in North East, Sargasso Sea 34° 00′ N. 48° 50′ W. (left); Equatorial Atlantic 02° 00′ N., 41° 20′ W. (centre) (after Seiwell, 1935); and Indian Ocean 33° 50′ S. 28° 45′ E. (right) (*Discovery Rept.*, 1950).

Dissolved organic phosphorus may form a large proportion (up to 50%) of the total phosphorus in the surface layer of the oceans. However, this proportion decreases rapidly with depth and is generally insignificant at depths greater than 1,000 m (Ketchum et al., 1955).

The average composition of phytoplankton can be roughly designated by the empirical formula $(CH_2O)_{106}(NH_3)_{16}H_3PO_4$, and the oxidative regeneration of nitrate and phosphate from it can be represented by the equation

$$(CH_2O)_{106}(NH_3)_{16}H_3PO_4 + 138\ O_2 \rightarrow$$

$$106\ CO_2 + 122\ H_2O + 16\ NO_3^- + PO_4^{3-} + 19\ H^+$$

This implies that the consumption of 276 atoms of oxygen would lead to the formation of 16 nitrate ions and 1 phosphate ion (Richards, 1965). On this basis, Redfield (1942) has used the *apparent oxygen utilization* (p. 108) to calculate the amounts of nitrate and phosphate regenerated in bodies of water which have left the surface and sunk below the euphotic zone. He has termed nitrate and phosphate produced in this way *nutrients of oxidative origin*, in distinction to the *preformed nutrients* which were present in the water before it left the surface. This technique has been extensively used to study the biological transformations occurring in oceanic water masses during horizontal transport (Redfield, 1942; Riley, 1951; Redfield *et al.*, 1963). However, its value in water mass analysis is only just being appreciated (Pytkowicz, 1968; Pytkowicz and Kester, 1966; Culberson and Pytkowicz, 1970).

IV. NITROGEN : PHOSPHORUS RATIO

Nitrogen and phosphorus are assimilated from sea water in an approximately constant proportion of 15 : 1 (by atoms) by phytoplankton as they grow. It is remarkable that ocean waters at all depths usually contain these elements in a similar ratio (Redfield, 1934). However, there are a number of exceptions. Thus, lower ratios have been found for South Atlantic water (Cooper, 1938). The ratio is often low in coastal waters and may show a seasonal effect, for example, English Channel water has been found to have a NO_3^-—N : PO_4^{3-}—P ratio varying from 10·5 : 1 in winter to 19 : 1 in summer (Armstrong and Butler, 1960).

V. SILICON

A. FORMS OF SILICON PRESENT IN THE SEA

Silicon is present in sea water both in solution and in particulate form. The soluble form of the element is probably orthosilicic acid $(Si(OH)_4)$ for which the first and second dissociation constants are $3·9 \times 10^{-10}$ and $2·0 \times 10^{-13}$ respectively at 25°C in a sodium chloride medium having a similar ionic strength to sea water (Ingri, 1959). This suggests that ca 5% is ionized at pH 8·2. Since polymeric forms of silicic acid are rapidly depolymerized in sea water (Burton *et al.*, 1970), they are unlikely to occur in the sea, and the evidence for their occurrence even in river water is contradictory.

Sea water contains in suspension a wide spectrum of finely divided siliceous materials. Many of these have been produced by the weathering of rocks on land and have been transported to the sea by rivers or by wind; they include quartz, feldspars and clay minerals (see p. 307ff.). As they sink to the ocean floor and contribute to the sediments some of these minerals react with the sea water forming secondary minerals. In some parts of the oceans the surface waters abound with organisms such as diatoms and radiolarians which have skeletons composed of a non-crystalline form of hydrated silica–opal. When these organisms die, their siliceous skeletons slowly dissolve as they sink. Only the remains of the larger and more resistant species reach the ocean bed. In regions where these siliceous organisms proliferate their remains often constitute the major part of the sediment e.g. the diatom oozes of the Antarctic regions (see p. 376).

The concentration of suspended matter in the sea varies widely with geographical location, on average about half of it is inorganic, having a silicon content ranging between 15 and 60%. Highest concentrations occur in inshore waters for example, Armstrong (1958) has found 37–410 μg Si/l of suspended silicon in water from the English Channel. In contrast, water from the east North Atlantic contained only 20–60 μg/l the amount depending upon depth. High concentrations (up to 100 μg Si/l) of biogenic silicon are found in Antarctic surface waters during diatom blooms.

B. DETERMINATION OF SILICON IN SEA WATER

The determination of dissolved silicon in sea water depends on the formation of yellow β-silicomolybdic acid when the sample is treated with an acidic molybdate reagent. Only silicic acid and its dimer react at an appreciable rate, and the method therefore gives a measure only of "reactive" silicate, which is probably a reasonable measure of the silicate available to growing phytoplankton. Since the β-silicomolybdic acid is unstable and has only a low molar absorbance it is generally reduced to the stable and more absorbent molybdenum blue complex and measured spectrophotometrically at 812 nm. Although the reduction can be carried out with a variety of reducing agents, a reagent containing metol (p-methyl-aminophenol sulphate) and sodium sulphite is usually used for this purpose. Phosphate produces a similar blue complex, but the formation of this can be prevented by incorporating oxalic or tartaric acid in the reducing reagent. For practical details of this method the reader should consult the monograph by Strickland and Parsons (1968).

C. THE BALANCE OF DISSOLVED SILICON IN THE SEA

The solubility of silica in sea water is in the order of 50 mg Si/l. However, there is considerable disagreement about the precise value. Sea water is very much undersaturated with respect to the element, since the average concentration of dissolved silicate-silicon in the sea (ca 1 mg Si/l) is only about 1/50 of this value.

The principal sources of dissolved silicate to the oceans are river waters and the glacial weathering of rocks in Antarctica. These two processes are estimated to introduce respectively $2 \cdot 0 \times 10^{14}$ g and $4-5 \times 10^{14}$ g of dissolved silicon into the oceans annually (Livingstone, 1963; Schutz and Turekian, 1965). The growth and sedimentation of siliceous plankton is the principal biological process stripping silicon from the ocean, and accounts for the removal of ca $0 \cdot 9-1 \cdot 4 \times 10^{14}$ g Si annually. The great majority of this is deposited in the Antarctic Ocean as diatomaceous ooze. Thus, there appears to be a massive imbalance between the rate at which silicon enters the ocean and the rate at which it is removed biologically (Burton and Liss, 1968). If the silicon content of the oceans corresponds to a steady-state concentration it is necessary to seek other, presumably non-biological, mechanisms for the removal of $5-6 \times 10^{14}$ g of silicon per year. It is not known at present what these processes are, but deposition of silica cannot be involved as sea water is more than an order of magnitude undersaturated with respect to it. Bien and Thomas (1958) concluded from their work on the Mississippi that dissolved silicon was almost completely removed from the water by interaction with the suspended matter present in the estuary. However, Schink (1967) has suggested that their data were consistent with only partial (10–29%) removal. This is in agreement with the findings of other workers that less than 10% of the dissolved silicon present in river water is precipitated in this way (Kobayashi, 1967; Banoub and Burton, 1968). Reaction of dissolved silicon with oceanic suspended degraded aluminosilicates has been postulated by Burton and Liss (1968) as an alternative mechanism; similar reactions are thought to be important in maintaining the geochemical balances of the major elements in sea water (see p. 62 and Mackenzie and Garrels, 1966). Much further work will be needed before this difficult problem can be solved.

D. UPTAKE OF SILICON BY MARINE PLANTS AND ITS REGENERATION

The sea contains several groups of plants (e.g. diatoms and some Chrysophyta) and animals (e.g. radiolarians, pteropods and sponges) having silicified structures and some of them are sufficiently abundant to give rise to specialized deep sea oozes (see p. 297). Of these, diatoms are ecologically by far the most important.

Not much is known about the process by which dissolved silicate is taken up by diatoms and deposited as hydrated silica to form their elaborately patterned valves. It is known that energy is required for the deposition and that the sulphydryl (—SH) groups of proteins are involved in some way. Physico-chemical studies have shown that after silicic acid has been adsorbed on a protein monolayer it polymerizes and forms a rigid structure, probably because of hydrogen bonding between the imino groups of the protein and the hydroxyl groups of the silicic acid (Clark *et al.*, 1957). A similar mechanism may be involved in deposition of silica in diatoms. Lewin (1962) has suggested that the deposition may take place on the cytoplasmic membrane and that this serves as a template for the final pattern. It has been established that silicification is a very fast process and that it spreads from particular centres. The silica content of diatoms differs greatly from one species to another in the range 1–50% (of the dry weight). The weight of silica per cell in any given species varies with the concentration of silicon in the medium and with rate of cell division. If diatoms are grown in a medium depleted of silicon, the cells become silicon-deficient. Such cells may be viable in the medium for several weeks in the dark. When they are supplied with silicate they assimilate it even in the dark when no cell division can take place. If the deficient cells are strongly illuminated they photosynthesize for a limited period but soon die.

Some mechanism must protect the silica in living siliceous organisms, such as diatoms, since it is insoluble in sea water as long as they are alive, but dissolves fairly rapidly when they die. This protection may perhaps be a thin organic skin as suggested by Cooper (1952), and/or a coating of insoluble silicates of aluminium or iron formed after adsorption of the ions or hydroxides of these element by the silica (it has been observed that treatment of recently killed cells with chelating agents, such as EDTA, very much accelerates the process of dissolution (Lewin, 1961)).

E. MARINE DISTRIBUTION OF DISSOLVED SILICA

The dissolved silicon content of coastal waters is generally comparatively high because of the effect of run-off from land. In regions where blooming of phytoplankton and particularly diatoms occurs, marked seasonal variations are found resembling those of phosphate. The spring outburst of phytoplankton growth causes a rapid decrease in the concentration of silicon (Fig. 7.11), although considerable amounts still remain in the water. Regeneration of silicon commences during the summer when phytoplankton growth slackens, and continues until the maximum value is attained in the early winter. In some areas a phytoplankton bloom in the autumn may cause a temporary break in the regeneration.

The concentration of dissolved silicon in the surface waters of the oceans is generally low, except in regions of upwelling. In deeper layers there is a rapid increase in the concentration. The general distribution pattern differs from one ocean to another (Fig. 7.12) and is determined

8

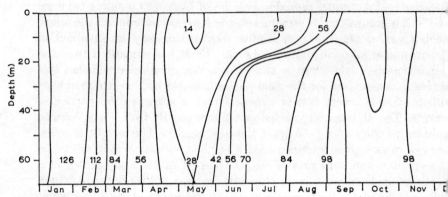

Fig. 7.11. Seasonal variation of silicon (as μg Si/l) at a station in the English Chan
1959 (from Armstrong and Butler, 1960).

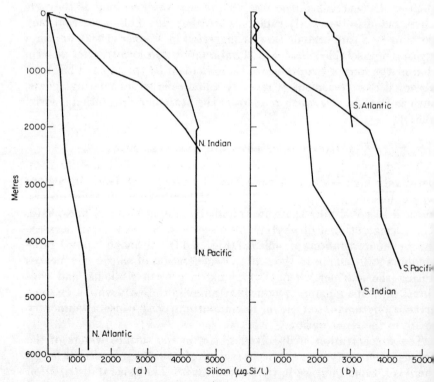

Fig. 7.12. Typical depth profiles of silicate in the oceans (after Armstrong, 1965).

by the water circulation pattern and by the supply of dissolved silicon from the Antarctic, and from diatoms dissolving as they fall from the surface. Grill (1970) has successfully developed a mathematical model for the vertical distribution of dissolved silicon in the oceans. This leads to the conclusion that the observed distribution is controlled by processes involving silicon utilizing-organisms rather than by aluminosilicate reactions.

REFERENCES

Armstrong, F. A. J. (1958). *J. mar. Res.* **17**, 23.
Armstrong, F. A. J. (1965). *In* "Chemical Oceanography" (J. P. Riley and G. Skirrow, eds), Vol. 1, p. 409. Academic Press, London.
Armstrong, F. A. J. and Harvey, H. W. (1950). *J. mar. biol. Ass. U.K.* **29**, 153.
Armstrong, F. A. J. and Butler, E. I. (1960). *J. mar. biol. Ass. U.K.* **39**, 299.
Armstrong, F. A. J. and Tibbitts, S. (1968). *J. mar. biol. Ass. U.K.* **48**, 143.
Banoub, M. W. and Burton, J. D. (1968). *J. Cons. int. Expl. Mer.* **32**, 201.
Bien, G. S. and Thomas, W. H. (1958). *Geochim. cosmochim. Acta*, **14**, 35.
Brand, T. von and Rakestraw, N. W. (1937–42). *Biol. Bull, Woods Hole* **72**, 165; **77**, 285; **79**, 231; **81**, 63; **83**, 273.
Brandhorst, W. (1959). *J. Cons. int. Expl. Mer.* **25**, 3.
Burton, J. D. and Liss, P. S. (1968). *Nature, Lond.* **220**, 905.
Burton, J. D., Leatherland, T. M. and Liss, P. S. (1970). *Limnol. Oceanog.* **15**, 473.
Butler, E. I., Corner, E. D. S. and Marshall, S. M. (1969). *J. mar. biol. Ass. U.K.* **49**, 977.
Carlucci, A. F. and McNally, P. M. (1969). *Limnol Oceanog.* **14**, 736.
Carlucci, A. F. and Strickland, J. D. H. (1968). *J. exp. mar. Biol. Ecol.* **2**, 156.
Chu, S. P. (1946). *J. mar. biol. Ass. U.K.* **26**, 285.
Clark, S. G., Holt, F. P. and Went, C. W. (1957). *Trans. Faraday Soc.* **53**, 1500.
Cooper, L. H. N. (1933). *J. mar. biol. Ass. U.K.* **18**, 677.
Cooper, L. H. N. (1937). *J. mar. biol. Ass. U.K.* **22**, 183.
Cooper, L. H. N. (1938). *J. mar. biol. Ass. U.K.* **23**, 179.
Cooper, L. H. N. (1948). *J. mar. biol. Ass. U.K.* **27**, 314.
Cooper, L. H. N. (1952). *J. mar. biol. Ass. U.K.* **30**, 511.
Craig, H. and Gordon, L. I. (1963). *Geochim. cosmochim. Acta.* **27**, 949.
Culberson, C. and Pytkowicz, R. M. (1970). *J. Oceanogr. Soc. Japan.* **26**, 95.
Discovery Reports (1950). **26**, 237.
Dugdale, R. C. (1967). *Limnol. Oceanog.* **12**, 685.
Dugdale, R. C. and Goering, J. J. (1964). *Limnol. Oceanog.* **9**, 507.
Dugdale, R. C. and Goering, J. J. (1967). *Limnol. Oceanog.* **12**, 196.
Dugdale, R. C., Menzel, D. W. and Ryther, J. H. (1961). *Deep Sea Res.* **7**, 297.
Fiadeiro, M. and Strickland, J. D. H. (1968). *J. mar. Res.* **26**, 187.
Fogg, G. E. and Than-Tun, (1958). *Biochim. biophys Acta*, **30**, 209.
Goering, J. J. and Dugdale, R. C. (1966a). *Limnol. Oceanog.* **11**, 113.
Goering, J. J. and Dugdale, R. C. (1966b). *Science, N.Y.* **154**, 505.
Goering, J. J., Dugdale, R. C. and Menzel, D. W. (1966). *Limnol. Oceanog.* **11**, 614.
Grant, B. R., Madgwick, J. and Dal Pont, G. (1967). *Austr. J. Mar. Freshw. Res.* **18**, 129.

Grill, E. V. (1970). *Deep Sea Res.*, **17**, 245.

Guillard, R. R. L. (1963). *In* "Symposium on Marine Microbiology" (C. H. Oppenheimer, ed.). Charles H. Thomas, Springfield, Illinois.

Hamilton, R. D. (1964). *Limnol. Oceanog.* **9**, 107.

Hargrave, B. T. and Geen, G. H. (1968). *Limnol. oceanog.* **13**, 332.

Harvey, H. W. (1940). *J. mar. biol. Ass. U.K.* **24**, 115.

Harvey, H. W. (1953). *J. mar. biol. Ass. U.K.* **31**, 477.

Harvey, H. W. (1955). "The Chemistry and Fertility of Sea Waters", 224 pp. Cambridge University Press, London.

Hutchinson, G. E. (1954). *In* "The earth as a Planet" (G. Kuiper, ed.). Chicago University Press, Chicago.

Ingri, N. (1959). *Acta chem. Scand.* **13**, 758.

Jones, K. and Stewart, W. D. P. (1969a). *J. mar. biol. Ass. U.K.* **49**, 475.

Jones, K. and Stewart, W. D. P. (1969b). *J. mar. biol. Ass. U.K.* **49**, 701.

Kester, D. R. and Pytkowicz, R. M. (1967). *Limnol. Oceanog.* **12**, 243.

Ketchum, B. H. (1939). *Am. J. Bot.* **26**, 399.

Ketchum, B. H., Corwin, N. and Keen, D. J. (1955). *Deep Sea Res.* **2**, 172.

Kittredge, J. S., Isbell, A. F. and Hughes, R. R. (1967). *Biochemistry*, **1**, 624.

Kittredge, J. S., Horiguchi, M. and Williams, P. M. (1969). *Comp. Biochem. Physiol.* **29**, 859.

Kobayashi, J. (1967). *In* "Chemical Environment in the Aquatic Habitat" (H. Golterman and J. Clymo, eds). North Holland Publishing Co. Amsterdam

Kreps, E. (1934). *In* "James Johnstone Memorial Volume" (R. J. Daniel, ed.), p. 193. Liverpool University Press, Liverpool.

Lewin, J. C. (1961). *Geochim. cosmochim. Acta.* **21**, 182.

Lewin, J. C. (1962). *In* "Physiology and Biochemistry of Algae" (R. A. Lewin, ed.) Academic Press, New York.

Livingstone, D. A. (1963). *Prof. Pap. U.S. Geol. Surv.* No. 440–G, 64.

MacIsaac, J. J. and Dugdale, R. C. (1969). *Deep Sea Res.* **16**, 45.

Mackenzie, F. T. and Garrels, R. M. (1966). *Am. J. Sci.*, **264**, 507.

Martin, J. H. (1968). *Limnol. Oceanog.* **13**, 63.

Murphy, S. and Riley. J. L. (1962). *Anal. Chim. Acta.* **27**, 31.

Murray, R. G. E. and Watson, S. W. (1963). *Nature, Lond.* **197**, 211.

Nutman, P. S. (1959). *Symp. Soc. exp. Biol.* **13**, 42.

Postma, H. (1964). *Netherlands J. Sea Res.* **2**, 258.

Provasoli, L. and McLaughlin, J. J. A. (1963). *In* "Marine Microbiology" (C. H Oppenheimer, ed.). Charles H. Thomas, Springfield, Illinois.

Pytkowicz, R. M. (1968). *J. Oceanogr. Soc. Japan.* **24**, 21.

Pytkowicz, R. M. and Kester, D. R. (1966). *Deep Sea Res.* **13**, 373.

Quin, L. D. (1965). *Biochemistry.* **4**, 324.

Rakestraw, N. W. (1936). *Biol. Bull. Woods Hole*, **71**, 133.

Ramamurthy, V. D. and Krishnamurthy, S. (1968) *Curr. Sci.* **37**, 21.

Redfield, A. C. (1934). *In* "James Johnstone Memorial Volume" (R. J. Daniel ed.). p. 176. Liverpool University Press, Liverpool.

Redfield, A. C. (1942). *Pap. phys. Oceanog.* **9**, 22.

Redfield, A. C., Ketchum, B. H. and Richards, F. A. (1963). *In* "The Sea" (M N. Hill, ed.), Vol. 2. p. 26. Wiley, New York.

Reid, J. L. (1962). *Limnol. Oceanog.* **7**, 287.

Richards, F. A. (1965). *In* "Chemical Oceanography "(J. P. Riley and G. Skirrow eds), Vol. 1, pp. 365. Academic Press, London.

Richards, F. A. and Benson, B. B. (1961). *Deep Sea Res.* **7**, 254.

Riley, G. A. (1951). *Bull. Bingham Oceanogr. Coll.* **13**, 1.
Riley, J. P. and Taylor, D. (1968). *Anal. Chim. Acta.* **40**, 479.
Schink, D. R. (1967). *Geochim. cosmochim. Acta.* **31**, 987.
Schreiber, E. (1927). *Helg. Wiss. meeresunters.* **16**, 1.
Schutz, D. F. and Turekian, K. K. (1965). *Geochim. cosmochim. Acta*, **29**, 259.
Seshappa, G. and Jayamaran, R. (1956). *Proc. Indian Acad. Sci.* **B43**, 287.
Seiwell, H. R. (1935). *Pap. phys. Oceanog.* **3(4)**, 1.
Solorzano, L. and Strickland, J. D. H. (1968). *Limnol. Oceanog.* **13**, 515.
Strickland, J. D. H. and Solorzano, L. (1966). *In* "Some Contemporary Studies in
    Marine Science" (H. Barnes, ed.). Allen and Unwin, London.
Strickland, J. D. H. and Parsons, T. R. (1968) "A practical Handbook of Sea
    Water Analysis." *Fish. Res. Bd Can. Bull.* **167**, 311 pp.
Sverdrup, H. U., Johnson, M. W. and Fleming, R. H. (1942). "The Oceans, their
    Physics, Chemistry and General Biology." Prentice Hall, New York.
Syrett, P. J. (1954). *In* "Autotrophic Micro-organisms" (B. A. Fry and J. L. Peel,
    eds), p. 126. Cambridge University Press, London.
Thomas, W. H. (1966). *Deep Sea Res.* **13**, 1109.
Tranter, D. J. and Newell, B. S. (1963). *Deep Sea Res.* **10**, 1.
Vaccaro, R. F. (1963). *J. mar. Res.* **21**, 284.
Vaccaro, R. F. and Ryther, J. H. (1960). *J. Cons. int. Explor. Mer.* **25**, 260.
Wattenberg, H. (1929). *Rappt. Proc. Verb. Cons. Perm. int. Expl. Mer.*, **53**, 90.
Wood, E. J. F. (1967). "Microbiology of Oceans and Estuaries." Elsevier,
    Amsterdam.

Chapter 8

Dissolved and Particulate Organic Compounds in the Sea

I. INTRODUCTION

The organic matter in the sea can be divided into two categories dissolved and particulate. The latter embraces material having a diameter greater than 0·5 μm, whereas the former includes true dissolved matter together with colloidal material passed by a 0·5 μm membrane filter. The amount of the dissolved organic matter in the sea usually exceeds the particulate organic fraction by a factor of 10–20 only 1/5 of the particulate matter, on average, consists of living cells. All this organic matter must ultimately have been produced by living organisms, and the range of compounds which it contains may therefore embrace the whole range of their cellular, metabolic and decay products These materials play a vital role in marine ecology since they provide part of the energy, food, vitamin and other requirements for bacteria plants and animals. Growth-promoting and inhibiting substances may have an important role in controlling the phytoplankton succession in the sea. For ease of treatment, dissolved and particulate organic matter will be discussed separately.

II. DISSOLVED ORGANIC MATTER IN THE SEA

A. ANALYTICAL

The total soluble organic-carbon content of ocean waters usually lies in the range of 0·3–3 mg C/l., although values as high as 20 mg/l may be found in coastal waters as a result of increased phytoplankton activity and pollution from land. In the method developed by Menzel and Vaccaro (1964) for the determination of dissolved organic carbon, the filtered sample is transferred to an ampoule, and acidified. Sparging with a stream of purified air removes the carbon dioxide associated with the carbonic acid equilibria. The sample is treated with potassium peroxydisulphate ($K_2S_2O_8$) and the ampoule is sealed. It is then heated

t 130°C in an autoclave for 1 hr. After cooling, it is opened in a closed
ystem and the carbon dioxide formed by oxidation of the organic
ompounds is removed with a current of helium or nitrogen, and
letermined by measurement of its infra red absorption (Menzel and
Vaccaro, 1964) or by adsorption chromatography. Alternatively, the
xidation can be carried out photochemically by irradiation with ultra-
violet radiation in the presence of a small amount of hydrogen peroxide
Armstrong *et al.*, 1966).

The determination of the dissolved organic nitrogen (5–300 μg N/l)
s most conveniently carried out by means of the method described by
Strickland and Parsons (1968). Organic nitrogen compounds are oxi-
lized to nitrate plus nitrite by irradiation with a powerful source of
ultra-violet radiation in the presence of hydrogen peroxide. The nitrate
s then reduced to nitrite using a cadmium reductor column, and total
nitrite-nitrogen is determined as described on p. 153. Organic nitrogen
s the difference between this figure and the nitrate+nitrite-nitrogen
oncentration of the original water sample. The determination of
lissolved organic phosphorus has been described on p. 168.

Although simple colorimetric procedures exist for the estimation of
ne or two *classes* of dissolved organic compounds in sea water (e.g. the
anthrone method for carbohydrates), the determination of individual
lissolved organic compounds in sea water is a much more difficult task.
The concentration of any individual compound will rarely exceed 10 μg/
and it is normally necessary to concentrate it from several litres of
sample in order to obtain sufficient for analysis; this also serves to
separate it from the large amounts of inorganic salts which might
otherwise interfere. The type of preconcentration technique to be used
will be determined by the class of compounds sought (Jeffrey and Hood,
1958). Among those which may be employed are solvent extraction
e.g. for fatty acids and insecticides), and adsorption on carbon (e.g. for
carbohydrates) and on polystyrene beads (e.g. for vitamin B_{12} and
Gelbstoff see Riley and Taylor, 1969). Highly sensitive methods of
determination must be used, and for this purpose thin-layer, or gas
chromatographic procedures are frequently chosen.

In a few instances, organic compounds can be determined directly
n sea water by bio-assay techniques (see e.g. Belser, 1963). These
lepend on the fact that some species of marine plants and bacteria
are unable to synthesize particular organic compounds which they
require for their growth (Provasoli, 1963). They therefore obtain
hem from the water, and will not thrive unless these compounds are
available. In practice, the filtered water sample or a concentrate from
t is enriched with micro-nutrients and chelated trace metals and

inoculated under aseptic conditions with an axenic culture of the assay organism (bacteria produce vitamin B_{12} and other essential compounds). The growth of the organism is then measured either photometrically or by cell-counting techniques. The increase in biomass when terminal growth has been attained is proportional to the amount of the essential organic compound. The procedure is standardized by examining another portion of the same sample which has been spiked with a known amount of the compound. The use of bio-assay procedures for the determination of vitamins is well established, and it may be possible to extend their use to other compounds if suitable assay organisms can be found or produced by mutation. Bio-assay techniques are extremely sensitive; thus vitamin B_{12} can be determined at concentrations down to 0·1 ng/l using *Euglena gracilis*.

B. ORIGIN OF DISSOLVED ORGANIC MATTER IN THE SEA

There are four main sources of dissolved organic compounds in the sea; (i) addition from the land; (ii) decay of dead organisms; (iii) addition of extracellular metabolites of algae, mainly phytoplankton; (iv) excretions of zooplankton and larger animals. These sources will now be considered separately.

(i) Organic compounds entering the sea from the land

Soluble organic materials from the land are transported to the sea by both wind and rivers, the latter being quantitatively the more important. The soluble organic content of river waters, which may be as high as 20 mg C/l, arises mainly from the leaching of humic materials and other decomposition products of vegetable matter from the soil. The humic materials are very resistant to oxidation and may constitute an appreciable fraction of the yellow-coloured material (Gelbstoff) present in coastal waters (see p. 190). In addition to organic matter added by natural agencies, ever-increasing amounts of soluble organic materials are being discharged into the sea in the form of sewage and industrial effluents. Most of these are readily oxidized, and are soon decomposed bacterially in the open waters of the sea. However, in restricted bodies of water, such as estuaries, their biological oxygen demand may be sufficient to cause anoxic conditions to be set up.

(ii) Production of soluble organic matter by the decay of dead organisms

The decay of dead organisms occurs by two mechanisms; autolysis and bacterial action. In nature, both mechanisms will act together, the extent of each varying according to the conditions of death and to the availability of the necessary enzymes and bacteria. In autolysis, the decomposition reactions are brought about by enzymes present in the

ead cell, and the reaction products are liberated into the surrounding
ater. It is uncertain what mechanisms are involved in the production
f dissolved organic compounds during bacterial decay of dead cells.
3acterial cells are known to autolyse very rapidly after death (see p.
59), and it seems probable that this process may account for some of
he dissolved organic carbon production. According to Johannes (1968),
xcretions of other micro-organisms, such as protozoa, which feed
hagotrophically on bacteria may be an important source of dissolved
rganic carbon compounds. Studies of the remineralization of nitrogen
nd phosphorus from dead organisms in sea water have shown that it is
iitially a rapid process. Thus, Waksman and his co-workers (1933,
938) have found that about half of the nitrogen present in dead
ooplankton was converted to ammonia in about 2 weeks and that
hosphate was liberated even more rapidly. During the initial period
f decay the bacterial count rose more than 50-fold. Little is known
bout the formation of soluble organic carbon compounds in the sea
uring the decay process. Skopintsev (1949) has observed that 70% of
ie particulate organic carbon in a dead algal culture was oxidized
) carbon dioxide after 6 months, and found that about 5% was con-
erted to comparatively stable soluble organic compounds.

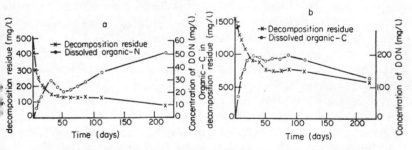

Fig. 8.1. Change in concentration of (a) dissolved organic nitrogen and (b)
ssolved organic carbon during the bacterial decay of dead cells of a fresh water
ecies of *Scenedesmus* (Otsuki and Hanya, 1968).

Our knowledge of decay processes occurring in fresh water media is
ore detailed. Otsuki and Hanya (1968) have examined the decomposi-
on of dead cells of a fresh water species of *Scenedesmus* by microflora
om a lake sediment. They found that the soluble organic carbon and
trogen concentrations increased rapidly during the initial decay of
e cells (Fig. 8.1). At its maximum the dissolved organic-carbon
ncentration represented ca 35% of the particulate carbon decomposed.
issolved organic nitrogen continued to be formed even after the rapid
itial phase of the decomposition of the phytoplankton, and at the

termination of the experiment amounted to ca 18% of the particulat nitrogen decomposed. The dissolved organic material remaining at the end of the experiment had an absorption spectrum similar to that of Gelbstoff (p. 190), and had the low C/N ratio characteristic of deep ocean waters. These observations were made on non-saline media, and it is uncertain whether they represent what would happen in saline waters. Very little is known about the nature of the soluble product formed during the decay of organisms in the sea, and much further work on the subject is required.

(iii) Excretion of extracellular products by algae

Algae liberate appreciable proportions of the compounds produced by photosynthesis into the surrounding water. These extracellular products are important as sources of energy for other marine organisms and may also exert some ecological control. The reader is referred to the excellent review by Fogg (1966) for more details of the subject of extra cellular metabolites than can be given in the compass of this section

The extent of the production of extracellular metabolites can be most readily investigated by a radiochemical method (Hellebust, 1965). Algal suspensions are treated with $NaH^{14}CO_3$ having a high specific activity. After a suitable period under the appropriate conditions the algae are filtered off using a membrane filter. A small aliquot of the filtrate is acidified and aerated to remove inorganic carbon. The extra cellular metabolites are then assayed by direct counting of the ^{14}C activity of the residue remaining after evaporation. Alternatively, the production of extracellular material during photosynthesis can be estimated from the difference between the gross productivity estimated from oxygen consumption, and the net productivity determined by the carbon-14 method (see pp. 232–235).

The quantity of extracellular material produced by marine phyto plankton grown in culture varies widely according to species. Hellebust (1965) studied 22 species of phytoplankton and observed that although the majority excreted 3–6% of the photosynthetically fixed carbon some liberated as much as 10–25%. Liberation of these materials takes place not only in the light but also in the dark. However, production in the dark, which may be much greater than in the light, appears to take place by a different mechanism (Fogg et al., 1965). The compound excreted are usually small molecules such as glycollic acid (Watt and Fogg, 1966), amino acids (Hellebust, 1965) and carbohydrates. Some times a single compound is excreted dominantly by a particular species e.g. Olisthodiscus sp. produces mainly mannitol (Hellebust, 1965). The extent to which these materials are produced by any particular specie is controlled by the chemical and physical conditions in the culture.

A knowledge of the occurrence of excretion in cultures, with their high nutrient concentrations and high cell populations, is of little value in estimating the extent to which this process occurs in the sea. Khailov and Burlakova (1969) have observed that large brown and red algae growing in natural sea water release up to 39% of their gross photosynthetic production as dissolved organic compounds. The only *in situ* demonstration of the excretion of extra cellular material in the sea has been carried out by Antia *et al.* (1963). These workers found that only 60–65% of the inorganic carbon removed from water by phytoplankton growing (in the absence of zooplankton) in a plastic sphere in the Pacific was incorporated into the cells; the remainder presumably being converted to extracellular metabolites. More recently, Vaccaro *et al.* (1968) have shown that the concentration of glucose in certain productive areas of the Atlantic could be correlated with phytoplankton productivity. At present, it has only been established that excretion of extracellular material does occur to an appreciable extent.* Much further work is required to establish the significance of material from this source in the oceanic budget of dissolved organic carbon.

(iv) Excretion by marine animals

The excretions of zooplankton and marine animals may be an important source of dissolved organic matter in the sea. The compounds introduced in this way are principally nitrogenous ones (Corner and Newell, 1967), such as urea, purines (e.g. allantoin and uric acid), trimethylamine oxide and amino acids, of which glycine, taurine and alanine usually predominate (Webb and Johannes, 1967). In addition to dissolved organic matter contributed in this way, marine animals make a highly significant indirect contribution to the budget of dissolved organic material through the bacterial decay of their faecal pellets.

C. THE NATURE OF THE DISSOLVED ORGANIC COMPOUNDS PRESENT IN THE SEA

The major fraction of the dissolved organic carbon in the sea consists of a complex material (Gelbstoff) which is very resistant to bacterial attack. However, from the standpoint of ecology the minor fraction composed of more labile compounds is more important. This fraction contains substances representative of the main biochemical classes (e.g. amino acids, carbohydrates, lipids and vitamins; see Table 8.1). The concentrations of these substances are low and must reflect a balance between the rates at which they are produced and that at which they are used by micro-organisms. The fact that the concentration of any particular compound is low does not necessarily imply that it has no importance as a nutrient, since it may have a high rate of turn-over.

---

* According to Anderson and Zeutschel (1970), the amounts of these materials are greatest relative to total production in oligotrophic areas, but in absolute terms they are greatest in eutrophic waters and near the surface.

In many instances the concentrations are probably maintained at the lowest levels at which the compounds can be assimilated by bacteria. Owing to analytical difficulties little is yet known about the distribution and seasonal variation of individual compounds in the sea. Further examination of this problem awaits the development of new and highly sensitive methods of analysis, which should preferably be suitable for use on board ship.

Carbohydrates appear to occur in sea water practically entirely as free saccharides, and to only a negligible extent as their derivatives e.g. amino sugars. Both hexoses and pentoses have been detected. However, in the light of the work of Degens *et al.* (1964) (Table 8.1) the older reports of large concentrations of pentoses and rhamnoses must be regarded with caution, even though these sugars are known to be present in the polysaccharides of some marine plants. The only dissolved carbohydrate that has been studied in any detail is glucose, which is readily assimilated by many species of marine bacteria down to concentrations of less than 8 μg/l (Vaccaro and Jannasch, 1966). Its concentration in the upper layers of a tropical trans-Atlantic section has been investigated by Vaccaro *et al.* (1968) using enzymatic and bioassay methods of analysis. The maximum concentration of glucose (ca 80 μg/l) was found in the Sargasso Sea where a fairly large phytoplankton crop was held in check by a shortage of nutrients. Such conditions are known to lead to the production of glucose as an extracellular metabolite of phytoplankton (see p. 186 and Hellebust, 1965). These high concentrations of glucose were probably sustained because the bacterial population used it only at a slow rate.

Dissolved amino acids occur in the sea mainly combined as peptides but also in the free form. The principal amino acids are α-alanine, serine, threonine, glycine and valine; subsidiary amounts of aspartic and glutamic acids can also usually be detected. Most of these compounds probably enter the sea as a result of decomposition of protein during the decay of organic tissue and excreta. However, maxima have been found in the free amino acid concentrations at the time of the spring and autumn plankton blooms in the Irish Sea. This suggests that they may also enter the water as extracellular metabolites of phytoplankton (Riley and Segar, 1970).

Several carboxylic acids have been detected in sea water, including acetic and glycollic and the Krebs-cycle acids—citric and malic acids (Creac'h, 1955). Owing to the difficulty of separating and determining such hydrophilic compounds not much is known about their concentrations. However, they may be major components of the dissolved organic matter in the sea since they are excreted by organisms and are

TABLE 8.1

Representative concentrations of some dissolved organic compounds in marine surface waters

Compounds or class of compound	Concentration $\mu g/l$	Locality	Reference
Methane	1	Trop. Atlantic	Swinnerton *et al.*(1969)
Paraffinic hydrocarbons[a]	400	South Pacific	Jeffrey (1966)
Pristane (2,6,10,14-tetramethylpentadecane)	tr	Cape Cod Bay	Blumer *et al.* (1963)
Pentoses	0·5	Pacific off California	Degens *et al.* (1964)
Hexoses	14–36	Pacific off California	Degens *et al.* (1964)
Glucose		Sargasso Sea	Vaccaro *et al.* (1968)
Malic acid	300	Atlantic Coast	Creac'h (1955)
Citric acid	140	Atlantic Coast	Creac'h (1955)
Triglycerides+fatty acids[b]	200	South Pacific	Jeffrey (1966)
Amino acids[c]	10–25	Irish Sea	Riley and Segar (1970)
Peptides[c]	10–100	Irish Sea	Riley and Segar (1970)
Vitamin B_{12}	0–0·010	Various	Provasoli (1963)
Thiamine (Vitamin B_1)	0·021	Long Island Sound	Vishniac and Riley (1959)
Biotin	0·01	G. of Mexico	Litchfield and Hood (1965)
Urea	0–80	English Channel	Newell, *et al.* (1967)
Adenine	100–1,000	G. of Mexico	Litchfield and Hood (1966)
Uracil	300	G. of Mexico	Litchfield and Hood (1966)
ρ-hydroxy-benzoic acid	1–3	Pacific off California	Degens *et al.* (1964)
Vanillic acid	1–3	Pacific off California	Degens *et al.* (1964)
Syringic acid	1–3	Pacific off California	Degens *et al.* (1964)

[a] C_{12}, C_{14}, C_{16}, C_{18}, C_{20} and C_{24} paraffins.

[b] A large number of saturated and unsaturated fatty acids were present, with oleic 18:1) and palmitic (16:0) predominating.

[c] Mainly alanine, serine, glycine, threonine and valine with lesser amounts of glutamic and aspartic acids.

common extracellular metabolites. Thus, Fogg (1966) has pointed out that alternation of sunny and cloudy periods may cause phytoplankton (particularly *Chlorella* spp.) to release considerable amounts of glycollic acid into the surface waters. Further evidence of the high concentrations of some of these acids is provided by the deposition of calcium citrate on the floor of the Weddell Sea (Bannister and Hey, 1936). In contrast, the concentrations of the higher fatty acids in the sea are extremely low.

Gelbstoff

In addition to these comparatively labile compounds, sea water contains considerably larger amounts of dissolved material which is resistant to bacterial attack. This material, which has been named *Gelbstoff* because of the yellow colour which it imparts to the water (Kalle, 1937, 1949), exhibits a blue fluorescence. Kalle (1966) recognized that it was not a single compound but a complex mixture. He concluded that it was formed in the sea and was the equivalent of the terrestial humic materials. Brown algae (Phaeophyta), which are the principal algal group growing in coastal waters of temperate and higher latitudes, excrete phenolic compounds (see p. 193 and Craigie and MacLachlan, 1964). At the pH of sea water these polyphenols are converted by secondary reactions with carbohydrates and proteins of algal origin into a brown polymer. The absorption and fluorescence spectra of this material and its behaviour in chromatography and gel filtration strongly suggest that it is identical with the Gelbstoff found in coastal waters (Sieburth and Jensen, 1969). Both of these substances soon precipitate in the sea in the form of organic aggregates indistinguishable from those described by Riley (1963) (see p. 209). Despite the very considerable amount of Gelbstoff liberated into the sea by the brown algae, its concentration level in the oceans, as a whole, is relatively low (av. ca 1 mg/l). It seems probable that removal takes place mainly by precipitation (Sieburth and Jensen, 1968, 1969), rather than by bacterial attack to which it is resistant because of its phenolic nature (Craigie and MacLachlan, 1964). Some humic material derived from the leaching of soils and having a composition somewhat different from that of Gelbstoff enters the sea in river water. However, its contribution to oceanic Gelbstoff is not thought to be large, since much of it is rapidly precipitated in contact with sea water (Sieburth and Jensen, 1968).

D. ECOLOGICAL EFFECTS OF DISSOLVED ORGANIC MATERIAL

During the course of evolution many marine organisms have come to depend on the dissolved organic material, not only as an energy source, but also to supply essential organic compounds which they are unable to synthesize. There are several degrees of dependence on dissolved organic materials. In auxotrophy, the organism has a requirement for one or two specific substances, such as vitamins, whereas in heterotrophy it depends entirely on dissolved and/or particle organic matter. Some organisms are able to transfer from one type of dependence to another according to the availability of food (see also p. 221).

For many years it has been realized that waters of the sea can be classified as "good" or "bad" according to their productivity. At first it was thought that this was controlled entirely by the availability of inorganic micronutrients, particularly nitrogen and phosphorus. Although this is frequently the case, it is not always so. For example, productivity in the waters around the British Isles is considerably higher than in the coastal waters off California, even though the latter waters are considerably richer in inorganic micronutrients. This, and the fact that many algae have a specific requirement for certain organic compounds, led to the conclusion that marine productivity might also be controlled by the presence of biologically active substances which Lucas (1947, 1955) has termed "ectocrines".

Laboratory experiments have demonstrated that many of the substances which are excreted by the lower forms of marine life may act as ectocrines, and either promote or inhibit growth. Johnston (1963) has concluded that of these, the growth-promoting factors are the more important in controlling the "quality" of sea water. Work on cultures has shown that different species frequently react to a particular ectocrine in different ways. This suggests that these substances may exert a controlling influence on the phytoplankton succession in the sea. However, there is, as yet, no direct evidence that they are important under natural conditions.

(i) *Growth-promoting factors* (*see also pp. 247–249*)

Thiamine, biotin and vitamin B_{12} are essential for the growth of many species of marine phytoplankton. These compounds occur in sea water at concentrations of the order of nanograms (10^{-9} g) or less per litre. Vitamin B_{12} (cyanocobalamin) is the most important of these substances as it is needed by the majority of algal auxotrophs. Droop (1968) has observed that the specific growth rate of *Monochrysis lutheri* is constant above 0·5 ng vitamin B_{12}/l. The vitamin is synthesized by many species of marine bacteria, by some seaweeds and also by a few comparatively scarce species of planktonic algae such as *Trichodesmium erythraeum* (which forms blooms in the Arabian Sea and elsewhere). The vitamin is probably not actively excreted by the organisms, but is released only after their death. There can be little doubt that bacteria are the principal source of this vitamin in the sea (Provasoli' 1963).

In addition to the vitamins, other types of organic compounds may act as growth promoters. There is considerable evidence that glycollic acid can act not only as a carbon and energy source for unicellular algae, but also that it can shorten the lag phase in the growth of these organisms in culture media (Nalewajko et al., 1963). Many species of phytoplankton will not grow in sea water which has been heated to 80°C, even if abundant inorganic nutrients are present—evidently some essential compounds are destroyed during the heating process. However, the addition of a sterilized extract of soil makes the medium suitable for growth, and this is the basis of many of the older culture media, e.g. Erd-Schreiber. This growth-promoting activity may, in part, be due to the chelating action of the soil humic acids, which improves the availability of trace metals. However, the presence in the extract of organic micronutrients such as vitamins and auxins may also play an important part. Entirely synthetic culture media, containing chelated trace metals and vitamins, have been developed in recent years which support the growth of many species of phytoplankton (see, e.g. Provasoli et al., 1957). Some types of algae (e.g. some dinoflagellates) also require the presence of low molecular weight humic acids before they will thrive. Prakash and Rashid (1968) consider that this growth-promoting property of humic acids may be ecologically important in coastal waters.

(ii) Growth-inhibiting factors

Growth-inhibitors which have been identified in culture medium include both antibiotics and toxins. Among the latter are the toxic substances secreted by many dinoflagellates, such as *Gymnodinium breve* and *Gonyaulax polyhedra*, which give rise to "red tide" blooms in certain regions where nutrient enrichment occurs (Brongersma-Sanders, 1957). When these organisms die following the peak of the bloom, their toxins are liberated into the water and mass mortalities of fish occur. Some species, e.g. *Gonyaulax catenella* do not kill marine organisms, but their toxins are concentrated by shell-fish into their digestive glands, which thus become toxic to man. The strongly basic nitrogenous poison of this last named species is one of the most toxic known (Schantz, 1960; Burke et al., 1960).

Many marine organisms are known to produce extracellular metabolites or decay products which are capable of retarding the growth of other organisms or even the same species. The activity of these substances has generally been demonstrated in culture medium. Because of their great dilution it seems unlikely that they would normally have any ecological significance in the open ocean. However they may have an effect in conditions where dilution is restricted, e.g. in algal blooms, estuaries and particularly in rock pools.

It is uncertain whether antibiotics produced by bacteria play any part in controlling the growth of marine organisms. No marine species of the actinomycetes and fungi imperfecti, which are the most important group producing antibiotics in the soil, are known to exist. However, some terrestial actinomycetes, such as *Streptomyces*, that are tolerant of saline conditions have been found on coastal sediments, nets, etc. Several of these have been shown to exhibit antibiotic activity in culture (Grein and Meyers, 1958). A few species of marine bacteria, e.g. *Bacillus* and *Micrococcus* have been shown to be bactericidal to non-marine species (Rosenfeld and Zobell, 1947, see also Burkholder, 1970). These may well be the cause of the well known germicidal action of sea water towards many pathogenic bacteria.

Many species of algae produce substances which have high activity against micro-organisms. Although the nature of the active substances is known only in a few instances, they appear to represent a wide range of compounds both simple and complex.

Polyphenols, related to catechol, which are produced by brown algae are known to inhibit the growth of many species of unicellular algae (Sieburth, 1968; McLachlan and Craigie, 1964). These substances may be ecologically important in suppressing the growth of epiphytes (Fogg, 1966). Proctor (1957) has demonstrated that *Chlamydomonas reinhardi* liberates a mixture of long-chain fatty acids which completely inhibits the growth of *Haematococcus fluvialis*. Although these observations were made with fresh water species, there is no reason to suppose that similar effects may not be produced by the fatty acids of marine algae including sea weeds.

At low concentrations, the tannin-like material excreted by the flagellate *Olisthodiscus luteus* is thought to stimulate the growth of the diatom *Skeletonema costatum*. However, at high concentrations it acts as an inhibitor and may in fact prevent the blooming of the diatom in Narragansett Bay (Pratt, 1966). *Monochrysis lutheri* (Chrysophyceae) has been shown to bind vitamin B_{12} with extracellular protein and render it inert. Droop (1968) has suggested that the organism takes up all the vitamin necessary for many generations in the early stages of the bloom. The gradual production of the inhibitor does not therefore affect the organism which produces it, but will inhibit the growth of others which require the vitamin.

In some instances inhibitor substances can be passed along the marine food chain. The reduced gastrointestinal flora of penguins and the antibiotic activity of their blood serum against Gram positive bacteria, has been attributed by Sieburth (1959, 1960) to acrylic acid. This is thought to have originated in a *Phaeocystis*-like alga which forms the diet for the crustacean *Euphausia superba*, which in its turn is the staple diet of the birds.

It has only been possible in this chapter to draw attention to a few aspects of the phenomena of growth promotion and inhibition. For further information the reader is referred to the review articles by Provasoli (1963), Shilo (1964), Fogg (1962, 1966) and Sieburth (1968) as well as to the classical papers by Lucas (1947, 1955, 1961). Active research in this field is continuing with a view to discovering the nature of the substances responsible and to determining their significance in marine ecology.

E. THE FATE OF DISSOLVED ORGANIC MATTER IN THE SEA

Dissolved organic matter is removed from the sea by several processes, the combined effect of which ensures that the total dissolved organic carbon level rarely rises above 3 mg C/l. Undoubtedly the most important of these processes are those in which the breakdown and respiration of this material provides energy and cellular carbon for living organisms, and brings about regeneration of carbon dioxide and nutrients.

It is probable that the heterotrophic bacteria are the principal users of dissolved carbon compounds. These micro-organisms occur widely throughout the oceans (Kriss *et al.*, 1964) where they are mainly associated with particulate matter, both organic and inorganic. Their most commonly used substrates are simple compounds such as acetate, lactate, citrate, amino acids and glucose (see, e.g. MacLeod *et al.*, 1954), although there are probably marine species capable of utilizing any organic substances existing in the sea.

Some species of phytoplankton are also able to subsist heterotrophically on dissolved organic substrates particularly at low light intensities, or in the dark (Lewin, 1963; Fogg, 1963; Hamilton and Presland, 1970). There is considerable selectivity among these with regard to the type of compounds which they use. Thus, *Chlorella pyrenoidosa* subsists in the dark best on glucose and not as well on acetate and galactose; it appears to be unable to utilize other organic compounds (Samejima and Myers, 1958). The role of phytoplankton in the removal of soluble organic material from the sea is probably quantitatively far less important than that of bacteria (Wright and Hobbie, 1965).

Radiochemical techniques, using ^{14}C-labelled compounds, have been developed to measure the uptake of these compounds by heterotrophic organisms (Parsons and Strickland, 1961; Wright and Hobbie, 1965, 1966). Assimilation of glucose and acetate by heterotrophic organisms present in coastal waters has been studied by Parsons and Strickland *loc cit.*, and Vaccaro and Jannasch (1966) who observed that their uptake is first order with respect to the substrate concentration and conformed with the Michaelis-Menten equation (p. 157). At substrate concentrations of 250 μg C/l assimilation rates of ca 10 μg C/l/hr were found. Williams and Askew (1968) have investigated the turnover time of organic substrates in sea water by measuring the rate of heterotrophic regeneration of carbon-14 dioxide from ^{14}C-labelled glucose. They observed that the turnover time at a station in the English Channel varied from over 60 days in winter to 1 day during July, presumably according to the numbers of heterotrophs available. Heterotrophic attack on the dissolved organic material in sea-water is thus of importance because: (i) it causes regeneration of carbon dioxide and nutrients, and (ii) it converts the dissolved compounds, which are probably of little value to marine animals, into bacterial tissue which can be utilized by them. Unfortunately, little is known of the food chain stemming from bacteria.

Generally, the amount of available organic carbon in the sea is rather limited, since most of the dissolved organic carbon present is resistant to bacterial decay (see p. 196). However, since heterotrophic activity can occur throughout the whole water column, heterotrophic carbon assimilation may amount to a significant fraction of the photosynthetic primary production (Parsons and Strickland, 1961).

Riley *et al.* (1964, 1965) have suggested that dissolved organic carbon is converted into particulate matter through the agency of the bursting of air bubbles at the sea surface (see p. 209). Experimental evidence for the existence of such a process is contradictory. Even if dissolved organic carbon is not removed from the sea by this mechanism it may be removed by adsorption onto suspended particulate matter both organic and inorganic. Thus, it is well known that vitamin B_{12} is readily adsorbed by marine detritus and clay minerals. Dissolved organic carbon may also be removed from the sea by being sprayed into the air as an aerosol when bubbles burst at the sea surface (see Blanchard, 1964).

F. DISTRIBUTION OF DISSOLVED ORGANIC CARBON IN THE SEA

Almost all the dissolved organic carbon in the sea originated ultimately from carbon dioxide fixed by phytoplankton. It would be expected therefore that its concentration in the euphotic zone at given locations would be related in a gross fashion to the prevailing rate of primary production. The actual concentration at any instant will depend on the balance between the rate at which dissolved organic carbon is formed by decay, excretion etc. and the rate at which it is removed by decomposition or utilization. The quantitative significance of these various processes in maintaining the balance is only poorly understood, as also is the mechanism by which dissolved organic carbon is distributed to the depths of the oceans. Practically nothing is known about the geographical or seasonal variations of individual dissolved organic compounds in the sea.

(i) Seasonal variations

Seasonal variations in the dissolved organic carbon content are usually restricted to the upper ca 100 m and correlate roughly with productivity. The results given by Duursma (1961) for a station in the North Sea are typical of productive coastal waters. The highest values (ca 1·8 mg C/l) were found in the spring and early summer, somewhat later than the period of maximum phytoplankton activity. The concentration then decreased slowly during the summer, and with the exception of a minor maximum in the month after the autumn plankton bloom, remained relatively constant in the range 0·6–0·9 mg C/l until the onset of the spring bloom. A similar trend was also shown by the dissolved organic nitrogen concentration. Analogous seasonal variations correlating with phytoplankton activity have been found for individual organic nutrients and combined (peptide) amino acids in coastal waters (Belser, 1959; 1963; Riley and Segar, 1970). Seasonal differences in the dissolved organic carbon content in the surface waters of the open oceans are usually much smaller than those in coastal waters.

(ii) Variation with depth

The depth distribution of dissolved organic carbon in all the oceans follows the same general pattern. In any particular water column the amount of organic carbon produced annually is only a small fraction of the total dissolved organic carbon. Most of the latter appears to be resistant to oxidation, and even in highly productive surface waters labile compounds usually account for less than half of the dissolved organic carbon. In waters deeper than a few hundred metres practically all the dissolved organic material is resistant to bacterial attack (Menzel, 1964; Menzel and Goering, 1966; Barber, 1968), and probably consists of Gelbstoff-type compounds (p. 190). The hypothesis that this organic material is resistant to biochemical attack has received added support from ^{14}C-age determinations. Williams et al. (1969) have observed that the dissolved organic carbon from 3,400 m in the north-east Pacific is of a very considerable age (ca 3,400 years), and is in fact on average probably more than twice as old as the inorganic carbon present. On the basis of their results these authors have estimated that if a steady state exists, then only ca 0·5% of the photosynthetically fixed carbon enters

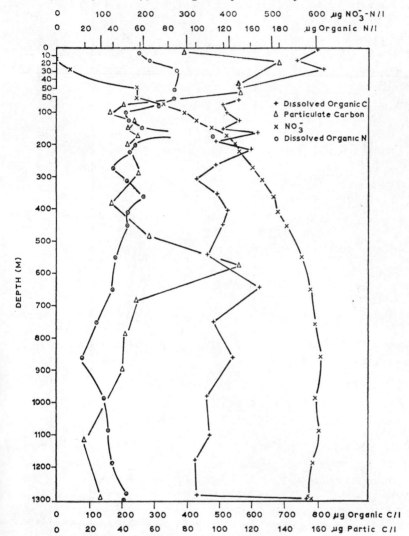

Fig. 8.2. Vertical profiles for concentrations of nitrate, dissolved and particulate organic carbon and dissolved organic nitrogen at a station off southern California (after Holm-Hansen et al., 1966).

the deep sea. Further confirmation of the stability of this dissolved organic material has been obtained from measurements of its $^{13}C/^{12}C$ ratio (Williams and Gordon, 1970).

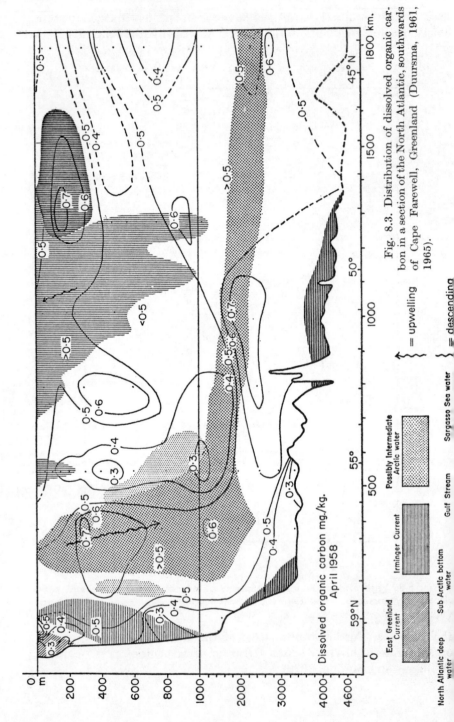

Fig. 8.3. Distribution of dissolved organic carbon in a section of the North Atlantic, southwards of Cape Farewell, Greenland (Duursma, 1961, 1965).

The highest concentrations of dissolved organic carbon and nitrogen are found in the surface layers (Fig. 8.2). It is only in this layer that significant seasonal variations occur. Around the base of the euphotic zone the concentrations of both begin to decrease with increasing depth at rates which differ from place to place, in accordance with the productivity, availability of heterotrophs and the hydrographic conditions. At depths greater than a few hundred metres the concentrations are relatively constant. Menzel and Ryther (1968) have concluded that the organic carbon content of deep water arises neither from decomposition of particles falling from the surface, as was once thought, nor by the *in situ* decay of entrained particulate matter, since this is now known to be resistant to oxidation. They have suggested that it is evolved during the sinking of surface water and its transport and mixing in the deep water circulatory system. For this reason, the dissolved organic carbon content of the deep waters shows small, but significant, variations with depth (see, e.g. Fig. 8.2). These reflect the origins of the

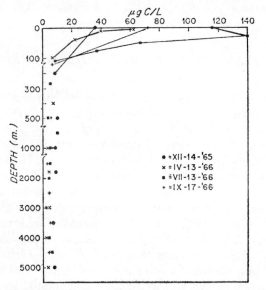

Fig. 8.4. Vertical distribution of particulate organic carbon at 36° 30′ N. 67° 50′ W. in the Sargasso Sea in 1966 (Menzel and Ryther, 1968).

various water masses and can be correlated with variations in temperature and other chemical properties such as salinity, silicate and nitrate.

The ratios of the concentrations of dissolved organic carbon to nitrogen in the near surface waters lie in the range 100 : 15 to 100 : 25, values not very dissimilar to that found for plankton. The values for deep waters have been found to vary considerably from one water mass to another. Thus, Holm-Hansen *et al.* (1966) have observed a ratio of 100 : 8 in the deep water off Southern California, whereas much lower ratios (up to 100 : 50) occur in the North Atlantic (Menzel and Ryther, 1964).

(iii) Geographical variations of dissolved organic carbon

The distribution of dissolved organic carbon in the world oceans has been intensively studied in recent years (Atlantic Ocean, Duursma (1961), Menzel and Ryther (1968), Barber (1968); Pacific Ocean, Menzel (1967), Ogura and Hanya (1967), Ogura (1970); Indian Ocean, Menzel (1964)). It has been found that the distribution patterns for these oceans are quite similar, that found by Duursma (1961) for the North Atlantic probably being fairly typical (Fig. 8.3). If near-shore areas are ignored, the dissolved organic carbon concentration varies only with narrow limits (0·3–1·2 mg C/l). The highest values are found at the surface during spring and summer. In deep waters the concentration is usually less than 0·5 mg C/l. Values of greater than 1 mg C/l are often found in productive shallow water regions.

III. PARTICULATE ORGANIC MATTER IN THE SEA

A. INTRODUCTION

The suspended material which can be removed from sea water by a filter having a pore size of ca 0·5 μm is customarily referred to as particulate matter. The nature and composition of the inorganic fraction of this material is discussed in Chapters 10 and 12. The particulate organic material in the upper layers of the sea consists mainly of detritus (dead organisms) and phytoplankton; even in the euphotic zone the latter rarely constitute more than 1/4 of the total particulate matter. Bacteria may also sometimes comprise a significant fraction of it. In addition, fungi and yeasts may be present as minor components. Zooplankton and fish, which are not considered in the present discussion, represent only a very small fraction of the particulate carbon present in the water column. The relative proportions of the various forms of particulate carbon found by McAllister *et al.* (1961) in the surface waters of the north-east Pacific are probably fairly characteristic of ocean surface waters as a whole. The concentration of detritus, phytoplankton, zooplankton and fish corresponded to about 125, 20, 2 and 0·02 μg C/l. These values should be compared with a dissolved organic carbon content of ca 1,000 μg/l.

The concentration of particulate organic carbon in the euphotic zone is usually considerably higher than in the water lying beneath it.

However, particulate matter, both organic and inorganic is present at all depths in the oceans.

B. ANALYTICAL CHEMISTRY

Particulate organic matter is an extremely important part of the marine food chain as it provides food for organisms at several trophic levels. In order to study the food balance in the sea, ecologists require information about the distribution and composition of this material. The acquisition of this data is a task in which the chemical ocean-ographer plays a key role.

(i) Elementary analysis

In the determination of particulate organic carbon or nitrogen the particulate matter is removed from a known volume of water by filtra-tion through a glass fibre filter. If it is desired to exclude zooplankton, the sample should first be strained through a 300 mesh stainless steel or nylon mesh. Carbon is determined most readily by heating the filter and its associated particulate matter with a solution of potassium dichromate in concentrated sulphuric acid. The amount of dichromate ion reduced by the organic matter is measured photometrically. The relative amount of dichromate reduced depends on the type of organic compound being oxidized. However, since the particulate matter does not vary much in composition, it is possible to obtain a reasonably realistic value for organic particulate carbon by use of an empirical factor (Strickland and Parsons, 1968). A more accurate measure can be obtained by combustion of the filter and sample in a current of oxygen and deter-mination of the carbon dioxide gravimetrically or with a non-dispersive infra red analyser. If inorganic carbon is present (e.g. in the tests of cal-careous organisms) it must first be removed by washing the filter with dilute hydrochloric acid. It should be remembered that these methods determine total particulate organic carbon, not just that which can be assimilated by marine organisms. For this reason there is some danger in using such figures for ecological purposes, since even in productive surface waters there are appreciable proportions of non-assimilable material.

Organic nitrogen is usually determined by the Kjeldahl method. In this, the filter with its associated matter is heated with concentrated sulphuric acid in the presence of a catalyst (e.g. selenium dioxide). Amino and certain other nitrogenous groups are quantitatively con-verted to ammonium-nitrogen which is determined photometrically using ninhydrin (Holm-Hansen, 1968). Alternatively, if milligram amounts of particulate material are present, organic nitrogen can with advantage be determined by the micro-Dumas method which gives a more accurate measure of total nitrogen.

In the determination of phosphorus in particulate matter, organophosphorus compounds are first decomposed by fuming with perchloric acid. The orthophosphate which is produced is then estimated photometrically by the molybdenum blue method (p. 168). It should be borne in mind that not only phosphorus in organic combination, but also particulate inorganic phosphate (e.g. $FePO_4$) will be determined by this procedure.

(ii) Determination of classes of compounds and individual compounds

Elementary analysis will give a rough estimate of the amount of organic matter which is available for assimilation or use as an energy source. However, it is often of greater value to ecologists to know the concentrations of each of the main classes of algal components. Simple spectrophotometric methods are available for the determination of carbohydrate, lipid, protein and plant pigments (Strickland and Parsons, 1968). In addition, for special purposes, information may be required about the concentration in the particulate matter of specific compounds, such as xanthophylls, component fatty acids, deoxyribonucleic acid and adenosine triphosphate. For the determination of these compounds it will probably be necessary to employ much more sophisticated techniques using, for example, thin-layer or gas-liquid chromatography or spectrofluorimetry.

The determination of algal pigments (viz. chlorophylls a, b and c and total carotenoids) is probably the most important one carried out on the particulate matter, since it provides a chemical measure of the living plant material in the water. Unfortunately, the estimate is only approximate since the pigment contents of marine algae vary considerably depend'ng upon species and their nutritional state (the ratio of chlorophyll a to total carbon may range from 1 : 20 to 1 : 100). In the method which is usually used (Richards and Thompson 1952; Strickland and Parsons, 1968), the pigments are extracted from the particulate matter using aqueous 90% acetone and the extract is examined photometrically at the absorption maxima of the various pigments. There is considerable overlap between the absorption bands and reasonably accurate results are obtained only for chlorophyll a. Chlorophyll degradation products, such as phaeophytins and phaeophorbides, may be a significant fraction of the total green pigments in the particulate matter of water below the euphotic zone, or in samples from areas where heavy grazing by zooplankton has taken place. These compounds absorb light strongly at the wavelengths of maximum absorption of the chlorophylls, and thus cause interference in the method.

Lorenzen (1967) has shown that chlorophyll a can be determined by measuring the decrease in absorption at 665 nm which occurs when the pigment extract is treated with hydrochloric acid to convert it to phaeophytin. The photometric method can only give very limited information about the carotenoid pigments present; if more detailed data are required they can be obtained by means of quantitative thin-layer chromatographic techniques (Jeffrey, 1968; Garside and Riley, 1969).

The red fluorescence which chlorophylls exhibit when irradiated with ultra violet radiation can be used at sea for the direct semiquantitative determination of chlorophyll (Lorenzen, 1966). In this method, the sea water is pumped through a flow-cell in a specially sensitive fluorimeter, and the red fluorescence is detected with a photomultiplier, the signal from which is fed to a recorder.

C. THE NATURE AND ORIGIN OF THE PARTICULATE MATTER IN THE SEA

The organic particulate matter in the sea comprises both living organisms (mainly phytoplankton) and detritus; the composition and relative proportions of these two classes varying with depth and location (Parsons, 1963). There is at present no entirely satisfactory method for determining the proportion of living material in the particulate organic matter. Although microscopical examination is of great value in studying the nature of the individual particles present (Krey, 1961), it is only of limited use for quantitative purposes. Since chlorophyll is rapidly degraded after the death of plant cells, it is common practice to determine it as a measure of the plant biomass present in the sea (see above p. 202). This method suffers from three disadvantages: (i) the chlorophyll a : carbon ratio in plant cells can vary by as much as five-fold; (ii) the degradation products of chlorophyll interfere; (iii) organisms which do not contain chlorophyll (e.g. bacteria and yeasts) are not determined. Recently, it has been suggested that determinations of adenosine triphosphate or deoxyribonucleic acid can be used to estimate *total* living matter in marine particulates (Holm-Hansen and Booth, 1966; Holm-Hansen et al., 1968). However, before these substances can be adopted for this purpose, it is necessary to be sure that non-living material does not contain them, and also that they occur in a constant, or predictable, ratio to cell carbon.

(i) *Living organisms*

Live phytoplankton, which comprise the bulk of the biomass in the sea, are confined to the euphotic zone and to the water lying immediately beneath it. The chemical composition of phytoplankton varies from species to species and according to the environmental conditions.

Table 8.2 shows the proportions of the major metabolites (expressed as ratios to carbon) in a variety of phytoplankton species grown under identical conditions (Parsons *et al.*, 1961). The ratios for the metabolites of diatoms from a natural bloom (McAllister *et al.*, 1961) lie in the same ranges. The principal amino acids of the proteins of phytoplankton are glutamic and aspartic acids, alanine and leucine; lesser amounts of about 20 other amino acids are usually present.

TABLE 8.2

Proximate analysis of algal cells (Parsons et al., 1961)

Species	Metabolites (percentage dry weight of cells)					
	Protein*	Carbo-hydrate	Fat	Total pigment	Ash	Total
Chlorophyceae						
Tetraselmis maculata	52	15·0	2·9	2·1	23·8	96
Dunaliella salina	57	31·6	6·4	3·0	7·6	106
Chrysophyceae						
Monochrysis lutheri	49	31·4	11·6	0·8	6·4	99
Syracosphaera carterae	56	17·8	4·6	1·1	36·5	116
Bacillariophyceae						
Chaetoceros sp.	35	6·6	6·9	1·5	28·0	78
Skeletonema costatum	37	20·8	4·7	1·8	39·0	103
Coscinodiscus sp.	17	4·1	1·8	0·5	57·0	81
Phaeodactylum						
tricornutum	33	24·0	6·6	2·9	7·6	73
Dinophyceae						
Amphidinium carteri	28	30·5	18·0	2·4	14·1	93
Exuviella sp.	31	37·0	15·0	1·1	8·3	92
Myxophyceae						
Agmenellum						
quadruplicatum	36	31·5	12·8	1·5	10·7	93

* Nitrogen × 6·25

The component amino acid distributions in the proteins of plankton grown under similar conditions usually differ little from one species to another (Chuecas and Riley, 1967), and little differentiation occurs when the amino acids are incorporated into the tissues of zooplankton and fish (Cowey and Corner, 1963). In contrast, the component fatty acid distributions of phytoplankton lipids are characteristic of particular Phyla or even classes. For example, the Bacillariophyceae (diatoms) are differentiated from other species by the virtual absence of octadeca-dienoic, -trienoic and -tetraenoic acids; the Cryptophyta are distingished by their content of eicosenoic acid (Table 8.3). Species belonging to the same genus frequently contain similar fatty acid arrays (Chuecas and Riley, 1969). The fatty acid composition of phytoplankton lipids is known to be influenced by the temperature at which the organisms live; low temperatures favour the production of the more unsaturated acids. The relative amounts of lipid in the organisms are influenced by the availability of nitrogen in the medium.

The usual plant pigments of phytoplankton are carotene and chlorophylls, together with xanthophylls of many types. The latter pigments can often be used for taxonomic purposes since particular classes of algae are frequently characterized by specific types of xanthophylls (Table 8.4). Thus, most members of the Chlorophyta contain, in addition to carotene and chlorophylls a and b, lutein, neoxanthin and violoxanthin.

The carbohydrates of phytoplankton occur in three principal forms: (i) as polysaccharide components of the cell wall which are often resistant to bacterial decay; (ii) as mono and oligosaccharides in the cell sap; and (iii) as polysaccharides which serve as food reserves. The saccharides which are involved in each of these roles differ considerably from one class of algae to another. Thus, the cell wall of *Phaeodactylum tricornutum* is composed of a sulphated glucuronomannan (Ford and Percival, 1965), whereas that of *Chlorella pyrenoidosa* consists of α-cellulose, glycoprotein and hemi-celluloses (Northcote et al., 1958). The cell saps may contain a wide variety of sugars, including sucrose, fructose, glucose, galactose and di- and tri-saccharides. A number of different polysaccharides are known to be used as energy reserves by phytoplankton. Starch is used for this purpose by many members of the Rhodophyceae and Chlorophyceae (Bursa, 1968), and laminarin is employed by Phaeophyeans. The reader should consult the reviews by Percival (1969) and Percival and McDowell (1967) for further information on the occurrence and chemistry of the algal carbohydrates.

In addition to these major metabolites, phytoplankton contain small amounts of other biologically important compounds, such as vitamins, nucleic acids, etc. These substances may play an important part in the food chain as many of them can be utilized directly by zooplankton to satisfy their essential requirements.

ii) *Detritus*

The non-living particulate organic material in the sea may be of many types. Dead organisms, both plants and zooplankton, constitute one of the most important sources. An appreciable part of their contribution is in the form of material which will decay slowly, e.g. the cell walls of phytoplankton and the chitinous exoskeletons of zooplankton. Handa

TABLE 8.3

Component fatty acids of the lipids of some phytoplankton species (as g of fatty acid/100 g of total fatty acids) after Chuecas and Riley (1969)

Class	Bacillariophyceae		Cryptophyceae	Chlorophyceae	Chrysophyceae	Haptophyceae
Order	Bacillariales	Pennales	Cryptomonadales	Volvocales	Chromulinales	Prymnesiales
Family	Coscinodiscaceae	Cymbellaceae	Cryptomonadaceae	Dunaliellaceae	Chromulinaceae	Coccolithaceae
Genus Species	*Skeletonema costatum*	*Phaeodactylum tricornutum*	*Hemiselmis brunescens*	*Dunaliella primolecta*	*Monochrysis lutheri*	*Coccolithus huxleyi*
*Fatty acid Double-bond position						
12:0	—	—	0·6	0·8	0·6	0·4
13:0	—	—	0·9	—	0·2	1·1
14:0	6·2	8·6	1·2	4·7	9·2	6·4
14:1	?	—	—	0·1	—	—
15:br	0·1	0·6	0·8	—	0·5	0·1
15:0	0·3	0·4	0·6	1·7	0·6	0·4
15:1	?	—	—	—	—	—
16:br	0·2	—	—	—	—	—
16:0	11·1	10·7	12·5	11·3	10·1	16·9
16:1 (9)	21·7	27·3	3·4	9·8	20·2	27·8
16:2 (6,9)	2·6	4·7	0·4	—	2·5	2·4
16:2 (9,12)	3·5	8·7	2·7	8·3	4·6	1·4
16:3 (6,9,12)	11·4	9·9	0·2	7·4	14·8	7·1
16:4 (4,7,10,13)	—	—	—	—	—	0·4
16:4 (6,9,12,15)	1·0	2·8	0·1	5·9	1·5	0·8

Fatty acid*						
17:br	—	—	—	—	—	—
17:0	0·3	0·2	0·2	0·3	—	—
18:0	—	0·1	0·4	0·1	0·3	0·7
18:1 (9)	1·8	4·7	0·1	5·8	0·4	10·0
18:2 (9,12)	2·1	0·5	2·3	5·8	0·4	2·1
18:3 (6,9,12)	—	—	0·1	2·1	5·7	—
18:3 (9,12,15)	—	0·2	0·2	10·4	1·6	1·3
18:4 (6,9,12,15)	0·8	0·1	8·4	6·7	—	1·1
19:0	—	—	30·5	—	0·6	—
19:1	?	0·2	—	0·1	—	—
20:0	—	—	0·1	0·2	0·2	—
20:1 (11)	—	—	0·2	0·8	0·3	0·2
20:2 (8,11)	0·1	—	18·0	1·3	—	0·3
20:3 (8,11,14)	—	0·2	0·1	0·4	1·7	0·4
20:4 (5,8,11,14)	1·9	0·5	0·2	1·7	—	—
20:4 (8,11,14,17)	2·0	18·2	0·3	9·7	0·5	0·5
20:5 (5,8,11,14,17)	30·2	—	13·9	0·3	18·9	17·1
22:0	—	0·1	—	—	—	—
22:1	?	—	—	—	—	0·3
22:2	?	—	—	—	—	—
22:3	?	—	0·3	0·2	0·4	—
22:4 (7,10,13,16)	—	0·4	—	—	—	0·2
22:5 (4,7,10,13,16)	1·7	0·6	—	3·9	0·5	0·1
22:6 (4,7,10,13,16,19)	0·6	0·3	0·3	0·2	3·3	0·1
24:0	—	—	—	—	0·4	0·4

* The following abbreviated nomenclature is employed for the fatty acids. The first figure is the number of carbon atoms in the molecule, the number following the colon is the number of double bonds and the numbers in parentheses denote the positions of the double bonds counting from the carboxyl group which is numbered 1. Thus, 18:2 (9,12) designates linoleic acid.

and Tominaga (1969) have provided evidence that the bulk of the particulate carbohydrate in subsurface water does consist of structural carbohydrates, and starch grains have been shown to be present, frequently in considerable amounts (Bursa, 1968). In some waters at certain times in the year zooplankton faecal pellets may be a major

TABLE 8.4

The principal plant pigments of the main phytoplankton classes

	Chloro-phyceae	Crypto-phyceae	Dino-phyceae	Chryso-phyceae	Bacillario-phyceae
Chlorophyll *a*	++	++	++	++	++
Chlorophyll *b*	++	—	—	—	—
Chlorophyll *c*	—	++	++	++	++
α-Carotene	—	++	—	—	—
β-Carotene	++	—	++	++	++
Xanthophylls					
Fucoxanthin	—	—	—	++	++
Lutein	++	—	—	—	—
Neoxanthin	+	—	—	—	—
Zeaxanthin	—	—	—	—	—
Violoxanthin	+	—	—	—	—
Dinoxanthin	—	—	+	—	—
Peridinin	—	—	++	—	—
Diadinoxanthin	—	—	+	+	+
Diatoxanthin	—	—	—	+	+
Alloxanthin*	—	++	—	—	—
Monodoxanthin*	—	+	—	—	—
Crocoxanthin*	—	+	—	—	—
Biliproteins†					
Phycoerythrin	—	++	—	—	—
Phycocyanin	—	++	—	—	—

++ Major pigment.
+ Secondary pigment in most species.
* Acetylenic pigment.
† Water soluble pigments.

component of the particulate matter. In coastal areas, the water may contain fragments of seaweed and other littoral and shallow water organisms. In addition, such waters will also contain appreciable amounts of wind-, or river-transported terrestial or industrial debris, such as wood, spores and carbonaceous combustion products. Although the influence of these materials is usually restricted to near-shore areas, currents may occasionally carry them into deeper waters. Thus,

McAllister *et al.* (1961) have observed waterlogged wood fibres in the deep water in the Pacific as far as 900 km from land.

Organic aggregates have been observed in the sea both by direct observation (Suzki and Kato, 1951) and by underwater photography (Inoue *et al.*, 1955; Nishizawa *et al.*, 1954). This "marine snow" as it has been called is not confined only to the surface layers, since Dietz (1959) has observed particulate matter of relatively large size at 1,100 m in the Mediterranean during a descent in the bathyscaphe Trieste. There is some doubt about the size range of this particulate material. Inoue *et al.* (1955) state that it may be as large as 1–3 mm. However, other workers have found it to be much smaller, the average size being ca 15–20 μm (Gillbricht, 1951; Jerlov, 1961). It seems likely that the size range of the particles may vary considerably with both locality and depth. Under the microscope the particles appear to be formed of yellowish amorphous material. Bacteria and diatom and other phytoplankton cells may be found embedded in the larger particles. The origin of the amorphous particulate material is at present uncertain. Riley *et al.* (1964, 1965) have suggested that dissolved proteinaceous organic material from the water becomes concentrated as a film on the surfaces of bubbles in the upper layer of the sea. When these bubbles burst at the surface this organic skin remains behind and forms a seed for the aggregation of further dissolved matter. In support of this hypothesis Riley (1963) has cited the fact that the concentration of particulate organic carbon in Long Island Sound could be correlated with turbulence, and hence with bubble formation. Additional evidence was supplied by Sutcliffe *et al.* (1963) who observed that the occurrence of particulate organic carbon was associated with windrow formation. Experimental evidence for the formation of such particles is contradictory. Riley *et al.* (*loc. cit.*) and Sutcliffe *et al.* (1963) have found that aeration of filtered sea water produced organic particles. More recently Batoosingh *et al.* (1969) have observed that particles in the size range 0·22–1·2 μm serve as nuclei in the formation of particles during bubbling. In contrast, Barber (1966) was unable to detect any formation of particulate matter when aeration was carried out using *purified* air. It is perhaps significant that the only occasion on which Barber found aggregates to be formed in a sea water medium, enriched with the cell material of a diatom (mol. wt $< 10^5$), was when bacteria were present. However, Batoosingh *et al.* (*loc. cit.*) consider that bacteria are not more important than non-living detritus in the formations of particulate matter. Obviously much more work is needed before it is possible to decide whether aggregation by bubble bursting is significant in the sea.

From a study of the distribution of particulate matter with depth in

9

the Strait of Georgia, British Columbia, Sheldon *et al.* (1967) were led to conclude that little of that present below 50 m had originated in the surface waters, and that it must have been formed *in situ.* They were able to demonstrate that particle formation occurred spontaneously in water samples which had been filtered through a $0 \cdot 22$ μm membrane filter. This only occurred under non-sterile conditions and was thought to be associated with the rapid multiplication of bacteria which gave rise to sizeable particles by growing in clumps, or under natural conditions, by stimulating other species to grow around them. Such clumps may then, perhaps, become aggregated into much larger particles (McAllister *et al.*, 1961).

It may be significant that marine organic particulate matter usually has a large inorganic content and may contain complex mineral assemblages (Chave, 1965). In part, this may arise from the growth of bacteria, or the adsorption of organic matter, on preexisting suspended minerals, e.g. clays and feldspars. However, Sieburth (1965) has suggested a novel way in which bacteria may participate in the formation of organic-rich particulate matter in the sea. When bacteria degrade proteins, ammonia is formed; in the micro-environment this will lead to an increase in pH and to the consequent precipitation of calcium carbonate and magnesium hydroxides. These highly adsorptive precipitates will scavenge organic matter from solution.

To summarize, several processes are probably involved in the production of organic particulate matter in the sea. However, we are still a long way from a complete understanding of their nature. Much further work is required in order to resolve the question of whether bubble formation plays any part in the production of organic aggregates in the sea, and to define the role of bacteria.

Organic particulate matter is not only distributed throughout the water column, but also occurs as surface-active films over extensive areas of the sea. These films often considerably modify the physical properties of the sea surface, including the surface tension and viscosity, and lead to the production of slicks, which are areas of wave-damping. In addition, they may inhibit evaporation and thus reduce the amount of heat lost from the sea surface (Jarvis *et al.*, 1962). They have been shown to be monolayers composed of fatty acids (C_8—C_{18}), methyl esters of fatty acids (C_{11}—C_{22}), higher aliphatic alcohols (C_{12}, C_{16} and C_{18}) and hydrocarbons (Garrett, 1964; Jarvis *et al.*, 1962). These film forming materials probably originate mainly from metabolites of phytoplankton and zooplankton which are gathered onto bubbles, which carry them to the surface and disperse them there when they burst. Other sources include discharges of oil from ships and pollution from land

D. THE CHEMICAL COMPOSITION OF MARINE DETRITUS

The composition of marine detritus varies considerably with depth. n the euphotic zone it consists to a large extent of the major biochemi- al metabolites (such as carbohydrates, proteins, plant pigments) and heir degradation products. The majority of these materials are readily xidized bacterially and are subject to predation by filter feeders. 'robably only the resistant parts of the organisms, such as the cell valls of phytoplankton and the exoskeletons of zooplankton, and recipitated humic substances will fall to depths greater than 100– 00 m. Thus, Parsons and Strickland (1962) have found that 70% of he carbohydrate existing in detritus at a depth of 400 m in the north ast Pacific was of a resistant nature. Furthermore, protein accounted or only about half of the organic nitrogen present. The majority of the rganic particulate material present in deep water appears to be ex- remely refractory to bacterial attack (Menzel and Goering, 1966). Its ow C : N ratio and the blue fluorescence which it emits suggests that t may be related to the Gelbstoff fraction of the dissolved organic arbon (see p. 190). Recent work by Gordon (1970a) has shown that a roportion of the organic particulate matter can be readily hydrolysed y digestive enzymes such as trypsin and chymotrypsin.

. ECOLOGICAL SIGNIFICANCE AND FATE OF PARTICULATE ORGANIC MATTER
) *Euphotic zone*

We have seen above that the organic particulate matter in the eu- hotic zone consists of phytoplankton and bacteria, together with etritus, a considerable proportion of which is refractory in nature. 'he key role of phytoplankton in the marine food chain has been known or a long time, but it is only in the last 20 years that their nutritional value and those of bacteria and detritus have been studied.

Many filter feeders are quite selective with regard to the species of hytoplankton which they are able to utilize. Not all species of zoo- lankton prefer the same species of phytoplankton, and often better rowth is obtained with a mixture of two or more species than with a ingle one. Under natural conditions, diatoms are probably the staple iet of copepods (Beklemishev, 1954), but coccolithophores and dino- agellates can also be used (Marshall and Orr, 1952). The phytoplankton equirements of oysters have been investigated by several workers and ound to be very exacting (Loosanoff, 1949). Thus, Davis and Guillard 1958) have found that of ten genera which they examined, only two pecies of Chrysophyceae provided satisfactory food (see also p. 272).

Under conditions of phytoplankton bloom, zooplankton may onsume considerably more phytoplankton than they are able to assimilate (Beklemishev, 1954, 1962). During this so-called *luxury* or *superfluous* feeding the animals excrete the phytoplankton in a semi-digested state. This contributes considerable quantities of nourishing and readily assimilable material to the detritus. Beklemishev has suggested that in some areas this may be an important source of food for filter feeders living below the euphotic zone. For example, he has estimated that ca 20 g of organic carbon/m^2 is produced annually as faecal pellets in the water over Georges Bank, Gulf of Maine. Experiments in which *Calanus hyperboreus* was fed with diatoms have shown that this species does not indulge in luxury feeding, which has led Conover (1966) to doubt whether this phenomenon does occur in the sea. Further work is obviously required to elucidate this problem.

While phytoplankton are abundant, detritus probably acts only as a supplement to the diet of zooplankton (Jørgensen, 1962). However, when the crop is sparse, it may form a major part of the food of these animals (Harvey, 1950; Fox, 1950; Riley, 1959). In shallow coastal waters, much detritus reaches the sea bed where it is digested by the organisms of the benthic community. MacGinitie (1932) has pointed out that detritus is much more important to bottom living animals than it is to the free swimming organisms. In confirmation of this, Mare (1942) has estimated that 75% of the benthic fauna in the English Channel off Plymouth depends on detritus and the organisms associated with it for their main source of food.

Bacterial decomposition is thought to be the principal mechanism by which detritus is removed from the water column and degraded. Bacteria use the particulate matter to supply both their energy and the materials for their protoplasm. During their respiration and metabolic processes carbon dioxide, ammonia and phosphate ions are regenerated into the water. Work by Johannes (1965) has suggested that much of the regeneration may actually be brought about by protozoa, such as ciliates, which feed on the bacteria; these organisms are known to take a considerable part in the breakdown of sewage. If this is the case, the bacteria would merely serve to assimilate the detritus. Whichever mechanism of regeneration is operative, its efficiency is high. This is shown by the low standing concentration of organic particulate matter in the sea, and by the fact that very little of it penetrates to depths below a few hundred metres. Species of bacteria are known which will break down materials which marine animals cannot digest or can digest only with difficulty, e.g. chitin (Hock, 1941) and cellulose (Waksman *et al.*, 1933).

It has been established that some species of benthic animals, e.g. *Mytilus californianus* can utilize bacteria as a food source (MacGinitie

1932; Zobell and Feltham, 1938), and that this may in fact be the principal item of diet of a few species of molluscs (Newell, 1965). However, in the upper parts of the water column it seems likely that their use as a food for zooplankton is of secondary importance in the marine organic cycle compared with their regenerative capacity (Parsons, 1963).

(ii) *Deeper waters*

Although much of the detritus found in the euphotic zone is relatively large in size (Fox, 1950), the efficiency with which it is decomposed by bacteria is so high that very little of it reaches a depth of 200–300 m. Most of that which does remain is unlikely to be consumed by deep water filter feeders, as it has little nutritive value (Harvey, 1955), and sinks to the ocean floor to be incorporated into the sediments, which on average contain ca 0·3% of organic carbon. It is therefore necessary to consider an alternative source of food for these animals. This role is probably fulfilled by the marine aggregates (see p. 209 and Sheldon *et al.*, 1967) which are rich in protein and likely to be nutritious. Recent work by Gordon (1970a) has shown that a significant fraction up to 25%) of the particulate organic carbon present in the deep water of the Atlantic can be readily hydrolysed by enzymes, such as trypsin and α-amylase, which are known to occur in zooplankton. He has postulated that this fraction may be an important food source for bathypelagic filter feeders (see also Gordon, 1970b). In this context it may be noted that Baylor and Sutcliffe (1963) have demonstrated that aggregates produced by bubbling can in fact be utilized by marine animals.

F. DISTRIBUTION OF PARTICULATE ORGANIC CARBON IN THE SEA

The distribution of organic particulate matter in the sea has been investigated by a number of workers (e.g. Parsons and Strickland, 1962; Menzel and Goering, 1966; Holm-Hansen *et al.*, 1966; Sheldon *et al.*, 1967), whose results show a similar pattern. The total organic particulate carbon in oceanic water columns is always at least an order of magnitude greater than the yearly primary productivity. The concentration occurring in the euphotic layer is relatively high and variable and can be roughly correlated with phytoplankton activity. For this reason it shows geographical and seasonal variations similar to that of primary productivity. Much of the organic particulate material in this layer is bacterially labile and consists of phytoplankton, both living and dead, and the excreta of zooplankton and other marine animals. However, according to Menzel and Goering (1966) a significant proportion of it is detritus, which is strongly resistant to bacterial attack.

Beneath the euphotic zone the concentration of particulate organic carbon decreases rapidly and below a depth of ca 200 m remains more or less constant to the bottom. According to Menzel and Goering (1966 and Menzel (1967) this material is remarkably constant in concentration (Fig. 8.4, p. 199), with depth and space in the water below 200 m. However, Holm-Hansen et al. (1966), Dal Pont and Newell (1963), Nakajima (1969) and others have found evidence that its concentration in deep water varies with depth, perhaps, according to the stratification in the water column (Fig. 8.2, p. 197). Geographical variations in the Pacific Ocean may be inferred from the range of concentrations which have been reported for different areas (Parsons and Strickland, 1962; Dal Pont and Newell, 1963; Hobson, 1967). In addition, temporal variations have been observed for a number of regions, such as the Indian Ocean (Newell, 1969), and the north eastern Pacific (Hobson, 1967). Further work is obviously necessary to determine the extent to which the water circulation pattern determines the distribution of organic particulate matter in the sea. The particulate matter from this sub surface water is very refractory towards bacterial attack (Menzel and Goering, 1966). Thus, Handa and Tominaga (1969) have found that the decrease of particulate carbohydrate with depth is associated with the disappearance of storage sugars (p. 205); that which is present in the deeper layers is almost entirely structural polysaccharide (p. 205). The C : N ratio in the particulate matter from deep water has usually been found to be ca 100 : 50 in contrast to values of ca 100 : 20 reported for surface waters (Dal Pont and Newell, 1963; Menzel and Ryther 1964). However, there is some indication that the ratio may depend on the history of the water mass, since Holm-Hansen et al. (1966) have found much lower ratios (ca 100 : 8) in particulate matter in the water column off southern California.

REFERENCES

Anderson, G. C. and Zeutschel, R. P. (1970). *Limnol. Oceanogr.* **15**, 402.

Antia, N. J., McAllister, C. D., Parsons, T. R., Stephens, K. and Strickland J. D. H. (1963). *Limnol. Oceanog.* **8**, 166.

Armstrong, F. A. J., Williams, P. W. and Strickland, J. D. H. (1966). *Nature Lond.* **211**, 479.

Bannister, F. A. and Hey, M. H. (1936). *Discovery Rep.* **13**, 60.

Barber, R. T. (1966). *Nature, Lond.* **211**, 257.

Barber, R. T. (1968). *Nature, Lond.* **220**, 274.

Batoosingh E., Riley, G. A. and Keshwar, B. (1969). *Deep Sea Res.* **16**, 213.

Baylor, E. R. and Sutcliffe, W. H. (1963). *Limnol. Oceanog.* **8**, 369.

Beklemishev, C. W. (1954). *Zool. Zhurn. Inst. Oceanol., Akad Nauk. SSSR.* **33** 1210.

Beklemishev, C. W. (1962) *Rappt. Proc. Verb. Cons. Perm. int. Expl. Mer.* **153**, 108

Belser, W. L. (1959). *Proc. nat. Acad. Sci.*, *Wash.* **45**, 1533.

Belser, W. L. (1963). *In* "The Sea" (M. N. Hill, ed.), Vol. 2. Interscience, New York.

Blanchard, D. C. (1964). *In* "Progress in Oceanography" (M. Sears, ed.), Vol. 1. Pergamon Press, New York.

Blumer, M., Mullin, M. M. and Thomas, D. W. (1963). *Science, N.Y.* **140**, 974.

Brongersma-Sanders, M. (1957). *In* "Treatise on Marine Ecology and Palaeoecology" (I. W. Hedgpeth, ed.), Vol. 1. *Geol. Soc. Am. Mem.* **67**, 941.

Burke, J. M., Marchisotto, J. McLaughlin, J. J. A. and Provasoli, L. (1960). *Ann. N.Y. Acad Sci.* **90**, 837.

Burkholder, P. R. (1970). *In* "Encyclopedia of Marine Resources" (F. E. Firth, ed.). Van Nostrand, New York.

Bursa, A. S. (1968). *J. Fish. Res. Bd Can.* **25**, 1269.

Chave, K. E. (1965). *Science, N.Y.* **148**, 1723.

Chuecas, L. and Riley, J. P. (1967). *J. mar. biol. Ass. U.K.* **47**, 543.

Chuecas, L. and Riley, J. P. (1969). *J. mar. biol. Ass. U.K.* **49**, 117.

Conover, R. J. (1966). *Limnol. Oceanog.* **11**, 346.

Corner, E. D. S. and Newell, B. S. (1967). *J. mar. biol. Ass. U.K.* **47**, 113.

Cowey, C. B. and Corner, E. D. S. (1963). *J. mar. biol. Ass. U.K.* **43**, 495.

Craigie, J. S. and MacLachlan, J. (1964). *Can. J. Bot.* **42**, 23.

Creac'h, P. V. (1955). *C. r. Acad. Sci., Paris*, **240**, 2551; **241**, 437.

Dal Pont G. and Newell, B. S. (1963). *Australian J. Mar. Freshw. Res.* **14**, 155.

Davis, H. G. and Guillard, R. R. L. (1958). *Fish. Bull., Fish and Wildlife Service*, **58**, 293.

Degens, E. T., Reuter, J. H. and Shaw, K. N. F. (1964). *Geochim. cosmochim. Acta.* **28**, 45.

Dietz, R. S. (1959). *Limnol. Oceanog.* **4**, 94.

Droop, M. R. (1968). *J. mar. biol. Ass. U.K.* **48**, 689.

Duursma, E. K. (1961). *Netherl. J. Sea Res.* **1**, 1.

Duursma, E. K. (1965). *In* "Chemical Oceanography" (J. P. Riley and G. Skirrow, eds), Vol. 1, p. 433. Academic Press, London.

Fogg, G. E. (1962). *In* "Physiology and Biochemistry of Algae" (R. A. Lewin, ed.), p. 475. Academic Press, New York.

Fogg, G. E. (1963). *Br. phycol. Bull.* **2**, 195.

Fogg, G. E. (1966). *Oceanog. mar. Biol. Ann. Rev.* **4**, 195.

Fogg, G. E., Nalewajko, C. and Watt, W. D. (1965). *Proc. R. Soc.* B **162**, 517.

Ford, C. W. and Percival, E. (1965). *J. Chem. Soc.* 7042.

Fox, D. L. (1950). *Ecology*, **31**, 100.

Garrett, W. D. (1964). U.S. Naval Research Laboratory Report 6201.

Garside, C. and Riley, J. P. (1969) *Anal. Chim. Acta.* **46**, 179.

Gillbricht, M. (1951). *Kieler. Meeresforsch.* **8**, 173.

Gordon, D. C. (1970a). *Deep Sea Res.* **17**, 233.

Gordon, D. C. (1970b). *Deep Sea Res.* **17**, 175.

Grein, A. and Meyers, S. P. (1958). *J. Bact.* **76**, 457.

Hamilton, R. D. and Presland, J. E. (1970). *Limnol. Oceanogr.* **15**, 395.

Handa and Tominaga (1969), *Mar. Biol.* **2**, 228.

Harvey, H. W. (1950). *J. mar. biol. Ass. U.K.* **29**, 97.

Harvey, H. W. (1955). "The Chemistry and Fertility of Sea Waters." Cambridge University Press, Cambridge.

Hellebust, J. A. (1965). *Limnol. Oceanog.* **10**, 192.

Hobson, L. A. (1967). *Limnol. Oceanog.* **12**, 642.

Hock, C. W. (1941). *J. mar. Res.* **4**, 99.

Holm-Hansen, O. (1968). *Limnol. Oceanog.* **13**, 175.

Holm-Hansen, O. and Booth, C. R. (1966). *Limnol. Oceanog.* **11**, 510.

Holm-Hansen, O. Strickland, J. D. H. and Williams P. M. (1966). *Limnol. Oceanog.* **11**, 549.

Holm-Hansen, O., Sutcliffe, W. H. and Sharp, J. (1968). *Limnol. Oceanog.* **13**, 507.

Hutchinson, G. E. (1954). *In* "The Earth as a Planet", (G. P. Kuiper, ed.), Vol. II. University of Chicago Press, Chicago, Illinois.

Inoue, N., Nishizawa, S. and Fukuda, N. (1955). *Proc. U.N.E.S.C.O. Symp. Phys. Oceanog. Tokyo*, p. 53.

Jarvis, N. L., Timmons, C. O. and Zisman, W. A. (1962). *In* "Retardation of Evaporation by Monolayers." Academic Press, New York.

Jarvis, N. L., Garrett, W. D., Scheiman, M. A. and Timmons, C. O. (1967). *Limnol. Oceanog.* **12**, 88.

Jeffrey, L. M. (1966). *J. Am. Oil Chem. Soc.* **43**, 211.

Jeffrey, L. M. and Hood, D. W. (1958). *J. mar. Res.* **17**, 247.

Jeffrey, S. W. (1968). *Biochim. Biophys. Acta.* **162**, 271.

Jerlov, N. G. (1961). *Goteborg Kungl. Vetensk. Samh., Handl.*, Ser. B, **8**, 1.

Johannes, R. E. (1965). *Limnol. Oceanog.* **10**, 434.

Johannes, R. E. (1968). *Adv. Microbiol. Sea.* **1**, 203–210.

Johnston, R. (1963). *J. mar. biol. Ass. U.K.* **43**, 427.

Jørgensen, C. B. (1962). *Rapp. Cons. Expl. Mer.* **153**, 99.

Kalle, K. (1937). *Annln. Hydrog. Berl.* **65**, 276.

Kalle, K. (1949). *Dt. hydrogr. Z.* **2**, 117.

Kalle, K. (1966). *Oceanog. mar. Biol. Ann. Rev.* **4**, 91.

Khailov, K. M. and Burlakova, Z. P. (1969). *Limnol. Oceanog.* **14**, 521.

Krey, J. (1961). *J. Cons. int. Expl. Mer.* **26**, 263.

Kriss, A. Y., Mishustina, I. Y., Mitskevich, I. N. and Zemtsovo, E. V. (1964). "Microbial Population of the Ocean and Seas (Species Composition, Geographical Range)." Nauka Press, Moscow, 296 pp.

Lewin, J. C. (1963). *In* "Marine Microbiology" (C. H. Oppenheimer, ed.). C. H. Thomas, Springfield, Illinois, 229 pp.

Litchfield, C. D. and Hood, D. W. (1965). *Appl. Microbiol.* **13**, 886.

Litchfield, C. D. and Hood, D. W. (1966). *Appl. Microbiol.* **14**, 145.

Loosanoff, V. L. (1949). *Science, N.Y.* **110**, 122.

Lorenzen, C. J. (1966). *Deep Sea Res.* **13**, 223.

Lorenzen, C. J. (1967). *Limnol. Oceanog.* **12**, 343.

Lucas, C. E. (1947). *Biol. Rev.* **22**, 270.

Lucas, C. E. (1955). *Deep Sea Res.* **3** (Suppl.), 139.

Lucas, C. E. (1961). *Symp. Soc. exp. Biol.* **15**, 190.

McAllister, C. D., Parsons, T. R. and Strickland, J. D. H. (1961), *Limnol. Oceanog.* **6**, 237.

MacGinitie, G. E. (1932). *Science, N.Y.* **76**, 490.

MacLeod, R. A. Onofrey, E. and Norris, M. E. (1954). *J. Bacteriol.* **68**, 680.

MacLachlan, J. and Craigie, J. S. (1964). *Can. J. Bot.* **42**, 287.

Mare, M. F. (1942). *J. mar. biol. Ass. U.K.* **25**, 517.

Marshall, S. M. and Orr, A. P. (1952). *J. mar. biol. Ass. U.K.* **30**, 527.

Menzel, D. W. (1964). *Deep Sea Res.* **11**, 757.

Menzel, D. W. (1966). *Deep Sea Res.* **13**, 963.

Menzel, D. W. (1967). *Deep Sea Res.* **14**, 229.

Menzel, D. W. and Goering, J. J. (1966). *Limnol. Oceanog.* **11**, 333.

Menzel, D. W. and Ryther, J. H. (1964). *Limnol. Oceanog.* **9**, 179.

Menzel, D. W. and Ryther, J. H. (1968). *Deep Sea Res.* **15**, 327.

Menzel, D. W. and Vaccaro, R. F. (1964). *Limnol. Oceanog.* **9**, 138.

Nakajima, K. (1969). *J. oceanog. Soc. Japan*, **25**, 239.

Nalewajko, C., Chowdhuri, N. and Fogg, G. E. (1963). *In* "Studies on Microalgae and Photosynthetic Bacteria." Japanese Society of Plant Physiologists, Tokyo, 171.

Newell, R. (1969). *Proc. zool. Soc. Lond.* **144**, 25.

Newell, B. S., Morgan, D. and Cundy, J. (1967). *J. mar. Res.* **25**, 201.

Nishizawa, S., Fukuda, M. and Inoue, N. (1954). *Bull. Fac. Fisheries Hokkaido Univ.* **5**, 36.

Northcote, D. H., Goulding, K. J. and Horne, R. L. (1958). *Biochem. J.* **70**, 391.

Ogura, N. and Hanya, T. (1967). *Int. J. Oceanol. Limnol.* **1**, 91.

Ogura, N. (1970). *Deep Sea Res.* **17**, 221.

Otsuki, A. and Hanya, T. (1968). *Limnol. Oceanog.* **13**, 183.

Parsons, T. R. (1963). *In* "Progress in Oceanography", Vol. 1, p. 203 (M. Sears, ed.).

Parsons, T. R., Stephens, K. and Strickland, J. D. H. (1961). *J. Fish Res. Bd Can.* **18**, 1011.

Parsons, T. R. and Strickland, J. D. H. (1961). *Deep Sea Res.* **8**, 211.

Parsons, T. R. and Strickland, J. D. H. (1962). *Science, N.Y.* **136**, 313.

Percival, E. (1969). *Oceanog. mar. biol. Ann. Rev.* **6**, 137.

Percival, E. and McDowell, R. H. (1967). "Chemistry and Enzymology of Marine Algal Polysaccharides." Academic Press, New York, 219 pp.

Prakash, A. and Rashid, M. A. (1968). *Limnol. Oceanog.* **13**, 598.

Pratt, D. W. (1966). *Limnol. Oceanog.* **11**, 447.

Proctor, V. W. (1957). *Limnol. Oceanog.* **2**, 125.

Provasoli, L. (1963). *In* "The Sea" (M. N. Hill, ed.), Vol. 3, p. 165. Interscience, New York.

Provasoli, L., MacLaughlin, J. J. A. and Droop, M. R. (1957). *Arch. Mikrobiol.* **25**, 392.

Richards, F. A. and Thompson, T. G. (1952). *J. mar. Res.* **11**, 156.

Riley, G. A. (1959). *Bull. Bingham oceanog. Coll.* **17**, 83.

Riley, G. A. (1963). *Limnol. Oceanog.* **8**, 372.

Riley, G. A., Wangersky, P. J. and Van Hemert, D. (1964). *Limnol. Oceanog.* **9**, 546.

Riley, G. A., Van Hemert, D. and Wangersky, P. J. (1965). *Limnol. Oceanog.* **10**, 354.

Riley, J. P. and Segar, D. (1970), *J. mar. biol. Ass. U.K.* **50**, 713.

Riley, J. P. and Taylor, D. (1969). *Anal. Chim. Acta.* **46**, 307.

Rosenfeld, W. D. and Zobell, C. E. (1947). *J. Bact.* **54**, 393.

Samejima, H. and Myers, J. (1958). *J. gen. Microbiol.* **18**, 107.

Schantz, E. J. (1960). *Ann. N.Y. Acad. Sci* **90**, 843.

Sheldon, R. W. Evelyn, T. P. T. and Parsons, T. R. (1967). *Limnol. Oceanog.* **12**, 367.

Shilo, M. (1964). *Verk int. Verein. theor. angew. Limnol.* **15**, 782.

Sieburth, J. McN. (1959). *J. Bact.* **77**, 521.

Sieburth, J. McN. (1960). *Science, N.Y.* **132**, 676.

Sieburth, J. McN. (1965). *J. gen. Microbiol*, 41, XX.

Sieburth, J. McN. (1968). *Adv. mar. Microbiol.* 1, 63.

Sieburth, J. McN. and Jensen A. (1968). *J. exp. mar. Biol. Ecol.* 2, 174.

Sieburth, J. McN. and Jensen A. (1969) *J. exp. mar Biol. Ecol.* 3, 275.

Skopintsev, B. A. (1949), *Tr. Vses. Gidrobiol. Obshchestva Akad. Nauk. SSSR.* 1, 34.

Sorokin, Y. (1961). *Mikrobiologiya*, 30, 289.

Strickland, J. D. H. and Parsons, T. R. S. (1968). "A Practical Handbook of Sea Water Analysis." *Fish. Res. Bd Can. Bull.* 167, 311 pp.

Sutcliffe, W. H., Baylor, E. R. and Menzel, D. W. (1963). *Deep Sea Res.* 10, 233.

Suzki, N. and Kato, K. (1951). *Bull. Fac. Fish. Hokkaido Univ.* 4, 132.

Swinnerton, J. W., Linnenbom, V. J. and Cheek, C. H. (1969). *Environ. Sci. Tech.* 3, 836.

Tsujita, T. (1955). *J. oceanog. Soc. Japan*, 11, 199.

Vaccaro, R. F. and Jannasch, H. W. (1966). *Limnol. Oceanog.* 11, 596.

Vaccaro, R. F., Hicks, S. E., Jannasch, H. W. and Carey, F. G. (1968). *Limnol. Oceanog.* 13, 356.

Vishniac, H. S. and Riley, G. A. (1959). *Proc. Intern. Oceanog. Congr. N.Y.* 1959, 942.

Waksman, S. A. Carey, C. L. and Reuszer, H. W. (1933). *Biol. Bull. Woods Hole*, 65, 57.

Waksman, S. A., Hotchkis, M., Carey, C. L. and Hardman, Y. (1938), *J. Bact.*, 35, 477.

Watt, W. D. and Fogg, G. E. (1966). *J. exp. Bot.* 17, 117.

Webb, K. L. and Johannes, R. E. (1967). *Limnol. Oceanog.* 12, 376.

Williams, P. J. Le B. and Askew, C. (1968). *Deep Sea Res.* 15, 365.

Williams, P. M. and Gordon, L. I. (1970). *Deep Sea Res.* 17, 19.

Williams, P. M., Oeschager, H. and Kinney, P. (1969). *Nature, Lond.* 224, 256.

Wright, R. T. and Hobbie, J. E. (1965). *Ocean Sci. Ocean. Eng.* 1, 116.

Wright, R. T. and Hobbie, J. E. (1966). *Ecology*, 47, 447.

Zobell, C. E. and Feltham, C. B. (1938). *J. mar. Res.* 1, 312.

Chapter 9

Primary and Secondary Production in the Marine Environment

I. Introduction

The growth of marine plants is of paramount importance since it provides the basis of the marine food chain which culminates in fish and marine mammals. During this photosynthetic process, which is known loosely as *primary production,** the plants remove dissolved carbon dioxide and micronutrients from the water and, using solar energy, convert them to complex organic compounds of high potential energy. Quantitatively, the group of microscopic organisms known as *phytoplankton*, accounts for by far the greatest production in the sea. Although the large attached algae may be responsible for a considerable proportion of the production in some shallow water areas, their contribution to the total productivity of the sea as a whole is probably insignificant since they are restricted to the shallow water (< 50 m deep) of the Continental Shelf. For this reason, discussion of primary production will be confined practically entirely to that brought about by phytoplankton. The growth and distribution of marine plants is controlled in a complex fashion by a combination of numerous factors, which may be physical (e.g. light, temperature, current velocity), biological (e.g. the intrinsic growth rate or interaction between organisms) and chemical (availability of nutrients or specific types of growth-promoting substances). The uptake of the latter substances during a plankton bloom may modify the composition of the sea water to such an extent

* For a more rigid definition of primary production see p. 231. In the literature, the word *productivity* is frequently considered to be synonymous with *production*. However, these terms are used more specifically by some ecologists (e.g. Davis, 1963); production being considered to be the actual rate in nature, whereas productivity is taken to imply maximum possible rate in a particular water mass under optimum conditions of illumination and hydrography. The terms production and productivity will be used synonymously throughout the present chapter with the wider meaning.

as to limit the further growth of the organism. Furthermore, release by the plankton of growth-inhibiting substances may prevent the growth of other species of plankton and thus influence the species succession.

Under the conditions normally prevailing in the sea a large proportion of the phytoplankton is consumed by herbivorous zooplankton which in their turn may be eaten by carnivorous zooplankton.* Although the main purpose of this chapter is the consideration of primary productivity some attention will be given to herbivorous zooplankton since many economically important fish rely on them more or less directly for food. For a detailed account of primary productivity the reader should consult the reviews by Strickland (1965), Raymont (1963, 1966), Goldman (1965) and Conover (1968). Production by zooplankton has been reviewed by Raymont (1966) and Mullin (1969), and that in the food chain as a whole has been summarized in a short but informative article by Ryther (1969).

II. Phytoplankton

Phytoplankton are free-floating microscopic plants, which as they have only limited mobility, are distributed by the ocean currents. They may be divided on a basis of their size into *microplankton*, *nanoplankton* and *ultraplankton*, which have lengths in the ranges $50–500 \mu m$, $10–50 \ \mu m$ and $0 \cdot 5–10 \ \mu m$ respectively. Those species which live primarily by photosynthesis are described as *phototrophic*. Some species of phytoplankton (*autotrophs*) are able to satisfy their dietary requirements from purely inorganic sources; some others (*auxotrophs*) are unable to grow without a supply of specific organic compounds, e.g. vitamin B_{12}, which they are unable to synthesize for themselves. In addition, others (*heterotrophs*) obtain the carbon, and perhaps nitrogen, phosphorus, and sulphur which they require from dissolved or particulate organic†

* Wood (1967) has pointed out that non-photosynthetic micro-organisms (such as bacteria, colourless flagellates and ciliates) are in fact the second trophic level. These organisms, which often greatly outnumber the photosynthetic plankton, and sometimes surpass them in biomass, feed heterotrophically on the living or dead phytoplankton cells, or on the dissolved organic matter which the latter produce. They are in all probability ingested along with the phytoplankton by the herbivores which should therefore be considered as the third trophic level. These organisms are of particular importance in the deep sea food chain as they are able to breakdown particulate matter, such as zooplankton faeces, falling from the euphotic zone, and convert it into a form in which it can be digested by deep sea fauna.

† Heterotrophism is obviously an extremely important factor in the survival of phytoplankton which have fallen beneath the euphotic zone. In this connection it may be noted that the presence of viable members of the Dinophyceae and Chrysophyceae has been observed by Bernard (1958a, 1961) in deep waters. The cell densities of these organisms, which appear to be associated with sub-surface currents, may sometimes be only slightly less than those in the surface layer at depths of several thousands of metres (see also p. 249).

sources; those heterotrophs which feed on phytoplankton or detritus are frequently referred to as *phagotrophs*. Some of these algae (*facultative chemotrophs*) possess photosynthetic pigments and are able to grow either photosynthetically, or without light; others (*obligate chemotrophs*) have no photosynthetic pigments and cannot photosynthesize.

These divisions should not be considered hard and fast since many algae are able to change from one mode of living to another according to the conditions in which they find themselves, for example, many which are autotrophs in the light become heterotrophs in the dark. A discussion of the general biology of the marine planktonic algae lies outside the scope of the present book; for further information about this interesting topic the reader should consult Round (1965), Wood (1967) and Prescott (1969). The book by Fraser (1962) can be recommended as a good general introduction to the subject. Physiological and biochemical aspects of the algae are treated in the volume edited by Lewin (1962).

TAXONOMY

In the internationally accepted Code of Biological Nomenclature the marine algae are classified into the following categories; suffixes which are appended for the various categories are given in brackets.

DIVISION or PHYLUM (-phyta)
  CLASS (-phyceae)
    SUB–CLASS (-phycidae)
      ORDER (-ales)
        SUB–ORDER (-inales)
          FAMILY (-aceae)
            SUB–FAMILY (-oideae)
              TRIBE (-eae)
                GENUS usually a Greek name
                SPECIES usually a Latin name

Until comparatively recently the taxonomy of marine planktonic algae has been based largely on the morphology revealed under the optical microscope. Since the amount of detail which can be seen in this way is limited and because unrelated species frequently resemble one another, classification is a subjective and highly controversial matter. However, considerable progress towards a more systematic classification has been made possible in the last decade by the employment of the electron microscope and by comparative study of the life histories of the organisms. Chemistry can be an important aid to

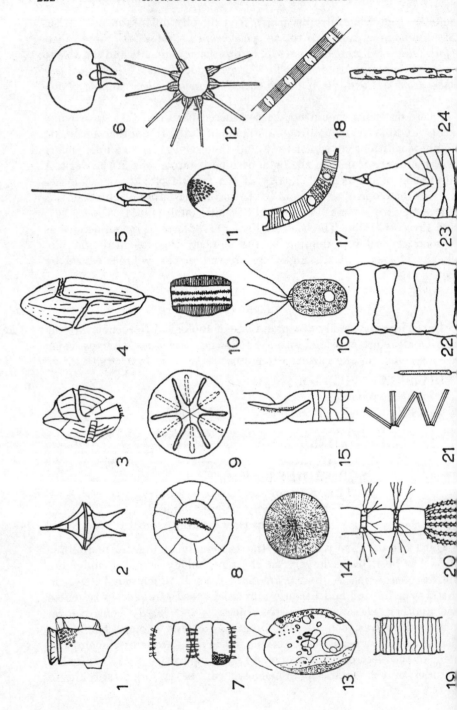

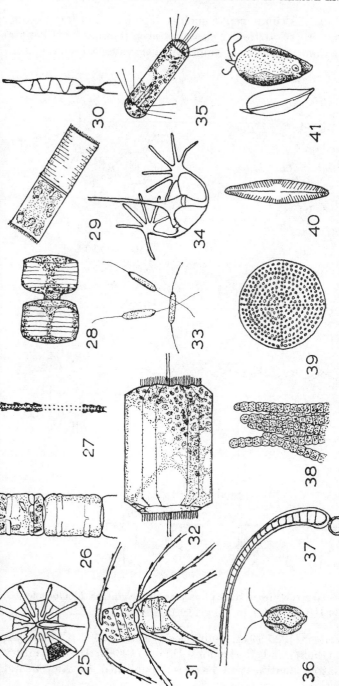

Fig. 9.1. Some examples of marine planktonic algae. 1, *Dinophysis*; 2, *Peridinium*; 3, *Gonyaulax*; 4, *Gyrodinium*; 5, *Ceratium furca*; 6, *Ceratium cephalotum*; 7, *Stephanopyxis*; 8, *Planktoniella*; 9, *Asterolampra*; 10, *Pseudoeunotia*; 11, *Actinocyclus*; 12, *Asterionella*; 13, *Chroomonas placoidea*; 14, *Thalassiosira nordenskiöldii*; 15, *Rhizosolenia alata*; 16, *Carteria oliveri*; 17, *Eucampia zodiacus*; 18, *Skeletonema costatum*; 19, *Guinardia*; 20, *Bacteriastrum*; 21, *Thalassionema*; 22, *Climacodium*; 23, *Rhizosolenia*; 24, *Leptocylindrus*; 25, *Asteromphalus*; 26, *Bacteriosira*; 27, *Thalassiothrix*; 28, *Porosira glacialis*; 29, *Detonula confervacea*; 30, *Rhizosolenia*; 31, *Chaetoceros convolutum*; 32, *Ditylum brightwellii*; 33, *Nitzschia closterium*; 34, *Ceratium ranipes*; 35, *Corethron criophilum*; 36, *Dunaliella*; 37, *Calothrix*; 38, *Trichodesmium*; 39, *Coscinodiscus*; 40, *Navicula*; 41, *Prorocentrum*.

classification since cell components, such as pigments and fatty acids, may be characteristic of particular classes or even families (see Chapter 8, pp. 203–208). Comprehensive accounts of algal systematics have been

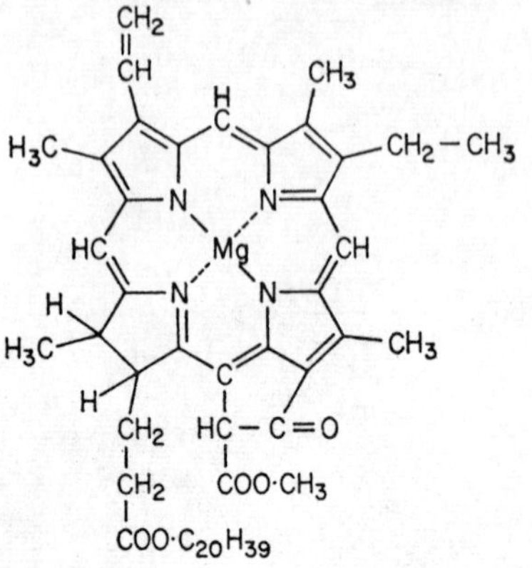

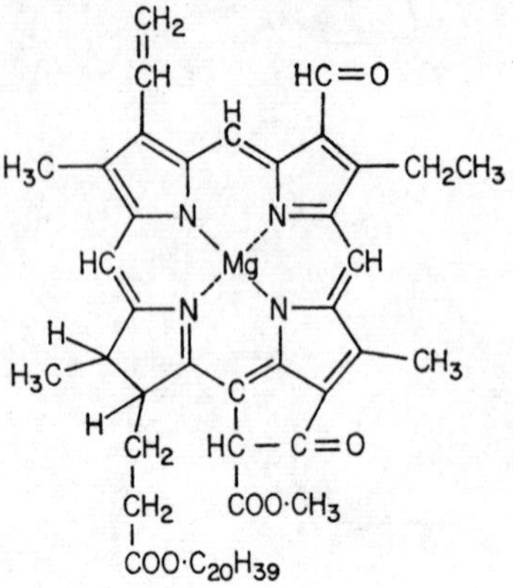

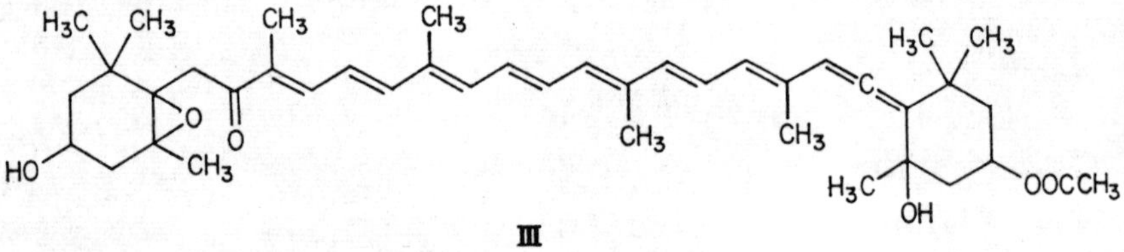

Fig. 9.2. Molecular structures of some phytoplankton pigments. I, chlorophyll *a*; II, chlorophyll *b*; III, fucoxanthin; IV, phycoerythrobilin.

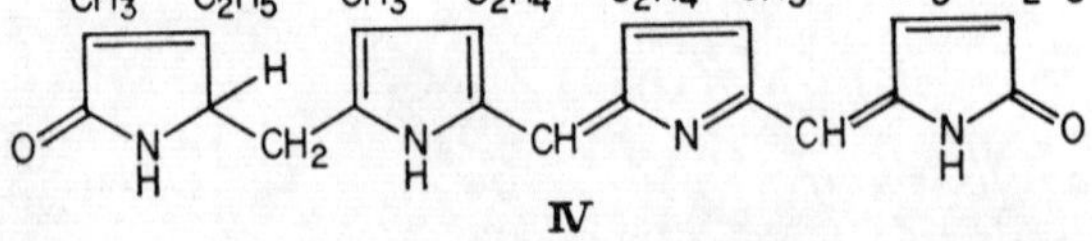

published by Fritsch (1935, 1945), Smith (1951), Lewin (1962), Christensen (1964), Silva (1962), and Parke and Dixon (1968). An elementary guide to plankton identification has been published by Newell and

Newell (1963). In order to assist general readers and non-biologists an abridged classification is provided below showing some of the more important species, some of which are illustrated in Fig. 9.1. In this classification the organisms have been classified up to Class rank according to Parke and Dixon (1968) and Classes have been grouped into Phyla according to the scheme of Round (1965).

Algae have been divided into eight Phyla by Round (1965). However, marine phytoplankters are mainly restricted to five of these. Of the remaining Phyla, the Rhodophyta consists principally of the attached red algae, but also contains a few planktonic forms. The Phaeophyta comprises mainly the large brown seaweeds. The marine forms of the Euglenophyta are mainly estuarine types, but are occasionally reported in ocean waters. They are important in surface sediments, but probably contribute little to the total biomass in the oceans.

1. CYANOPHYTA

Cyanophyceae

Chroococcales, e.g. *Microcystis, Chroococcus, Anacystis*
Nostocales, e.g. *Trichodesmium, Microcoleus, Calothrix, Oscillatoria*

The Cyanophyceae or blue green algae are probably the most primitive organisms, and show many affinities with the bacteria. They are motile by a creeping movement. Pigments are not limited within a sharp boundary membrane. Reproduction is by a variety of mechanisms, such as production of endospores, exospores or unencysted fragments of filaments. They are commonly found in great abundance in shore line and estuarine environments. Oceanic species are rarer, but *Trichodesmium* spp. occur in large numbers in certain tropical seas, e.g. the Indian Ocean, where their ability to fix nitrogen enables them to thrive when nutrient concentrations are low (see p. 156). In addition, some genera are known which are epiphytic on marine algae and corals.

2. CHRYSOPHYTA

Chrysophyceae

Ochromonadales, e.g. *Pavlova, Ruttnera*
Chromulinales, e.g. *Monochrysis, Pseudopedinella, Chromulina*
Dictochales, e.g. Silicoflagellates (*Dictyocha*)

The chrysomonads are motile organisms with one to three unequal terminal flagella. They possess an elastic cuticle which may be naked or have lightly silicified or calcified plates. Silicoflagellates have a siliceous endoskeleton. They may be phototrophic, heterotrophic or phagotrophic according to circumstances. Most species contain chlorophyll a and phototrophic species have in addition chlorophyll c and fucoxanthin. Reproduction is by fission or budding. Their pigments and other attributes suggest that they may have a common origin with the bacillariophyceae. Although they occur widely in the oceans they are important only in coastal waters.

Haptophyceae

Isochrysidales, e.g. *Isochrysis*

Prymnesiales, e.g. *Apistonema, Chrysochromulina, Prymnesium, Phaeocystis, Syracosphaera, Coccolithus* and other coccolithophorids

Members of the Haptophyceae have many of the attributes of the Chrysomonads, e.g. similar pigments, flagella and mode of nutrition. The coccolithophores may constitute the majority of the planktonic biomass in parts of the open oceans. This class may also contain the many unidentified species of ultraplankton which are abundant in the oceans in tropical and sub-tropical regions.

Bacillariophyceae

Centrales (centric diatoms), e.g. *Biddulphia, Chaetoceros, Coscinodiscus, Cyclotella, Ditylum, Lauderia, Melosira, Planktoniella, Rhizosolenia, Skeletonema, Thalassiosira, Triceratium*

Pennales (pennate diatoms), e.g. *Asterionella, Bacillaria, Cylindricotheca, Fragilaria, Navicula, Nitzschia, Phaeodactylum,* Rhabdonema

The diatoms, which are divided into two Orders, are classified on the basis of the structure and shape of their silicified frustules; much of the structure is so fine that it can only be resolved using an electron microscope. The degree of silicification is very variable varying from a few percent to forty or more (as SiO_2). Some species, particularly the Raphinideae (Order Pennales) are motile. All are phototrophic but many show increased growth in the presence of organic compounds such as glycine. Reproduction occurs both asexually and sexually by fusion of gametes or by autogamy. The diatoms are usually the main component of the blooms in spring and autumn in temperate coastal waters and in the summer in Arctic and Antarctic Seas.

3. CHLOROPHYTA

Chlorophyceae

Volvocales, e.g. *Carteria, Chlamydomonas, Dunaliella, Platymonas*

Chlorococcales, e.g. *Ankistrodesmus, Chlorella, Oocystis, Scenedesmus*

The chlorophyceae are green algae. The marine species range from uni-cellular to macroscopic types. The marine planktonic forms are

---

* This organism which has been extensively studied was formerly called *Nitzschia closterium forma minutissima*. It is probably not a true diatom, but rather the sole member of a sub-order of the Bacillariophyceae (Lewin *et al.*, 1958).

mainly in the ultra- and nano-plankton range and occur free or in colonies. Most species are phototrophic. Members of the Order Volvocales are flagellates with two, or rarely more, flagella and reproduce mainly by simple cell division. Members of the Order Chlorococcales are non-flagellate cells, reproducing asexually by means of zoospores or autospores. They occur mainly in coastal waters of temperate latitudes, particularly in late summer and autumn.

4. CRYPTOPHYTA

Cryptophyceae

Cryptomonadales, e.g. *Hemiselmis, Cryptomonas, Chroomonas*

The cryptophyceaens are a small group of naked monads having two rather unequal flagella which may be lateral or terminal. Reproduction is asexual, by longitudinal division. Those species which possess chlorophyll are phototrophic, but may also be phagotrophic. The taxonomy of this class is uncertain. Originally they were classed with the Pyrrophyta, but work on their pigments has shown that in common with the Rhodophyta and Cyanophyta they contain pink water-soluble photosynthetic biliproteins. However, in contrast to these they also possess several unique mono and diacetylenic carotenoids, e.g. alloxanthin (Mallams *et al.*, 1967; Riley and Segar, 1969). They are common members of the ultraplankton in coastal, estuarine, and ocean waters, particularly in warm conditions. Their quantitative significance in the plant biomass is uncertain.

5. PYRROPHYTA

Dinophyceae

Prorocentrales, e.g. *Exuviella, Prorocentrum, Dinophysis*
Peridiniales, e.g. *Amphidinium, Gymnodinium, Gyrodinium, Noctiluca Glenodinium, Peridinium, Gonyaulax, Ceratium*

The flagellates are a large group of very diverse pigmented or colourless unicellular organisms possessing two flagella which differ in structure and position according to species. The most extensively studied ones are the dinoflagellates, and little is known about the occurrence of the other flagellates in the marine environment. Although some dinoflagellates (e.g. members of the Gymnodiniaceae) are naked, most are encased in a cellulose wall which is often sculptured into patterned plates. Reproduction is by cell division. The majority of the dinoflagellates contain chlorophyll and are phototrophic. However, some of these can also live phagotrophically. Others, particularly the colourless species have lost the power to photosynthesize and exist phagotrophically or symbiotically. Dinoflagellates, and perhaps other flagellates make up a considerable part of the phytoplankton population at some times of the year. Particular species can often be used as "indicators" for water masses. The massive blooming of certain species, e.g. *Gonyaulax polyhedra* and *Gymnodinium breve* gives rise to "red tides" which occur mainly in tropical or sub-tropical waters, e.g. Walvis Bay and the Gulf of Mexico; many of the species involved give rise to toxins or growth inhibitors (see p. 192). The considerable amounts of particulate organic matter associated with these blooms are probably extra-cellular metabolites of these organisms.

III. PHOTOSYNTHESIS AND RESPIRATION

The most important stage in the marine food chain is photosynthesis. During this process, solar energy is absorbed by the phytoplankton cells and converted into biological energy, which is stored in the form of organic compounds of high potential energy. Much of the organic carbon fixed in this way is consumed by organisms at higher trophic levels. These, and the phytoplankton themselves, obtain their energy requirements by breaking down these compounds to those of lower energy; this process is known as respiration. It is only possible to give a very brief account of these two complicated processes here; for further information the reader is referred to textbooks of biochemistry.

A. PHOTOSYNTHESIS

The photosynthetic process can be considered conveniently in three stages.

(i) Photons of light are absorbed by the photosynthetic pigments which are contained in the chromatophores. The principal of these pigments are the chlorophylls (see Fig. 9.2 and p. 208) in which the resonating system of double bonds that stabilizes the molecule also provides π electrons that are easily excited to orbitals of higher energy when light energy is absorbed. Light energy absorbed by certain other pigments, e.g. fucoxanthin (Fig. 9.2) in diatoms and Phaeophyta, and complexes formed between globulin and phycobilin (Fig. 9.2) in Rhodophyta and Cryptophyta, is also transferred to the chlorophyll; the processes involved in this energy transfer are in many cases uncertain, but may include direct electron transfer and fluorescence.

(ii) Part of the energy of these excited electrons is converted to chemical energy through a cyclic series of enzymatic reactions, involving Cytochrome I, which leads to the production of high energy adenosine

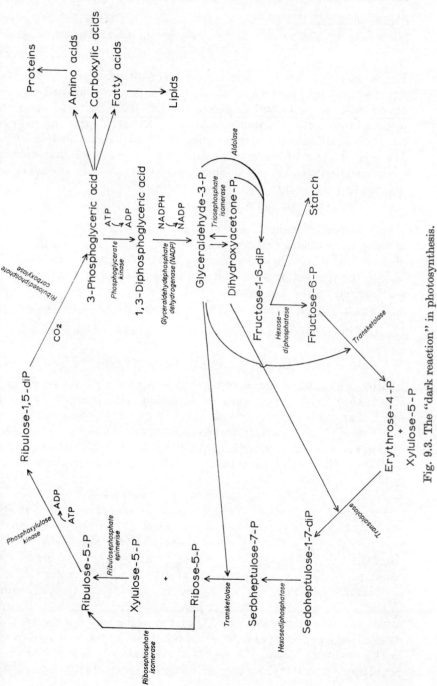

Fig. 9.3. The "dark reaction" in photosynthesis.

triphosphate (ATP) from adenosine diphosphate (ADP) and orthophosphate ions (P). The overall reaction may be summarized by the equation:

$$ADP + P \rightarrow ATP \tag{1}$$

The remainder of the energy of the electrons is used in another series of cyclic enzymatic reactions involving riboflavin phosphate and nicotinamide adenine dinucleotide phosphate (NADP). In these reactions protons from water, excited electrons and $NADP^+$ form the reduced form of nicotinamide adenine dinucleotide phosphate (NADPH). The OH^- moiety of the water yields molecular oxygen and donates electrons to chlorophyll *via* the cytochrome I chain. The overall equation for this reaction is:

$$4NADP + 2H_2O + 2ADP + 2P \rightarrow 4NADPH + O_2 + 2ATP \tag{2}$$

(iii) Carbon dioxide is assimilated in a cyclic series of reactions utilizing the reducing action of NADPH and the phosphorylating power of ATP (Fig. 9.4). These reactions, which can take place in the dark lead to the production of carbohydrate, represented as (CH_2O), and are summarized by the equation:

$$CO_2 + 4NADPH + ATP \rightarrow (CH_2O) + H_2O + 4NADP + ADP + P \tag{3}$$

Combination of equations (1), (2) and (3) leads to the basic equation of photosynthesis:

$$CO_2 + H_2O \rightarrow (CH_2O) + O_2 - 112 \text{ kcal} \tag{4}$$

Photosynthetic experiments using carbon dioxide labelled with ^{14}C have shown that fixed carbon finds its way very quickly into compounds other than carbohydrates. It has been shown that these compounds (e.g. fats and amino acids) are synthesized from intermediates in the carbon cycle in addition to being produced from photosynthetic carbohydrates. The pathways leading to some of these compounds are indicated in Fig. 9.4. The treatise of Calvin and Bassham (1962) should be consulted for a detailed account of photosynthesis.

Because of the production of these other classes of compound during photosynthesis, the *photosynthetic quotient* (PQ = molecules of O_2 liberated during photosynthesis/molecules of CO_2 assimilated) differ from the value of unity corresponding to the synthesis of hexose (see equation 4). For example, a value of 1·4 would be expected if only lipid is formed. If protein alone is produced the PQ will vary, according to the oxidation state of the nitrogen assimilated, from 1·05 with ammonia to 1·6 with nitrate. The PQ of any algal cells will therefore be determined by their fat and protein content and by the form of nitrogen in the medium. Values of 1·20–1·33 are commonly found for natural populations provided that there are adequate nutrients (Ryther, 1956a).

At low intensities, light may play a part in the heterotrophic utilization of organic compounds, such as acetate (Pringsheim and Wiessner, 1961; Fogg, 1963). This may be advantageous to the organism concerned since it can thus obtain high potential energy compounds using less energy than if it had to fix carbon dioxide.

B. RESPIRATION

The energy required by the algae for their metabolic processes is obtained by the oxidation of the photosynthetically-produced complex molecules to compounds of lower potential energy, and ultimately to carbon dioxide. This process, which is known as respiration, involves reaction pathways which are probably similar to those used during the respiration of other organisms. The overall reaction may be crudely represented by the equation:

$$(CH_2O) + O_2 \rightarrow CO_2 + H_2O$$

Although the algal cells contain fats and protein, in addition to carbohydrate, the latter appears to be the substrate for most respiratory processes and the *respiratory quotient* (RQ = molecules of CO_2 liberated/ molecules of O_2 assimilated) therefore generally lies close to unity. Respiration takes place both in the light and in the dark. The rate in the dark is usually about 5–10% of the maximum rate of photosynthesis exhibited by the same cell growing under optimal conditions (Steeman Nielsen and Hansen, 1959a). The rate of respiration in the light is generally considered to be the same as in the dark (Van Norman and Brown, 1952; Brown, 1953), although at high light intensities it may increase (Brown and Weis, 1959). However, there is still some controversy about the matter.

IV. PHYTOPLANKTON PRODUCTION

A. SOME DEFINITIONS

Primary production in the sea is of extreme importance since it is the initial stage in the marine food chain, which terminates with fish and marine mammals, such as whales. In considering the quantity of plankton potentially available to the food chain in a particular location, we must take into consideration not only the amount of phytoplankton available at any instant, but also the rate at which it is being produced. The amount of phytoplankton is termed the *standing crop* and is usually expressed as the amount of living phytoplankton (as mg of carbon) present in 1 m^3 of sea water, or beneath 1 m^2 of sea surface. The rate of primary production is usually defined as the weight of inorganic carbon fixed photosynthetically in unit time per unit volume, or under unit area of sea surface (i.e. mg. C/m³/hour or gC/m²/day). It is necessary to distinguish between the *gross* rate of production, which is the rate at which the plant carbon is produced photosynthetically, and the *net* rate, which is the gross rate less the rate of loss of carbon through respiration by plants.

B. MEASUREMENT OF THE RATE OF PRIMARY PRODUCTION

Lack of fast and accurate methods for direct determination of the low concentrations of plant material (see Lovegrove, 1970) or organic carbon produced during photosynthesis in the sea, makes it necessary to use indirect procedures for the estimation of the rate of primary production. A consideration of the overall equation of photosynthesis

$$CO_2 + H_2O \rightarrow (CH_2O) + O_2 \tag{5}$$

suggests that the gross rate of photosynthesis could be determined by measuring (i) the rate of liberation of oxygen, (ii) the rate of removal of carbon dioxide, (iii) the rate at which carbon is converted into plant tissue. In addition, since the $N : C$ and $P : C$ ratios in phytoplankton are fairly constant it should be possible to estimate primary production indirectly by measuring the rate of removal of nitrate or phosphate from the water. Only the first and third of these approaches have been widely employed. The procedure depending on the rate of consumption of micro-nutrient elements has been used to obtain integrated values of primary productivity over fairly lengthy periods (e.g. by Cooper, 1933 and Steele, 1956, 1958). However, since part of the nitrogen and phosphorus may be regenerated and recycled several times a year (Harris, 1959; Ketchum, 1962; Vaccaro, 1963) these values will tend to be minimal. Furthermore, difficulties may arise from the introduction of nutrients by horizontal or vertical exchange of water (see, e.g. Cooper, 1957).

In the determination of primary production by measurement of the rate of oxygen liberation a series of 300 ml glass stoppered bottles is filled with the sea water being studied. These bottles are suspended at various depths throughout the euphotic zone, together with corresponding light-tight bottles filled with the same water. After a suitable period (3–8 hours) the oxygen content of the water in the bottles and that of the initial water sample is determined by the Winkler method. The increase in oxygen concentration occurring in the "light" bottles is a measure of the mean net photosynthesis which has taken place. The loss of oxygen which occurs simultaneously, because of respiration by plant cells, bacteria and zooplankton, is assessed from the decrease in the oxygen content of the water in the "dark" bottles. The difference between the final oxygen concentration in the light and dark bottles gives an estimate of the mean gross productivity.

For experimental details the Handbook of Seawater Analysis by Strickland and Parsons (1968) should be consulted.

The oxygen bottle method suffers from several drawbacks, the most important of which are:

(i) it is insufficiently sensitive (sensitivity ca 3 mgC/m³/hr) to be used with waters of low productivity (i.e. *oligotrophic* waters).

(ii) it is unsuitable for use in highly polluted eutrophic waters, particularly if the bacterial population is high.

(iii) if appreciable amounts of lipids or proteins are photosynthesized as well as carbohydrate, as with the diatoms, results will be low. Allowance can be made for this by dividing the calculated productivity by the photosynthetic quotient (p. 230) of the particular phytoplankton population; somewhat arbitrarily this is frequently assumed to be 1·2.

(iv) bacteria tend to proliferate on the bottle walls and this will lead to an unusually high bacterial density in the water.

(v) the growth of the phytoplankton may be influenced by its confinement in the more or less static water in the bottle.

A more direct approach to the determination of the rate of photosynthesis is to measure the rate at which carbon dioxide is taken up from the water. Direct measurement of the rate of uptake by means of a gas analyser, such as a Van Slyke apparatus (Saruhashi, 1953; Strickland and Parsons, 1968) is generally feasible only with fairly dense cultures, because of the difficulties of measuring, with sufficient precision, the small differences in total carbon dioxide concentrations which occur. The most satisfactory method for measuring the uptake of carbon dioxide in both oligotrophic and eutrophic waters is that developed by Steeman Nielsen (1952, 1954) using carbon–14 (see also Steeman Nielsen and Jensen, 1957; Steeman Nielsen, 1964). In this technique a series of 300 ml clear glass stoppered bottles is almost filled with samples and treated with 2 ml of bicarbonate solution in which the carbon is labelled with ^{14}C (1–25 μCi depending on the productivity expected). The bottles are then suspended in the sea at the depths at which it is desired to measure the productivity. Less satisfactorily, they may be placed in a thermostatic bath aboard ship and illuminated with light from fluorescent lamps which is filtered to give intensities and spectral distributions corresponding to those at the desired depths in the sea. After 2–6 hours the contents of the bottles are filtered through a 0·45 μm membrane filter, using only gentle suction in order to avoid the destruction and loss of delicate flagellates. The filters are washed with sea water, and if the algae are calcareous (e.g. if coccolithophores are dominant) they are exposed to hydrochloric acid vapour to remove any *inorganic* carbon. The amount of radiocarbon fixed is measured by counting the filters with an end-window Geiger counter, or preferably a scintillation counter (Jitts and Scott, 1961; UNESCO, 1967) to determine the activity. Even in complete darkness some uptake of carbon dioxide occurs with both photosynthetic and non-photosynthetic organisms, mainly by exchange. Normally this only amounts to 1–2% of the total photosynthesis, but it may be more than 10% in tropical waters. In order to compensate for carbon dioxide taken up non-photosynthetically in this way a *dark blank* must be carried out in a similar fashion, using a sample of the same water contained in a light-tight bottle. Then

$$\text{Productivity (mgC/m}^3\text{/hr)} = \frac{1 \cdot 05(C_S - C_D)W}{C.N.}$$

where W is the total carbon dioxide content of the water (as mgC/m^3). This can be calculated from the alkalinity and pH of the waters (see p. 135). N is the number of hours for which the sample is exposed to light.

The factor $1 \cdot 05$ compensates for the slower uptake by the algal cells of $^{14}CO_2$ compared with $^{12}CO_2$.

C_S and C_D are the normalized counting rates of the sample and dark blank respectively.

C is the normalized counting rate given by the amount of ^{14}C which was added to the light and dark bottles. The accurate determination of this factor is a matter of considerable difficulty, and the reader is referred to Strickland and Parsons (1968) for an account of its determination and for experimental details of the whole method.

If it is desired to know the productivity beneath unit area of sea surface measurements are made at a series of depths covering the whole of the euphotic zone and the results are integrated.

Although the radiocarbon method is probably the most satisfactory procedure available for the determination of productivity in the sea, there are a number of factors which detract from its reliability. The most important of these are:

(i) the uncertainty as to whether the method measures gross, or net production, or some intermediate value, since, during the experiment some of the carbon fixed may be used in respiration. Many workers incline to the view that it gives results approximating to the net production. Steeman Nielsen (1964) considers that in temperate waters of moderate productivity net production probably amounts to ca 90% of the gross production. However, the difference between the net and gross production varies geographically; thus in oligotrophic tropical waters the net production may be only 60% of the gross value.

(ii) photosynthesis may give rise to soluble extra-cellular metabolites, e.g. glycollic acid, which will not be retained by the membrane filter, and hence the productivity will tend to be underestimated. However, during the productivity measurement, bacteria, or the algae themselves may utilize some of this material to synthesize particulate matter, and this will tend to compensate for this error. The amount of material lost as extra-cellular metabolites is probably usually small although under some circumstances more than 30% of the carbon fixed may be excreted to the water (see also pp. 186–7).

(iii) as with the oxygen-bottle method, errors will arise from the enclosure of the sample. It is known that under such confined conditions large changes can occur in the species composition of the phytoplankton population in quite short periods.

(iv) as with other methods of production measurement, since the photosynthetic activity of the phytoplankton varies diurnally, the time of day at which the determination is made can be important.

(v) it should be remembered that the radiocarbon method measures only the fixation of inorganic carbon and does not take into account any photo-assimilation of organic compounds (see p. 231).

Comparative studies of methods for determining primary production in the sea have been carried out by McAllister et al. (1961) and by Antia et al. (1963). They have suggested that the ^{14}C method provides an estimate of *net* production, and have emphasized the necessity to take the photosynthetic quotient into account when using the oxygen production method. Laboratory experiments with cultures have shown reasonable agreement between estimates of primary production based on oxygen production, carbon dioxide assimilation (as measured by pH changes), uptake of phosphate, and the volume of packed cells obtained after centrifugation of the culture (Ansell et al., 1963, 1964). In conclusion, it may be stated that present methods for determination of primary productivity are rather in the nature of *faute de mieux*. Ideally what is required is a method of measuring the rate at which *planktonic material* is synthesized. Even if rapid and precise procedures for the determination of dry organic matter or organic carbon were to become available, they would only provide a partial solution to the problem since (a) they would take no account of the production of soluble extracellular metabolites (see (ii) above), and (b) no distinction would be made between plant material and detritus or animal tissue.

For an account of recent work on the general subject of measurement of production the book edited by Vollenweider (1969) should be consulted.

V. Factors Influencing the Growth of Phytoplankton

A. LIGHT

In considering the influence of light on primary production in the sea, two factors must be taken into account. (i) The various factors controlling the intensity and spectral composition of light at any level

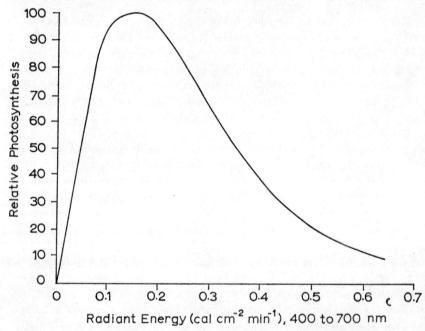

Fig. 9.4. The relationship between light intensity and photosynthesis of phytoplankton (after Ryther, 1956b; from Currie, 1962, by permission of the Royal Society.)

in the sea. (ii) The way in which the rate of primary production of any particular organism depends on the intensity and wavelength of the light.

(i) *Light in the sea*

The amount of solar energy falling on the sea surface depends on the sun's altitude and the extent of cloud cover. In temperate latitudes, the maximum total energy with a clear sky will be about 1·4 ly/min 1 ly/min = 1 g.cal./cm²/min) at midsummer falling to ca 0·5 ly/min in December. Cloud cover may reduce these figures by 70% or more. Only about 50% of this radiation lies in the photosynthetically useful wavelength range (370–720 nm), the exact proportion depending on the sun's height and the atmospheric conditions. Not all this radiation penetrates the sea surface since some of it is lost by reflection. The amount lost in this way is greatest when the sun is low in the sky, or when the sea surface is rough.

Radiation is attenuated in the sea by absorption and scattering by water molecules, ions and particulate matter, and also by absorption by Gelbstoff-like substances (p. 190). The attenuation of light of wavelength λ follows the logarithmic relationship $\log I_{0\lambda}/I_{\lambda} = kl$ where $I_{0\lambda}$ and I_{λ} are the intensities of the incident and attenuated radiation respectly, and l is the thickness of the attenuating medium (expressed in metres). The absorption coefficient, k, varies with the wavelength of the light and with the concentrations of absorbing and scattering material in the water column. Typical transmission spectra for sea water are shown in Fig. 2.11 (p. 34). For a detailed account of solar radiation and its attenuation in sea water reference should be made to Strickland (1958), Jerlov (1968) and Holmes (1957).

Even in clear ocean waters only ca 20% of the total solar radiation penetrates to a depth of 10 m (practically all the infra red radiation of wavelength greater than 750 nm is in fact absorbed in the first 2–3 m). As the depth increases the spectral range of the residual light becomes restricted to the blue-green region (ca 500 nm). The attenuation of light in the sea is determined principally by the amount of suspended matter present. Thus, in clear tropical oceans 1% of the surface visible light energy penetrates to a depth of ca 120 m, whereas in moderately turbid coastal waters this degree of attenuation may occur at a depth of 10–20 m.

The upper layer of the sea in which gross primary productivity exceeds respiration (i.e. the net production is positive) is termed the *euphotic zone* and its depth is determined mainly by the mean amount of solar radiation at the surface and the attenuation of light. The depth at which the net production is zero is termed the *compensation point*. This depth probably corresponds with a mean intensity of ca 0·005 ly/min, but will vary somewhat according to the light-adaptation of the algae concerned; for summer conditions it is frequently assumed to be the depth to which 1% of the surface visible light penetrates. There is increasing evidence that the effects of acclimatization to changing light levels and variations in the photosynthesis : respiration ratios make this assumption of doubtful validity. It should be noted

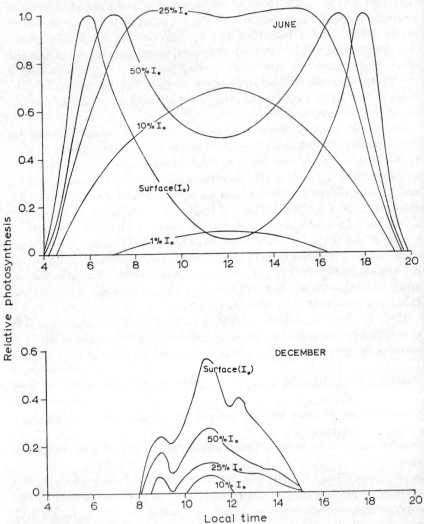

Fig. 9.5. Daily relative photosynthesis at various depths in the sea at temperate latitudes in summer and winter. Depths are expressed as those at which light is reduced to the specified percentage of its surface intensity, I_0 (after Ryther, 1956b).

that since the phytoplankton cells are usually in continual movement up and down the water column, viable cells may occur well below the compensation point (see also p. 249).

(ii) Effect of light on growth of phytoplankton

(a) Light intensity. At low light intensities the rate of photosynthesis of phytoplankton increases in proportion to light intensity. At moderate intensities the rate of increase becomes less and light saturation is attained. As the light intensity is increased still further inhibition sets in, perhaps as a result of inhibition of the production of chlorophyll, and the rate of photosynthesis decreases (Fig. 9.4, p. 236). This phenomenon probably accounts for the fact that the depth of maximum productivity in the sea often lies a few metres below the surface in the middle of the day in the summer at temperate latitudes and throughout the year in the tropics (Fig. 9.5, p. 238). At very high light intensities (>1.5 ly/min) photo-oxidation of the enzyme systems and of the chlorophyll results in permanent inhibition of photosynthesis.

Saturation intensities vary considerably from one class of algae to another. Thus, Ryther (1956b) has found that the saturation intensities for members of the Chlorophyta, diatoms and Dinoflagellates lie in the ranges 5–7.5 k.lux, 10–20 k.lux and 25–30 k.lux respectively. In each instance a further increase of 10 k.lux caused inhibition. There is considerable evidence that in many parts of the oceans it is possible to differentiate between "sun" plankton which are adapted to live at high light intensities, and which occur in the surface layer, and "shade" plankton which live near the base of the euphotic zone (see, e.g. Steeman Nielsen and Hansen, 1959b). As part of their adaptation to living at low light intensities these "shade" organisms are packed with chromatophores and tend to have higher chlorophyll a : carbon ratios and greater proportions of active accessory carotenoids than the "sun" forms. The mean minimum light intensity required by these shade species for their growth is probably ca 0.005 ly/min. Ryther and Menzel (1959) have noted that differentiation of phytoplankton into these two types occurs particularly in clear, highly stratified tropical waters, e.g. the Sargasso Sea. It should be noted that some species can adapt themselves quite rapidly to living under different lighting conditions (Jorgensen and Steeman Nielsen, 1965; Steeman Nielsen and Park, 1964; Harvey, 1955). This adaptability is probably of considerable importance in the sea. Thus, during phytoplankton blooms the growing cells will cause shading of the deeper layers and thus the phytoplankton at these levels must photosynthesize at lower light intensities. Another example of adaptation is the existence of diatoms which are capable of photosynthesis under Arctic ice (Smayda, 1963).

The relationship between light intensity and primary productivity should not be considered in isolation since it is influenced in a complex way by both nutrient concentration and temperature (Curl and McLeod,

1961; Steele and Baird, 1962; Maddux and Jones, 1964). For example, impoverishment of the water with micronutrients will lead to the production of cells deficient in chlorophyll, and therefore having a lower productivity.

(b) *Spectral composition of light.* Only light having wavelengths lying in the range ca 370–720 nm can be used by phytoplankton for photosynthesis. At wavelengths shorter than 380 nm the efficiency of photosynthesis falls rapidly and complete inhibition with irreversible damage to the cells occurs with ultra-violet radiation of wavelength 300 nm or less (McLeod and Kanwisher, 1962). There is some evidence that 750 nm infra red radiation causes inhibition of the photosynthesis of some algae (Rabinowitch and Thomas, 1960).

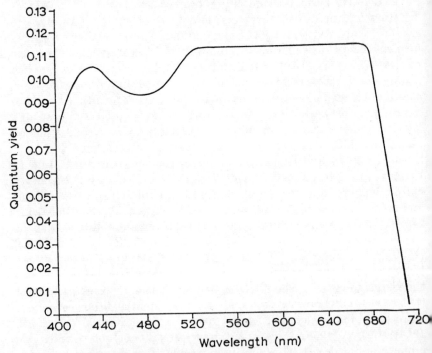

Fig. 9.6. Quantum yield of photosynthesis by *Navicula minima* as a function of wavelength (after Tanada, 1951).

The efficiency with which algae are able to use light of various wavelengths has been investigated in the following way (see Haxo, 1960). The algal cells are irradiated with monochromatic light, the wavelength of which can be varied. At each wavelength, measurements are made of the number of quanta of energy absorbed by the cells and of the corresponding amount of oxygen liberated photosynthetically (determined polarographically). The quantum yield of the process (number of molecules of oxygen liberated per quantum of energy absorbed) is then calculated and plotted as a function of wavelength. A typical *action spectrum* determined in this way for the fresh-water diatom *Navicula minima* is given in Fig. 9.6 (Tanada, 1951). This spectrum is very similar to the absorption spectrum of a pigment extract of the organism, if correction is made for shifts in the wavelength of the absorption bands which occur, because of the breakdown of their complexes with

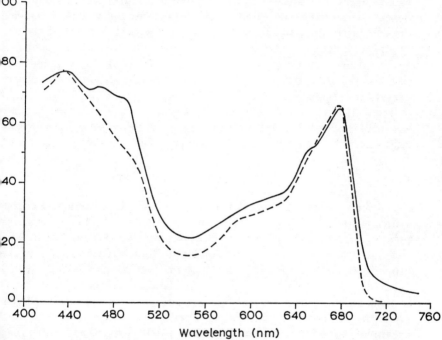

Fig. 9.7. Absorption and action spectra of *Ulva taeniata* (Chlorophyceae). Action spectrum broken line; Absorption spectrum of thallus continuous line (from Blinks, 1954).

lipids and protein, when the pigments are extracted. The horizontal line between 520 nm and 680 nm indicates that the light absorbed by fucoxanthin (λ_{max} 540 nm) is used for photosynthesis with a similar efficiency to that absorbed by chlorophyll *a*, which is responsible for most of the absorption above 560 nm and between 400 and 450 nm. This is because the energy absorbed in this way is transferred to the chlorophyll. The correspondence between action and absorption spectra is particularly striking for certain of the larger algae (see, for example, Fig. 9.7 and Blinks, 1954). Not all the carotenoids are able to act as

10

transferrers of energy in photosynthesis. Thus, of the pigments of the Rhodophyta only phycobilin was effective in photosynthesis (Haxo and Blinks, 1950). The dip at 480 nm in the action spectrum of *N. minima* (Fig. 9.6) probably results from non-utilization of light absorbed by the subsidiary pigments which are responsible for the absorption of most of the light of this wavelength.

The ability of some carotenoids (e.g. fucoxanthin and peridinin) and phycobilin-protein complexes to act as transferrers of energy has a very important implication for marine primary productivity. Since light of the wavelengths absorbed by chlorophyll penetrates only a few metres beneath the surface, it is only in the uppermost layer that it makes a direct contribution to photosynthesis. In the deeper parts of the euphotic zone light utilization is carried out by these accessory pigments which absorb the blue-green light (λ_{max} 500–550 nm) which penetrates to these depths, and most of the species adapted to living in the deeper parts of the euphotic zone have high ratios of accessory pigments : chlorophyll *a*. It seems likely that the carotenoids of the marine plants have a secondary function and serve to protect chlorophyll against photo-oxidation at high light intensities, perhaps by acting as oxygen acceptors.

B. TEMPERATURE

Efficient photosynthesis takes place over the whole range of temperatures encountered in the marine environment (i.e. from $-2°C$ in the polar regions to $>30°C$ in tropical lagoons or on mud flats). The species thriving in any location will have adapted themselves not only to the prevailing temperature but also to light conditions existing at the depth at which they live. There is considerable evidence for the interdependent effect of light and temperature on photosynthesis. Phytoplankton will only tolerate quite limited changes of temperature and are rapidly killed at temperatures of $10°-15°C$ above that at which they are adapted to live. However, the effect of decreasing temperature is less dramatic, and provided that the temperature is lowered sufficiently slowly, the organisms often become acclimatized to low temperatures. It is a curious, and so far unexplained, fact that most species of phytoplankton grow best in culture at temperatures $5°-10°C$ higher than those at which optimum growth occurs in their natural habitat.

Temperature changes of $4°-5°C$ produce little effect on the rate of photosynthesis at light intensities at which the photosynthesis/intensity curve is linear. However, under conditions of light saturation an increase in the temperature brings about an appreciable increase in the photosynthetic rate. For example, Wimpenny (1958) has observed that the

ate of carbon uptake by *Rhizosolenia* illuminated with 16 k.lux in-
reased by 40% when the temperature was raised from 10° to 15°C.
ncrease in temperature will raise the intensity at which light saturation
akes place (Sorokin and Krauss, 1958). The decrease in the rate of
hotosynthesis which occurs at supraoptimal light intensities is accent-
ated by elevation in temperature. It seems likely that temperature
ffects in photosynthesis involve Stage 3 (dark reaction) of the photo-
ynthesis process (see p. 229) in which many enzyme systems are
nvolved, rather than the second stage which is known to be compara-
ively insensitive to temperature changes.

The combined effects of temperature and light on photosynthesis and
ell division may help to determine the pattern of succession of species
n the sea, especially at temperate latitudes. Mechanisms of this type are
ertainly operative in some lakes where the ability of the various
rganisms to thrive or germinate under particular combinations of high
nd low temperature, and strong and weak light, controls the succession
f phytoplankton throughout the year (Vollenweider, 1950). There is
ess evidence for a control of this type in the sea. However, it seems
ikely that the succession of *Thalassiosira nordenskioldii* by *Skeletonema
ostatum* in spring in the temperate waters of the North Atlantic off
America can be explained by the different requirements of the two
diatoms (Conover, 1956). *T. nordenskioldii* appears to be favoured by
he low water temperature and relatively low light intensities of the
arly spring, whereas *S. costatum* prefers warmer water and stronger
ight. This may be only part of the explanation since Braarud (1962) has
ointed out that the high nutrient requirements of the former could
nly be satisfied at the beginning of the spring bloom.

C. SALINITY

Although variations in salinity have some effect on the rate of photo-
ynthesis, most truly marine phytoplankters will grow well at salinities
f $15^0/_{00}$ or even less. Indeed, many of them grow best at low salinities;
hus the optimum salinity for the growth of *Skeletonema* is $15-20^0/_{00}$
Curl and McLeod, 1961; see also, e.g. Braarud, 1961). Some species,
articularly diatoms, grow poorly at salinities greater than $35^0/_{00}$ and
his perhaps explains their preference for coastal waters (Hulburt and
Rodman, 1963). Specialized organisms have adapted themselves to
iving in brackish areas such as estuaries or inland seas, as the Baltic.
Some of these, which are termed *stenohaline*, will only thrive over a
imited range of salinities, e.g. *Peridinium balticum*, for which the
ptimum salinity ranges from $8-12^0/_{00}$. Others (*euryhaline* ones) can
ive over the wide range of salinities encountered in estuaries; these

include oceanic species such as *S. costatum* and also brackish water
forms such as *Exuviella*. The influence on productivity of the small
variations of salinity encountered in the oceans is likely to be insigni
ficant. Salinity rather than temperature appears to be the controlling
factor in estuarine waters (Wood, 1964). According to Braarud (1962)
the density changes produced in coastal waters by variations of temper
ature and salinity may indirectly affect the phytoplankton specie
succession by causing certain species to sink.

D. MICRONUTRIENTS AND TRACE METALS

For their healthy growth phytoplankton require a supply of both
nitrogen and phosphorus (see also Chapter 7). Although members of the
Cyanophyeae are able to fix dissolved molecular nitrogen (see p. 156)
members of other Phyla must satisfy their needs for this element by
using either dissolved inorganic forms of the element (e.g. ammonia
nitrite or nitrate ions), or, if they are heterotrophs, organic nitrogen
compounds. Dissolved inorganic phosphate ion is utilized readily by al
species of phytoplankton. However, some species can also satisfy their
needs for phosphorus heterotrophically by using dissolved or par
ticulate organic forms of the element. Although there is an upper limi
to the concentrations of phosphorus and nitrogen which can be tolerated
most species can grow well at concentrations of >20 times the maximum
encountered in the oceans. Growth of phytoplankton will be prevented
if the concentrations of these micronutrient elements in the medium fal
below critical levels. Thus, Ketchum (1939) has shown that the growth
of *Nitzschia closterium* was not affected provided that the medium con
tained more than 16 μg PO_4^{3-}—P/l, and even down to 5 μg PO_4^{3-}—P/
growth continued at a somewhat reduced rate. However, below the
latter concentration the rate of cell division decreased rapidly. Curl and
McLeod (1961) have found that for nutrient–deficient cells of *Skeletonem*
growing in a nutrient-depleted medium there is a linear relationship
between the photosynthetic rate and the nitrogen and phosphorus
concentration in the medium. Although, because of experimenta
difficulties, the nutrient requirements of only a few species of phyto
plankton have been determined there is considerable evidence that
various species differ considerably in their needs. For example, it i
probable that the green algae *Stichoccus* and *Nannochloris* requir
relatively high nitrogen concentrations (Ryther, 1954). On the othe
hand, *Isochrysis* appears to require higher concentrations of phosphat
than many diatoms (Kain and Fogg, 1958).

These differences in micro-nutrient requirements may have an import
ant influence on the distribution and succession of phytoplankton

species in the sea. Thus, the high nutrient concentrations found in spring favour the growth of the diatom *Thalassiosira nordenskioldii* in the North Atlantic. Similarly, the relative abundance of nutrients in many inshore waters encourages the proliferation of those species of phytoplankton which have high requirements for these elements. This may, at least in part, account for the restriction of certain species to coastal areas. There is reason to believe that the sizes of the cells of phytoplankters may be related to their nutrient needs. Those species which are adapted to low nutrient concentrations are generally considerably smaller than those which need high concentrations; their relatively greater surface : volume ratios favouring a more efficient uptake of the nutrients. As the nutrient concentration decreases the cells tend to become smaller and then to sink.

Deficiency of nutrients, especially nitrate, is probably the main factor limiting marine primary production. In temperate latitudes, the development of the seasonal thermocline prevents replenishment of nutrient elements assimilated during the spring plankton bloom. The nutrient deficiency which results reduces the production during the summer.* In tropical and subtropical areas the production is generally low because the waters are kept permanently depleted in nutrients by the thermocline which exists at a depth of 100–200 m. In contrast, production is high in waters which are enriched in nutrients by turbulent mixing or upwelling (see p. 268).

The pollution of restricted bodies of water, such as estuaries, with large amounts of nitrogen and phosphorus (e.g. from sewage, effluents or fertilizer run-off) can lead to heavy blooms of phytoplankton. This process is known as eutrophication. The decomposition of the algae after their death frequently strips the water of oxygen and leads to the mass mortality of fish and all other organisms which cannot live anaerobically. Eutrophication also takes place in a few continental shelf areas (e.g. Walvis Bay off South Africa) where nutrients are enriched in the euphotic zone because of upwelling. A supply of combined nitrogen is not essential for eutrophication since in many instances the organisms concerned are nitrogen-fixing Cyanophyceae which only require a source of phosphorus. This is present in abundance in many effluents because of the increasing employment of condensed phosphates in detergents and as water softening agents.

Silicon is essential for the growth of those organisms, such as diatoms and silicoflagellates which possess siliceous frustules or skeletons (see also Chapter 7, Section VI). Limitation of growth of diatoms because of

---

* In this connection it is of interest to note that *Phaeodactylum tricornutum* can reduce the concentration of phosphorus in media to levels of 0.002 μg PO_4^{3-}—P/l.

lack of silicon has been demonstrated in cultures and is preceded by the production of silicon-deficient cells. The limiting concentration varies from species to species, but generally lies somewhat below 50 μg Si/l. Because the concentration of the element is generally relatively high, and since it is regenerated rapidly it is likely that shortage of silicon rarely limits production in the sea. However, this may be the case in the impoverished sub-tropical areas of the oceans. (Thomas, 1958; Ryther and Guillard, 1959.)

It is well known that marine plants require for their healthy growth a number of trace elements, in addition to the micronutrient elements discussed above. These include iron, manganese, molybdenum, zinc, copper, cobalt and vanadium. The roles of many of these elements in algal metabolism have how been established. Thus, iron as a non-haem protein complex, called *ferredoxin* takes an essential part in the light reaction of photosynthesis. Manganese is known to be present in the enzyme co-factors involved in photosynthesis, and also in nitrate reduction, in which co-factors containing molybdenum are important. Copper-complexes serve as co-factors in oxidation-reduction cycles.

Since not all species of an element can be utilized with equal facility by phytoplankton, the concentration of the element in readily available species, rather than its total concentration, may be important in controlling growth (see Spencer, 1957). Phytoplankton can readily assimilate chelated forms of many elements, and in work with cultures it is usual to add trace metals as chelates with, for example, ethylene diamine tetraacetic acid (EDTA). This permits fairly high concentrations of the metals to be present, in relatively non-toxic forms, without risk of precipitation of their hydroxides. There is some evidence that, in the sea, the chelating action of the dissolved organic compounds, particularly humic acids, may play an essential role in maintaining trace elements in a form in which they can be utilized by algae. The addition of EDTA alone to sea water has been shown by Johnston (1963) to be as valuable as the addition of chelated trace metals. In addition to dissolved species of trace elements, algae are able to make use of some particulate and colloidal forms. This is particularly important for iron which has a very low ionic concentration. The mode of assimilation of trace metals from sea water by the organisms is thought to involve competitive chelation of the elements on the surface of the cell. The chelates are then decomposed and the trace elements are transported across the cell membrane as uncharged or singly charged chloro– or hydroxy–complexes. After passage through the membrane they are again complexed and transported to the site at which the coenzyme is produced.

Very little is known about the concentrations of trace elements required by phytoplankton. The requirements probably vary from species to species (Johnston, 1964). For example, chrysomonads need only very low concentrations (Pinter and Provasoli, 1963). It seems likely that the concentrations of trace elements present in most parts of the oceans are more than sufficient to sustain the optimum rate of photosynthesis (see, e.g., McAllister et al., 1961). However, there is some indication that shortage of iron may be significant in the open ocean far from land, e.g. the Sargasso Sea, or off Australia (Menzel and Ryther, 1961b; Menzel et al., 1962; Tranter and Newell, 1963). Hulburt and Rodman (1963) have postulated that Skeletonema is confined predominantly to neritic waters because of its high requirements for iron as well as micronutrients, and that its occasional blooms in the open sea are induced by the presence of unusual concentrations of iron, or other trace metals. In contrast, Strickland (1965) has suggested that the growth of some oceanic species of phytoplankton in coastal waters may be inhibited by the relatively higher trace element concentrations of these waters.

E. ORGANIC FACTORS

It has been known for many years that even when algal cultures are supplied with micronutrients and trace elements the growth of the organisms will often not take place unless minute traces of specific organic compounds are present. In contrast, some organic compounds which occur naturally in sea water have been found to inhibit the growth of algae. The origins and ecological effects of some of these ectocrines have been discussed in Chapter 8, Section II D. Vitamin B_{12}, B_1 (thiamine) and biotin are probably the most extensively investigated of the growth promoting compounds, although several other essential compounds such as ascorbic acid and cystine have also been detected in sea water (see Belser, 1963; and Wood, 1963a). A supply of vitamins is necessary for the growth of some species of phytoplankton which are unable to synthesize these compounds. Those present in the sea are synthesized principally by bacteria, but may also be produced by a few algae. However, it is thought that the relative abundance of these compounds in many of the red algae arises from their accumulation rather than from synthesis (Provasoli, 1963).

Class requirements for the above vitamins are summarized in Table 1 (Provasoli, 1963), from which it can be seen that almost all the species of Euglenineae, Cryptophyceae, Dinophyceae and Chrysophyceae examined require one or more of these vitamins. In contrast, about half the species of Chlorophyceae and diatoms studied and 9 out of 10 of

TABLE 9.1

Summary of vitamin requirements of fresh-water and marine algae (from Provasoli, 1963)

Algal group	No. of species	No. vitamins	Require vitamins	B_{12}	Thiamine	Biotin	B_{12}+ thiamine	Biotin+ thiamine	B_{12}+biotin +thiamine
Chlorophyceae	68	24	44	10	8		26		
Euglenineae	9	0	9	2	1		6		
Cryptophyceae	11	0	11	2	1		7	1	4
Dinophyceae	17	1	16	11			0	1	2
Chrysophyceae*	22	1	21	2	5		9		
Bacillariophyceae	39	21	18	11	3		4		
Cyanophyceae	10	9	1	1					
Rhodophyceae	4	0	4	4					
Totals	180	56	124	43	18		52	2	6
Totals for single vitamins				103	78	10			

* Two chrysomonads require B_{12}+biotin.

the species of Cyanophyceae were presumably able to synthesize the vitamins which they needed. Many of the species requiring vitamin B_{12} do not thrive on analogues of this vitamin.

Little is known about the concentrations of vitamins required by auxotrophs. Shortage of vitamin B_{12} appears to limit rates of photosynthesis at levels ranging from 0·1 to 50 ng/l according to species and growth conditions (Droop, 1961). Further reduction in the supply of the vitamin stops cell division and may cause loss of pigmentation and an increase in the size of the algal cells (Guillard and Cassie, 1963). In general, neritic species appear to require higher concentrations of the vitamins than oceanic types. The concentration levels of this vitamin in the surface waters of the sea vary both seasonally and geographically. In temperate latitudes the highest concentrations (5–10 ng/l) are found in coastal waters in winter. In spring the level falls drastically to about 1/10 of its previous value, presumably as a result of uptake by the plankton bloom (Droop, 1954; Cowey, 1956). The concentration of vitamin B_{12} in the euphotic zone in the more productive regions of the open oceans usually lies in the range 0·2–2 ng/l. However, it is much lower in barren parts of the oceans, such as the Sargasso Sea in which Menzel and Spaeth (1962) found less than 0·1 ng/l. It seems likely that the availability of this, and other essential compounds, may play an important role in controlling geographical and seasonal aspects of the phytoplankton succession. However, much more work is needed to decide whether this is the case.

There is considerable evidence for the existence of large healthy populations of photosynthetic organisms (mainly coccolithophores, but also flagellates and diatoms) at depths far below the euphotic zone in some parts of the ocean (Bernard, 1953, 1961; Kimball et al., 1963). It seems likely that these algae live heterotrophically on the dissolved organic compounds (0·5–1·0 mg C/l) present in the sea water (Wood, 1963a,b; Kimball et al., 1963). They are probably of considerable importance in the food chain in the deeper waters, particularly in the less fertile regions of the oceans where there is only a restricted supply of particulate matter from the surface.

VI. INFLUENCE OF GROWTH CONDITIONS ON CHEMICAL COMPOSITION OF PHYTOPLANKTON

Changes in the physical and chemical conditions of growth can considerably affect the chemical composition of phytoplankton. The influence of variations in light and temperature has been little studied. It is known that the chlorophyll content of the cells tends to decrease as the light intensity increases; at intensities more than ca 10 k.lux over saturation, photodecomposition of chlorophyll sets in (see also p. 239). In contrast, carotenoids are comparatively resistant to photochemical degradation. Decrease in growth temperature leads to the production of lipids containing relatively greater amounts of the more unsaturated fatty acids, mainly eicosapentaenoic and octadecatrienoic acids (P. R. Hinchcliffe, private communication).

The effect of changes in the micronutrient concentration of the medium on the composition of phytoplankton has been extensively studied. The phosphorus content of algal cells growing in a medium containing adequate nutrients is considerably greater than the minimum amount required by a viable cell, the excess amount being roughly proportional to the concentration in the medium. When the external nutrient concentration is reduced sufficiently to reduce the rate of photosynthesis, nutrient-deficient cells will be produced. If, as is normal in the sea, nitrogen is the limiting nutrient, the proportions of carbohydrate and lipid photosynthesized will increase relative to protein or chlorophyll. When the nutrients in the medium become exhausted each successive cell division will halve the amounts present in the parent cell. After three or four divisions the cell will abruptly cease to function because of the lack of crucial amounts of essential metabolites, such as enzymes, RNA and ATP. As a result, it will die or pass into a resting state or spore. Lack of trace elements, such as manganese and molybdenum, connected with the assimilation of nitrate can also give rise to symptoms of nitrogen deficiency. When nutrient-starved cells are transferred to a medium enriched with nitrogen (NO_3^- or NH_3) and phosphorus, these elements are taken up rapidly even in the dark. However, some time elapses before cell division recommences.

VII. PHYTOPLANKTON GROWTH AND DISSIPATION

A. MECHANISMS OF REPRODUCTION

Phytoplankton reproduce by a wide variety of mechanisms, which may be vegetative, asexual or sexual. Many species can employ more than one of these mechanisms according to circumstances. It is the vegetative process which occurs during the exponential growth phase, (see below) and its simplest form is seen in methods which do not directly involve division of the protoplast, but where the individual cells or cell aggregates are separated. Thus, the Dinophyceae generally divide diagonally along the longitudinal axis; although initially one daughter is smaller than the other, this difference in size does not persist for long. In contrast in the diatoms, one daughter is smaller than the other and scarcely grows at all, cell division being therefore accompanied by a progressive reduction in size. This process continues until either auxospore formation or death ensues. In asexual reproduction, the protoplast(s) which are released from the algal cells as either motile zoospores or non-motile aplanospores germinate to form the new plant. The number of zoospores released per cell varies widely from species to species, and each of them is usually a miniature edition of the flagellate vegetative cell of the parent. In sexual reproduction nuclear material and often cytoplasm from two cells of the same species unite. Although this process most commonly involves union of two morphologically identical gametes, other sexual mechanisms are also known.

B. KINETIC MEASUREMENTS ON PHYTOPLANKTON

For ecological purposes the kinetics of the growth of phytoplankton is usually considered in terms of changes in the number of living cells per unit time and volume. Cell counting and sizing is customarily carried out by microscopic examination, and this also enables the species distribution to be determined. In recent years electronic instruments, such as the Coulter Counter, have been widely adopted in oceanographic laboratories for the counting and sizing of particles in the sea and particularly in cultures (see, e.g. Sheldon and Parsons, 1967). However, although they are rapid in operation, they suffer from the disadvantage that they do not distinguish between living cells and detritus, or between species.

When considering growth kinetics from the standpoint of primary production it is of greater value to express growth in terms of the increase of the plant biomass. In practice the growth rate may be calculated directly from the change in plant biomass* with time, or by using a single measurement of the rate of primary production (see p. 232 ff.).

It should be pointed out that studies carried out on unialgal cultures subjected to alternating periods of light and dark have shown that for many species the division of all the cells in the culture tends to become synchronized. In such a synchronous culture the time of cell division will occur at a particular time of day for each species. Thus, division of *Dunaliella tertiolecta* takes place during the early hours of darkness (Eppley and Strickland, 1968). In addition, there is some indication that the composition of the cells may change throughout the day. Thus, during the light period, the doubling of the cell components of *Euglena gracilis* occurred consecutively in the order DNA, chlorophyll, carotenoids and finally weight; net synthesis of pigments ended in the final

---

* Although the biomass should ideally be measured by determining organic carbon, practical difficulties connected with the low cell densities and the presence of detrital carbon render this technique useless for natural populations. Instead, measurements of chlorophyll *a* are usually used (see p. 202 and UNESCO, 1966).

part of the light period (Edmunds, 1965). These observations can have important implications in kinetic studies in the sea, since if as seems probable, these rhythms also affect natural populations, measurements of cell numbers and chlorophyll are only truly comparable if made at the same hour each day. Certainly, diurnal variations have been observed in the chlorophyll content of the sea (see, e.g. Lorenzen, 1963).

C. GROWTH KINETICS

Experience suggests that, regardless of which reproductive mechanism is employed, the growth kinetics of the unicellular algae will closely

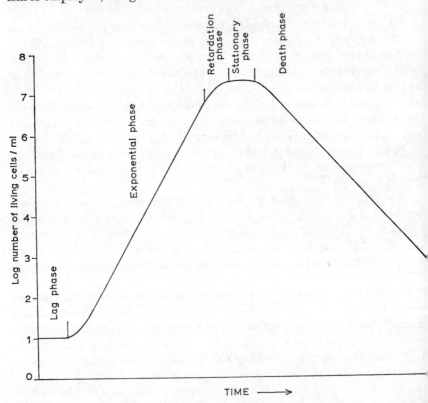

Fig. 9.8. Idealized phytoplankton growth curve.

resemble those of bacteria. The growth patterns of the latter have been very intensively studied, and mathematical models describing the response of the cells to variations in the conditions of their environment have been set up. The most successful of these is that developed by Hinshelwood (1946) who has interpreted bacterial growth in terms of a series of interacting enzyme systems separated by cellular membranes. Although this treatment can satisfactorily explain many of the phenomena observed in cultures, it is too complicated to be considered here, and the present discussion will be restricted to a purely qualitative account of the subject. For further information, the reader should consult Hinshelwood (1946, 1952), Strickland (1960) and Eppley and Strickland (1968).

Under carefully controlled conditions the growth pattern of a unialgal culture will generally resemble that shown in Fig. 9.8. In the sea, of course, the growth pattern will usually deviate widely from this ideal behaviour for a variety of reasons. These include not only the physical and chemical factors discussed in the previous sections, but also others which are biological (e.g. interaction of other species and predation; see p. 275), and hydrographic (e.g. horizontal and vertical water movements, and the stability of the water column).

For ease of discussion it is advantageous to divide the growth curve shown in Fig. 9.8 into five sections (i) the lag phase (ii) the exponential or logarithmic phase (iii) the retardation phase (iv) the stationary phase (v) the death phase; these will now be discussed individually.

(i) Lag phase

When a medium is inoculated with a culture of algal cells a variable and frequently lengthy period often elapses before vigorous growth takes place. This lag phase may originate from a lack of balance in the enzymatic reactions in the cells, brought about by the shortage of an essential growth promoting substance in the new medium. It will persist until this shortage is made up either by the organism itself, or by other organisms, e.g. bacteria, or alternatively, until the algal enzyme systems have adjusted themselves to cope with the reduced concentrations of ectocrine. Other factors may also contribute to the lag phase, for example, it may be necessary for the organism to deactivate a growth-inhibiting substance. The age and condition of the cells in the inoculum may be important. Thus, an apparent lag phase would result if only a small proportion of the cells in the inoculum were highly active and the remainder were more or less senescent or in a resting stage. In this instance, the lag phase would end when the active cells became an appreciable fraction of the population (see, e.g. Riley, 1943).

The combined effects of physical, chemical and hydrographic factors make it difficult to determine whether the lag phase occurs in the sea. However, the delayed blooming of phytoplankton in spring or summer, when growth conditions appear optimal, may well be a manifestation of this phenomenon (see Fogg, 1957).

(ii) Exponential phase

At the end of the lag phase, there ensues a period of growth in which the number of cells increases exponentially, viz

$$N = N_0 e^{kt} \quad \text{or} \quad kt = \log_e N/N_0$$

where N_0 and N are the numbers of cells initially and at time t, and k is a constant. The constant k, which has the dimensions of time^{-1}, depends on the species of organism, and also on environmental conditions, such as nutrient concentration, light intensity and temperature. Some authors prefer to express their results as the mean time elapsing between the formation of a cell and its division, this is termed the generation or doubling time, and is designated t_g where $t_g = \ln 2/k = 0.69/k$. Values reported for k for marine phytoplankton grown under near-optimal conditions in culture usually lie in the range 0.01—0.10 hours^{-1} (see Eppley and Strickland, 1968). Only a few attempts have been made to measure the generation times of the various species of algae in the sea, and these have usually given values in the range 8–48 hours ($k = 0.09$–0.015 hours^{-1}). Many of these measurements are open to doubt owing to the effects of predation and water movements; and even when the water is strained to remove predators and placed in a plastic bag errors will arise from attenuation of light by the bag (McAllister *et al.*, 1961). Using measurements of biomass and the rate of primary production McAllister *et al.* (1960) have arrived at a *mean* value for t_g of 4–5 days in the whole euphotic zone in the east Pacific Ocean. In fertile coastal waters at the time of the spring bloom this value may be less than 2 days (Eppley and Strickland, 1968). It seems likely that in temperate seas truly exponential blooming is restricted to the spring outburst because it is only at this time that all the various factors concerned are optimal. These include high and fairly uniform production, low levels of predation and the stability of the water column. During the spring bloom the number of cells may increase by 1,000 to 10,000 fold in somewhat less than a week.

(iii) Retardation phase

The period of exponential growth is of limited duration both under laboratory conditions and in the sea. Eventually the growth rate is reduced by a number of factors, which include:

(a) depletion of micro-nutrients, which appears to be the usual cause of the ending of the exponential phase in the sea (see also p. 242). When this is the limiting factor, the maximum cell population at the end of the exponential period will be proportional to the initial nutrient concentration. For an account of kinetic aspects of nutrient limitation the reader should consult Eppley and Strickland (1968).

(b) inhibition because of toxic or growth-inhibiting substances produced either by the cells themselves or by other organisms (see Chapter 8, Section II D).

(c) reduction in photosynthesis which occurs when the number of cells increases sufficiently to cause self-shading.

(iv) Stationary phase

The retardation phase leads into the stationary phase when, for example, supplies of nutrients are completely exhausted. During this period, which is only of very limited duration, there is no net increase in the number of living cells. However, this does not of necessity preclude metabolic activity, since it has been observed that nitrogen-starved diatoms can maintain themselves for a short time by photosynthesizing actively during the day and respiring the carbohydrate so formed during the following night (Ryther et al., 1958).

(v) Death phase

Unless the conditions during the stationary phase improve, most of the cells will soon die, the death rate being often a logarithmic function of time. It should be pointed out that there is also some cell mortality during the exponential phase of growth, but this is usually insignificant at this stage. Except under extremely drastic conditions, not all the cells die, since at all stages in the growth cycle a small proportion of them usually exists as a cyst or resting form. In these dormant forms the algae can survive for long periods under highly adverse conditions, e.g. under extremes of temperature and shortage of micro-nutrients. When circumstances again become favourable, those resting cells which have been brought into, or retained, in the euphotic zone by mixing will germinate and the cells will resume active growth. The resting forms are thus obviously of vital importance for the survival of the species in the sea.

D. DISSIPATION OF THE PHYTOPLANKTON BLOOM

In the sea, losses of the growing phytoplankton crop from the euphotic zone occur not only through natural mortality, but also from a number of other causes. These include:

(i) grazing, mainly by zooplankton which is probably the chief mechanism regulating the size of the phytoplankton population in spring and summer in mid and high latitudes (see, e.g. Harvey et al., 1935; Braarud, 1935; Wimpenny, 1936, 1938; Holmes,

1956). It seems likely that the dramatic reduction in the phytoplankton crop following the blooms in these seas is a result of
grazing by zooplankton rather than an effect of nutrient shortage
(Cushing, 1963, 1964). However, in highly productive waters
grazing may be insufficient to produce any appreciable effect on
the density of the phytoplankton crop (e.g. in Long Island
Sound; Riley, 1956, 1959). The effect of grazing on phytoplankton
growth dynamics has been described by Cushing (1959a) and will
be further discussed in Section IX.

(ii) sinking of the organisms below the photosynthetic zone. Practically all phytoplankton, except some flagellates tend to sink
when they have passed their highest metabolic efficiency.
Riley *et al.* (1949) have computed a mean value of ca 12 cm/hour
for the sinking velocity of a natural mixed population, but
particular populations may sink 5 or 10 times more rapidly or
slowly, according to the species involved and their metabolic
state (see p. 259).

(iii) the effect of eddy diffusion carrying phytoplankton cells below
the compensation point. In temperate and high latitudes this
process prevents the development of blooms during the winter.
However, it probably produces its maximum effect on the dissipation of phytoplankton blooms in the autumn when the
seasonal thermocline breaks down. Riley *et al.* (1949) have given
a mathematical treatment of the dissipation of phytoplankton
cells by turbulence.

(iv) the effect of advection transporting cells into and out of the area
under consideration.

(v) effects of toxins which may be produced by succeeding crops of
alga (see p. 192). There is considerable evidence that organisms
are immune to their own toxins.

VIII. Growth and Distribution of Phytoplankton in the Sea

A. PHYTOPLANKTON BLOOMS AND THE SUCCESSION OF SPECIES

The rapid proliferation or blooming of phytoplankton is a phenomenon mainly associated with coastal waters of temperate and higher
latitudes (but see pp. 245 and 268). In winter in these regions although
there is always some growth this is usually only slight. This arises partly
because of the lack of sufficient light and the absence of predators, but
also because turbulence resulting from the lack of stability of the upper
water column carries the phytoplankton cells below the euphotic zone
before they can divide. With the coming of spring the depth of the

euphotic zone (D_E) increases and the depth of the mixed layer (D_M) decreases as a result of the development of the seasonal thermocline. The latter tends to confine the algal cells to the euphotic zone which is now rich with nutrients as a result of winter mixing. If the necessary growth-promoting factors are also present, conditions are optimal for the proliferation of phytoplankton from "seed" stock which may be either the plankton cells themselves, or their resting stages. The initiation of the spring bloom has been studied mathematically by Sverdrup (1953), who has derived a useful formula for the prediction of blooms. This states that a bloom will occur when

$$D_M \ll [1 - \exp(-k \cdot D_M)] \cdot \frac{I_{\text{surf}}}{I_{\text{comp}} \, k}$$

Where k is the vertical extinction of blue-green light (420–520 nm) and

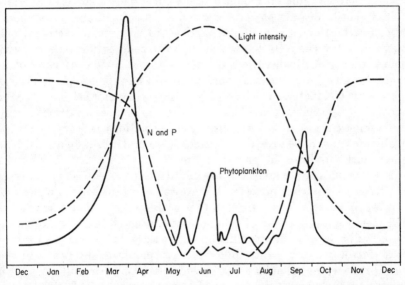

Fig. 9.9. Seasonal variation of phytoplankton, nutrients and light in a typical northern temperate sea (Raymont, 1963).

I_{surf} and I_{comp} are respectively the surface and compensation light intensities for light of this range of wavelengths. I_{surf} may be approximated as 20% of the total surface radiation (in ly min^{-1}); I_{comp} is frequently assumed to be 0·005 ly min^{-1} (see p. 239). Riley (1946) has shown that there is an inverse relationship between the rate of increase of phytoplankton in the Gulf of Maine and the depth of the mixed layer.

Year to year differences in the time of the commencement of the spring bloom may be, at least in part, attributable to variations in the time at which a stable stratification of the water column is attained. This in its turn depends on climatic factors, such as the rate of solar heating, and the amount of wind-created turbulence. In temperate regions, e.g. the North Sea, the onset of the bloom takes place in March. As higher latitudes are approached, the bloom generally occurs progressively later, except in regions where melting ice produces stable water conditions (see, e.g. Hart, 1934, 1942, for the Antarctic, and Zenkevitch, 1963, for the Arctic). At the highest latitudes there may be only one short-lived plankton bloom which takes place in mid-summer.

In many mid-latitude areas a succession of phytoplankton blooms occurs throughout the late spring, summer and early autumn (Fig. 9.9). The species succession from one year to another is not of necessity exactly the same in any locality. Little is known about the factors which cause phytoplankton to bloom or determine species succession. They will obviously include physical ones, such as light intensity, temperature, and the absence of too much vertical or horizontal advection which may transport the cells below the euphotic zone or carry them away before they can divide. Micro-nutrient and trace element concentrations, salinity, and the availability of growth-promoting substances may be crucial factors. The blooming of one species may set the stage for another by optimizing the concentrations of a micro-nutrient or by providing essential metabolites from their excretions or decay. Succeeding blooms may have to await the bacterial remineralization of sufficient micro-nutrients from the previous crop.

Even quite small changes in the physical and chemical parameters can have a significant effect on the growth constants of the algae (see p. 254). When it is borne in mind that a difference in doubling time of 25% between two fast-growing organisms can lead to one outnumbering the other by 15 to 1 in a week, it will be appreciated how large a bearing these factors have in controlling the phytoplankton succession in the sea. However, they should not be considered in isolation, but along with other biological and hydrographic factors, some of which will be discussed below.

Biological factors obviously play a key part in determining the time of a bloom. Before a bloom can commence, a sufficiently high concentration of seed stock of the particular organism must be present in the water. In many instances this must involve the germination of auxospores or resting stages. In off-shore waters these encysted forms of the organisms are probably present in the mixed layer throughout the winter, but it is likely that shallow waters are often seeded by spores stirred up from the sea floor by turbulence (Wood, 1959). Algal cells, existing heterotrophically in deeper waters (see p. 249), may be responsible for seeding in regions of upwelling.

Sinking of phytoplankton below the euphotic zone under the influence of turbulence, or through lack of buoyancy of the cells, can exert an important influence on the bloom and its eventual dissipation. Many species of phytoplankton, representing all the Phyla except the Cyanophyta, are motile because they possess flagella. This enables them both to overcome the gravitational force and through phototaxis to remain at an appropriate level in the euphotic zone. Other species which are not motile will eventually sink below the compensation point, although turbulence will tend to delay this process. Sinking rates vary considerably from one species of algae to another according to their size, shape, and density;* healthy diatoms probably sink at 1–2 m per day. The sinking rate also depends on the nutritional state of the organisms since nutrient-deficient cells because of their density may sink as much as ten times as fast as normal cells. If such cells sink into nutrient rich water they may recover their buoyancy (Steele and Yentsch, 1960). Differential sinking is probably an important factor controlling species succession in the sea. For a further account of the sinking of phytoplankton the review by Margalef (1961) should be consulted. Motility may also enable an organism to live at lower nutrient concentrations than non-motile forms since its motion will reduce the thickness of the diffusion layer around the cell and thus enable the motile cells to utilize nutrients more efficiently. This may account for the frequent predominance of flagellates in the plankton crops of nutrient-depleted waters, e.g. in the summer in temperate coastal waters and in the barren tropical oceans.

B. SPECIES DISTRIBUTION AND SEASONAL VARIATION OF PHYTOPLANKTON

A detailed account of the distribution and species succession of phytoplankton in the sea lies outside the scope of the present chapter and it is only possible to present a much simplified picture here; for further information the reader should consult the reviews by Raymont (1963) and Margalef (1958). Since various phytoplankton species have different physical and chemical requirements, each water mass will have its own characteristic species composition, succession and productivity. However, many common genera occur to some extent in almost all oceanic regions. Because of the higher content of micronutrients and organic matter, the species of organisms occurring in coastal waters

---

* Mechanisms involved in the control of the buoyancy include the presence of gas in the tissue, reduction of the concentrations of calcium and magnesium in the vacuole fluid until its density is less than that of the surrounding sea water, and formation of excess lipid (see Smayda, 1970).

differ considerably from those in ocean waters which will frequently not grow well in the chemical environment of shallow water. *Neritic* diatoms, e.g. *Coscinodiscus*, are often more strongly silicified than the *oceanic* species. So specific are the requirements of some algae that they can serve as indicators of both large and small scale water masses. Thus, *Thalassiosira antarctica* is characteristic of the cold (—1·8°C to 3·5°C) waters around the Antarctic continent; a related species *T. hyalina* is restricted to the cool waters of the Arctic seas. In contrast, *Planktoniella sol* characterizes tropical waters with a high temperature (mean 19°C) and high salinity (mean 35·7⁰/₀₀); and can be used as an indicator for Gulf Stream Water, in which it can be found as far north as 60° N. On a smaller scale, Braarud *et al.* (1953) have shown that the North Sea can be divided into 16 areas on a basis of their plankton assemblages. These areas are closely related to oceanographic features such as currents or eddies, or to particular coastal regions.

Phytoplankton are often not distributed uniformly throughout the euphotic zone, and at a given depth marked patchiness may occur even over quite small areas (see, e.g. Colebrook, 1969). These variations arise because growth of the algae is influenced in a complex fashion by the various physical and chemical gradients in the sea. These factors include temperature, salinity, light intensity, transparency of the water, and availability of nutrients. Their significance has already been discussed in Section V. In addition, the seasonal thermocline may limit the phytoplankton as a whole to the upper layers. Grazing by zooplankton may also contribute in large measure to the patchiness.

The species composition of phytoplankton in the sea varies both seasonally and geographically, and large differences in population may be found over quite short distances. In temperate coastal waters the first and largest blooms will consist of diatoms, with a similar pattern of genera, but not necessarily of species, from year to year. The sequence is often *Thalassiosira* spp., together with *Skeletonema costatum* and finally *Chaetoceros* spp. At the time of these blooms the seasonal thermocline will be established and act as a barrier against replenishment of micronutrients taken up by the bloom. The concentrations of these elements in the euphotic zone will therefore fall to low levels. Although much of the diatom bloom will be grazed by zooplankton, the subsequent regeneration of the micronutrients, particularly nitrogen, from the metabolic products of the zooplankton may be a fairly slow process, except when grazing is intensive (Cushing, 1963). Much will be lost to the euphotic zone when the faecal pellets fall below the thermocline. For this reason the nutrient level in the euphotic zone remains low throughout the summer and the phytoplankton crop is rather sparse.

However, minor blooms of dinoflagellates and chrysophycean flagellates do occur during this period since these organisms can obtain nitrogen and phosphorus auxotrophically and so are able to tolerate low nutrient levels. In the autumn, the mixing which occurs after the breakdown of the seasonal thermocline enriches the euphotic zone with micronutrients. This may result in an autumn diatom bloom which is usually only short-lived because of the decreasing light and increasing turbulence. At all times of year there is a relatively constant population of ultra-plankton, and in winter these organisms may make up a large proportion of the total biomass (see, e.g. Gross et al., 1947, 1950). Figure 9.10 shows diagrammatically seasonal variations in the species composition of the phytoplankton in the two basins of the Gulf of Maine and illustrates the quite wide differences which occur over relatively small areas (Lillick, 1940). Further south, in Long Island Sound, the earliest bloom consisted of *Thalassiosira nordenskioldii* and *S. costatum*, the latter being favoured as the temperature increased. Other species of diatoms and *Peridinium trochoideum* followed the main spring bloom. In May, *Guinardia* sp. and *Rhizosolenia fragillissima* bloomed and these were succeeded in summer by dinoflagellates, although diatoms were also present. The autumn bloom consists mainly of *Chaetoceros* sp. and other diatoms. Species successions have also been noted in waters of higher latitudes (see, e.g. Digby, 1953).

In tropical oceanic regions, variations in light and temperature are small, and the thermocline keeps the micronutrients in the euphotic zone at low levels throughout the year. As a result, blooming of phytoplankton occurs only rarely in these waters, and changes in species composition are slow. Much of the biomass consists of unidentified ultraplankton. The majority of the phytoplankton lies in the nano-plankton range and for much of the year consists mainly of coccolitho-phores and unidentified flagellates. The larger phytoplankton, which are mainly dinoflagellates, probably constitute only a small fraction of the total biomass. In sub-tropical seas, where there are strong enough winds and sufficient cooling in winter, the stratification of the upper layers may be broken down. The mixing which ensues may produce a small bloom such as that observed by Menzel and Ryther (1960, 1961a) in April in the Sargasso Sea, near Bermuda, but coccolithophores dominate the phytoplankton over most of the year.

C. THE STANDING CROP

Although the algal standing crop in the sea is extremely important from the point of view of productivity, its accurate determination is a matter of great difficulty. In much of the earlier work the cells retained by fine

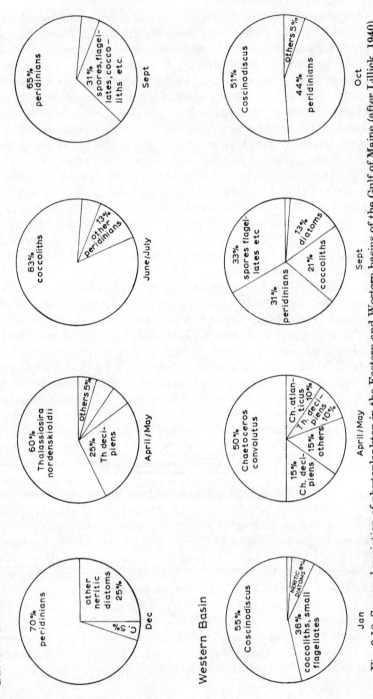

Fig. 9.10. Seasonal variation of phytoplankton in the Eastern and Western basins of the Gulf of Maine (after Lillick, 1940) (C = *Coscinodiscus*).

mesh silk nets were counted; this often led to a gross under-estimation since nanoplankton, because of their very small size, were not retained by the 40–50 μm meshes used. These organisms often constitute a highly significant and even dominant fraction of the phytoplankton crop, particularly in oligotrophic tropical waters (see, e.g. Hulburt, 1962; Wood, 1963a), in some of which they may contribute as much as 90% of the total photosynthetic activity (Teixeira, 1963). More recently the chlorophyll extraction method (Chapter 8 Section III B 2) has been widely adopted for the determination of the plant biomass. If, as is customary, the water sample is filtered through a 0·5 μm filter, nano-plankton and larger forms of plankton will be retained. However, owing partly to shortcomings in the photometric method (see p. 202) and to a great extent to the variability of the chlorophyll a : carbon ratio in the plankton, the results will only be of a moderate precision.

The standing crop is obviously closely related to the total production in the water column. However, its magnitude may frequently be less than that expected because of losses through grazing, mainly by zoo-plankton. Its concentration is very variable both geographically and seasonally. Expressed in terms of chlorophyll a values range from 10–40 mg/m^3 for blooms in fertile coastal waters to ca 0·05 mg/m^3 in barren tropical areas, with an average value for temperate oceans of 0·5 mg/m^3 (Strickland, 1960, 1965). The standing crop is often not distributed uniformly with depth, and the pattern may change from hour to hour where turbulence occurs. There is a tendency for sinking cells to be concentrated at the top of pycnoclines where they encounter water of increasing density. In temperate seas in spring and summer the maximum concentrations of phytoplankton often occur somewhat below the surface, probably because of the inhibiting effect of strong light on photosynthesis (Anderson, 1964; see also Fig. 9.5, p. 238). In tropical latitudes, the maximum concentration may occur near the bottom of the euphotic zone. Strickland (1965) has suggested values of 1–2 g C/m^2 for most oceanic areas, and 2–10 g C/m^2 in fertile coastal waters during the spring and summer. The highest of these values probably does not exceed one thousandth of the plant carbon/m^2 of fertile meadow.

D. PRIMARY PRODUCTION IN THE SEA

i) Introduction

When discussing primary production in the oceans, it is of more value to consider the production in unit time beneath unit area of sea surface (usually 1 m^2) than in unit volume. Because of the nature of the relationship between photosynthesis and light intensity (Fig. 9.4), and the way in which light of different wavelengths is attenuated in the seas, the theoretical maximum production rate beneath unit area of sea surface is not a linear function of surface light intensity but varies as shown in Fig. 9.11 (Ryther, 1959). The theoretical maximum production rate is ca 25 g C/m²/day with 600 ly/day, which is about the highest daily energy recorded in the tropics. Primary production values approaching the theoretical ones have been obtained with cultures of algae grown under optimum conditions, and values of 20 g C/m²/day have been found for beds of large attached algae. However, even in

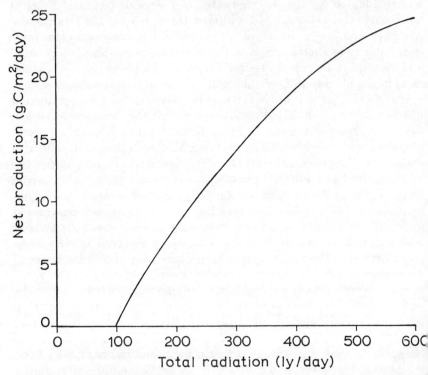

Fig. 9.11. Theoretical maximum primary production as a function of total daily solar radiation (after Ryther, 1959).

fertile seas the average net production by phytoplankton is probably only 0·5–1 g C/m²/day, although for short periods values of 5 g C/m²/day may be attained. In the barren parts of the oceans the net production will rarely reach 0·2 g C/m²/day. In spite of the great difference in the sizes of the standing crop these values are quite similar to those for the more productive land crops (1–2 g C/m²/day for wheat; 2–3 g C/m²/day for a young pine forest, see e.g. Westlake, 1963). Since the area of productive land is only about one tenth of the area of the sea it will be appreciated that the total annual production of the sea may be several times greater than that of the land which is ca 10^{10} tons C/year (see Riley, 1944; Ryther, 1959).

(ii) Variation of production with depth

The variation of primary production with depth is determined by a number of factors: (i) the intensity of light at the surface, (ii) the manner in which light is attenuated, not only by the water itself but also by the phytoplankton cells and by suspended matter, both organic and inorganic (it should be remembered that red and blue light are absorbed more readily than green), (iii) the nature of the relationship between photosynthesis and light intensity (Fig. 9.4). Typical curves showing the integrated daily photosynthesis as a function of depth in the summer and winter in clear temperate waters are shown in Fig. 9.12. It will be observed that in winter because of the low light intensities the compensation point lies very close to the surface and net production can only be significant in the surface layer; optimum production will occur if the plant cells are uniformly dispersed down the euphotic zone. Development of even a small winter surface bloom is usually inhibited by the instability of the water column which causes a high proportion of the dividing cells to be carried below the euphotic zone. In contrast, in summer the higher light intensity causes the rate of photosynthesis to be greater and the euphotic zone to be thicker. At this time of year the zone of maximum productivity lies some distance below the surface probably because of the inhibiting effect of strong light on photosynthesis (see p. 239). For this reason, production/m² will be greater if the algal cells are concentrated some distance below the surface rather than uniformly distributed throughout the euphotic layer. As the phytoplankton bloom develops the depth of the euphotic zone decreases as a result of self-shading, and the efficiency with which the light is used increases. Steeman Nielsen and Jensen (1957) have noted that at high cell densities the thickness of the euphotic layer tends to vary inversely with the rate of production per unit area rather than in direct proportion to it.

(iii) Geographical variation of primary production in the oceans

The fertility of the sea varies considerably geographically. However, the variation in production/unit area is not as great as perhaps might be anticipated since, in fertile areas, the depth of the euphotic zone is much reduced by the shading effect of the dense phytoplankton crop. Figure 9.13 gives a diagrammatic representation of the annual rate of primary production in the oceans, and in it a number of regions of high productivity can be recognized:

(1) Production, and also the standing crop, tend to be higher in in-shore areas and over submarine banks than in the open ocean. This is generally attributed to the more rapid replenishment of nutrients

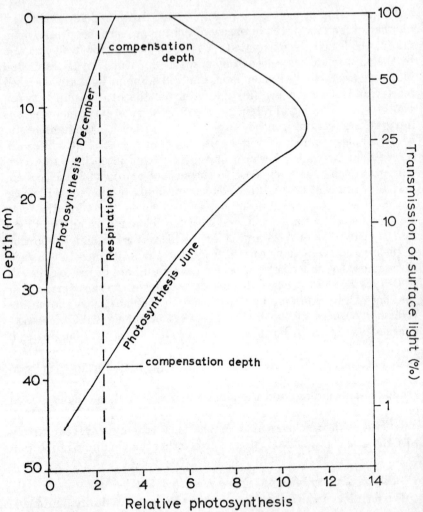

Fig. 9.12. Integrated daily rate of primary production as a function of depth (after Ryther, 1956b).

because of the greater turbulence in these regions. However, instances are known where coastal waters (e.g. those around Antarctica) display superior fertility even when adjacent oceanic waters are equally rich

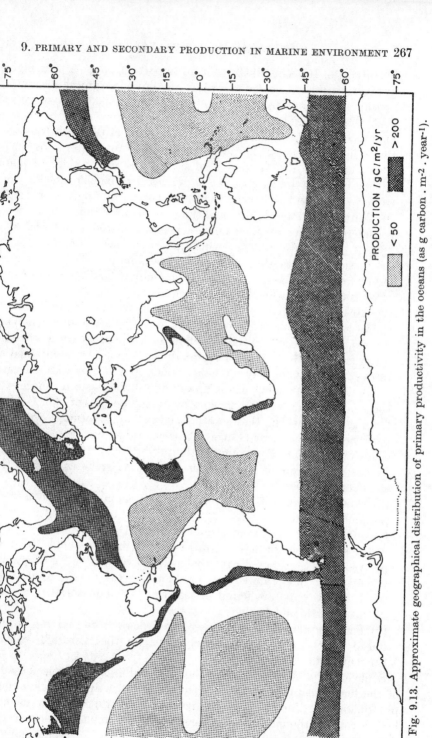

Fig. 9.13. Approximate geographical distribution of primary productivity in the oceans (as g carbon . m⁻² . year⁻¹).

in nutrients. It seems likely that an important cause of this higher productivity may be trace metals or organic growth-promoting factors carried out from the land. Strickland (1965) has suggested that the increased production in coastal waters may be caused mainly by an increase in the growth constant of the mixed population. This appears to be borne out by results of McAllister *et al.* (1960) who found that the rates of productivity per unit weight of plant carbon in the standing crop (actually measured as chlorophyll) were similar in nutrient-rich samples of both coastal and open water from the North East Pacific, although the sizes of the standing crops were considerably different. A phenomenon closely related to this *land mass effect* is the *island mass effect* which manifests itself as an increase in production and standing crop around oceanic islands. Thus, Jones (1962) found an increase in the plankton crop extending to a distance of ca 100 km from the Marquesas Islands.

(2) Areas of very high productivity occur on the continental shelf off the western margins of the continents (e.g. off Peru and South-West Africa), as the result of the upwelling of nutrient-rich intermediate water (see p. 243). In these areas productivity may be maintained at rates of 1·5–2 g $C/m^2/day$ for long periods. Because of seasonal variations in the prevailing winds upwelling off the eastern margins of the continents is much more spasmodic and periods of high and low fertility alternate in these waters. Upwelling associated with the passage of strong currents over submarine ridges can lead to regions of high production. For example, the productivity over the Iceland-Faroes Ridge can be as much as 5 times as great as that in the surrounding ocean.

(3) Slow upwelling of nutrient-rich water, combined with a very pronounced land mass effect over the continental shelf of Antarctica make the Antarctic Ocean probably one of the most fertile regions in the sea. Although active growth only occurs for about 100 days per year, the production over that period averages ca 1 g $C/m^2/day$.

(4) Areas of moderately high productivity (ca 0·3 g $C/m^2/day$ for much of the year) are found in the regions of equatorial upwelling in the Atlantic and Pacific.

(5) Fertility may be trebled in regions where strong mixing is produced by the meeting of currents, e.g. in the Kuroshio-Oyoshio Current region off Japan.

Some areas of the sea may have unexpectedly low fertilities because of the long generation times and thus, slow growth of the dominant phytoplankton. For example, the production of the Mediterranean is low, because although adequate amounts of micronutrients are available, the predominant phytoplankton consists of coccolithophores which have very much slower multiplication rates than diatoms (Bernard, 1958b).

(iv) Mathematical models of production

Several workers have tried to examine production in the sea in terms of mathematical models. Riley (1946, 1963) believes that it is possible to account for most of the phytoplankton variations in the water column in some areas in terms of illumination, temperature, depth, micro-nutrients and zooplankton grazing. In his model, the rate of productivity $(dP/dt) = P(P_h - R - G)$ where P is the total phytoplankton population per unit area, P_h is the photosynthetic coefficient, R is the coefficient of phytoplankton respiration and G is the coefficient of zooplankton grazing. He considers these values to be ecological variables rather than true constants. P_h is a function of light intensity (which, it will be remembered, decreases logarithmically with depth), the degree of micronutrient concentration, and the amount by which the depth of the euphotic zone exceeds that of the mixed layer. Taking these various factors into account he arrived at the equation

$$P_h = pI_0(1 - e^{-kz}NV)/kz$$

where p is a photosynthetic constant, found empirically to be ca 2·5 if the light intensity is expressed as $cal/cm^2/min$

I_0 is the incident light intensity

k is the extinction coefficient

z is the depth of the euphotic zone (assumed by him to be the depth at which the light intensity $= 0.0015$ g $cal/cm^2/min$)

N is a factor compensating for the effect of nutrient depletion

V is a factor compensating for loss of phytoplankton resulting from turbulence if the mixed layer is shallower than the euphotic zone.

The respiration rate R of the phytoplankton is a function of temperature. The grazing rate G, is assumed to be given by $G = gZ$, where g is the rate of consumption of phytoplankton per unit of herbivorous zooplankton (see p. 277), and Z is the amount of the latter present in the water column. Riley (loc. cit.) found satisfactory agreement between the experimentally determined production and that predicted by this comparatively simple model for seasonal variations in production in the Gulf of Maine and in the waters off Husan, Korea.

Other workers (e.g. Steele, 1956; Steele and Baird, 1962; Cushing, 1959a, 1963) have devised more complex models in which allowance is made for the effect of other variables, such as consumption of herbivores by zooplanktonic carnivores, or the replenishment of the euphotic zone with micronutrients either by regeneration or by eddy diffusion from deeper water. However, all these models are only of limited value

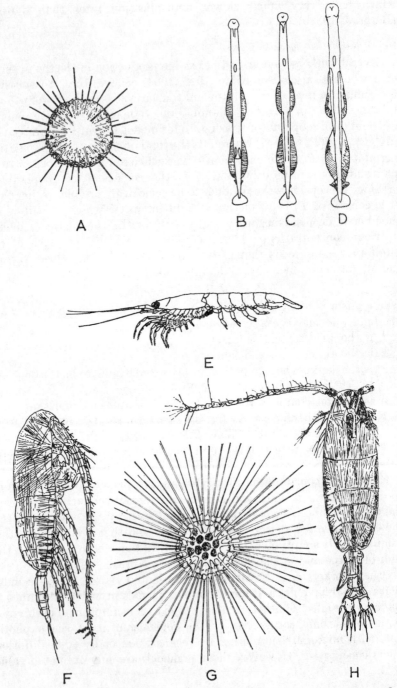

Fig. 9.14. Some typical zooplankton. A, a radiolarian (*Aulacantha scolymantha*); B, *Sagitta setosa*; C, *S. elegans*; D, *S. seratodentata*; E, *Euphausia superba*; F, *Calanus finmarchicus*; G, *Globigerina bulloides*; H, *Acartia discandata*.

since in most areas of the sea the physical and chemical conditions are extremely complex and because little is known of the factors affecting grazing and the rates of change of the zooplankton population.

IX. ZOOPLANKTON AS SECONDARY PRODUCERS

A. INTRODUCTION

In contrast to the terrestrial food chain, several steps usually occur between the plant plankton and the larger marine invertebrates. Phytoplankton serves as a food source for a floating population of small animals—the herbivorous zooplankton. This in its turn provides nutriment for carnivorous zooplankton and other animals. The higher stages of the food chain are frequently of considerable complexity; thus, carnivorous species of zooplankton may feed on other carnivorous species (e.g. sagittae serve as food for medusae and fish larvae). In many instances the specific food requirements of particular species of zooplankton are not known because, owing to the rapid digestive processes of these organisms, the guts of captured specimens usually prove to be devoid of identifiable food fragments. A full discussion of the higher levels of the food chain is clearly beyond the scope of the present chapter and for further details of this subject the reader should consult standard textbooks of marine biology.

Unlike the phytoplankton, the zooplankton are drawn from almost every phylum. Some of the species referred to in the text are illustrated in Fig. 9.14. Although many species are holoplankton and remain in the plankton throughout their existence, many other species (meroplankton) only exist in the plankton for part of their life cycle. The latter include the eggs and larvae of most fish, and the larval stages of echinoderms and bottom-living invertebrates. With a few exceptions, zooplankton are small in size, most being less than a centimetre in length and the majority not exceeding 2–3 mm. Almost all are capable of swimming by means of special appendages such as limbs, cilia, or flagella, or by muscular contraction. Since the majority of them are denser than their environment they must expend considerable energy swimming to maintain their level in the water (Vlymen, 1970). Many species are capable of swimming actively after their prey. The study of zooplankton is made difficult by our present inability to breed more than a very few species completely in the laboratory. Furthermore, there is evidence that results obtained under such conditions are often inapplicable to unconfined zooplankton in their natural environment. Thus, it has been found that the standard respiratory rate of laboratory specimens of *Calanus* may be only about 1/35 of the respiratory rate of the organism during its diurnal vertical migration in the sea (Petipa, 1966).

Zooplankton may be generally classified as herbivores or carnivores. However, some species are omnivorous and change their diet according to the type of food available, for example, *Calanus* will take animal food in times of winter scarcity. Among the carnivores are medusae, sagittae, ctenophores and larval stages of many fish and decapods. With the exception of fish larvae, which appear to feed mainly on copepods, little is known of the food organisms used by carnivores. The herbivorous group includes many copepods, euphausids and some other crustaceans, and the larvae of echinoderms and many bottom invertebrates. Of these the copepods (Crustacea) are probably the most important and have been most widely investigated. Although carnivorous copepods exist, particularly among the deep-living forms, the majority of these animals feed mainly on phytoplankton, especially diatoms. This is particularly the case in the Antarctic where the diatom-rich waters provide abundant food for euphausids and other copepods.

Our knowledge of the dietary requirements of even the commonest species of planktonic herbivores is very limited. It appears probable that many of these animals are quite selective in their diets. This may sometimes result from the inability of the organism to filter out the smaller species of phytoplankton, or to swallow the larger species.* However, size and shape are not the only factors involved. Many species of zooplankton will reject certain species or classes of phytoplankton; thus, *Calanus*, which can utilize a wide variety of diatoms and dinoflagellates as food will not take *Ceratium* (Marshall and Orr, 1955a). Walne (1963) has found that *Isochrysis galbana* and *Monochrysis lutheri* and low cell densities of *Phaeodactylum tricornutum* and *Dunaliella tertiolecta* are excellent foods for oyster larvae, whereas *Hemiselmis virescens* is only poor and *Chlorella* is practically useless. Work by Gibor (1956) has shown that there are considerable differences in food value between different species of phytoplankton, and studies by Shiriaishi and Provasoli (1959) and Walne (1956, 1963) suggest that for optimum growth it may be necessary for the diet to contain several species of algae. The trophic relationships of the zooplankton are further complicated by the fact that their various stages may differ considerably in dietary requirements. For further information on the complex subject of zooplankton feeding the reader should consult the excellent reviews by Raymont (1963, 1966).

---

* For example, Marshall and Orr (1952) have shown that the distance between the filtering setules of the copepod *Calanus helgolandicus* is 2–3 μm, which suggests that ultraplankton such as *Nannochloris sp.* will not be retained. In contrast, *Ditylum brightwelli* and *Coscinodiscus* appear to be too large to be ingested by young stages of *Calanus* (Marshall and Orr, 1958).

Phytoplankton cells usually enter the zooplankton digestive tract whole, and if they are to be digested at all, it is obviously essential that the gut should contain enzymes capable of breaking down the cell walls. Many species of zooplankton appear to be deficient in cellulase, although they contain amylase, and this may result in the cells of some organisms which have membranes of cellulose (e.g. Chlorophyceae and Dinophyceae) passing apparently unchanged through the gut.

When considering the marine food chain it is obviously of great importance to know the efficiency with which organic matter or biological energy is transferred from one trophic level to another. The efficiency can be expressed in a number of ways.*

(1) Assimilation efficiency $\left(= \frac{\text{assimilation}}{\text{ingestion}} \times 100\right)$ where assimilation = (ingestion—egestion). Maximum assimilation efficiencies of 40–100% have been found for zooplankton feeding on phytoplankton, under optimal conditions (see e.g. Marshall and Orr, 1955b; Monakov and Sorokin, 1961; Lasker, 1960). However, changes in the metabolic state of the zooplankton and in the environmental conditions can greatly influence the efficiency by affecting the proportion of the ingested food which is excreted or voided. For this season, the mean daily efficiency of an organism over its whole life span in the sea may be as low as 15%, or even less, although Steele (1965) has reported an average efficiency as high as 20–25% for zooplankton of the North Sea. Even higher efficiencies (up to 100%) have been reported for oligotrophic tropical and sub-tropical seas (Menzel and Ryther, 1961c; Grice and Hart, 1962).

(2) Growth efficiency, since it is a measure of the production of new tissue, is probably of greater relevance to marine production than assimilation efficiency. It may be expressed as either

$$gross \left(\frac{\text{growth}}{\text{ingestion}} \times 100\right) \text{ or } net \left(\frac{\text{growth}}{\text{assimilation}} \times 100\right) \text{ efficiency.}$$

Values for the net efficiency ranging from 5 to 90% are to be found in the literature, and even with the same species, it may vary 5 or more fold depending on the metabolic state of the animal and on the type and amount of food available.

B. ESTIMATION OF THE STANDING CROP OF ZOOPLANKTON

The quantitative estimation of zooplankton biomass and production is a difficult task, and at present there is no simple and reliable *in situ*

---

* The factors used in the efficiency calculations are usually expressed in terms of energy units, but other units (e.g. weights of carbon or nitrogen) have been used by some workers.

method which will give even an approximate estimate of zooplankton production. Most methods of estimating of the zooplankton crop involve trawling with fine mesh nets, followed by measurement of the displacement volume or dry weight of the unsorted catch. With such techniques much phytoplankton and detritus are also collected, and unless a time-consuming microscopical examination of the catch is made it is not possible to estimate the zooplankton volume accurately, or to distinguish between herbivorous and carnivorous forms. A further shortcoming of these procedures is that no single sampler will collect the whole range of zooplankton sizes owing to physical limitations of the nets. Thus, the fine nets necessary to retain the smaller forms filter slowly and clog before it is possible to filter a large enough volume to give a representative sample of the larger and more active zooplankton. However, even these nets are not sufficiently fine to catch the smallest protozoans. There is some prospect that it may eventually prove possible to estimate the zooplankton crop by chemical means; for example, Sutcliffe (1965) has used ribonucleic acid determinations for this purpose with some success (see also p. 203).

C. GEOGRAPHICAL DISTRIBUTION AND SEASONAL VARIATIONS IN THE ZOO-PLANKTON CROP

Over 30 years ago Hentschel (1933–36) working aboard the *Meteor* in the South Atlantic observed that there was a close correlation between the standing crops of phytoplankton and zooplankton. These observation have been confirmed by later workers who have pointed out that the zooplankton crop tends to be most abundant where primary production is high. This is very clearly demonstrated by the remarkable correlation found by Reid (1962) for the Pacific between the zooplankton biomass and the phosphate concentration of the surface waters which is a rough index of the primary production. The zooplankton crop is comparatively sparse in tropical and sub-tropical oceans ($0 \cdot 02$–$0 \cdot 07$ cm^3/m^3), but steadily becomes more abundant as high latitudes are approached (see, e.g. Foxton, 1956); values of $0 \cdot 5$–$1 \cdot 0$ cm^3/m^3 are found for the boreal northern Atlantic and Pacific. However, there is evidence that in the extreme north and south the zooplankton biomass is much reduced (see, e.g. Zenkevitch, 1963; Johnson, 1963). The zooplankton crop is usually greater in neritic waters than in the adjacent ocean waters at all latitudes, and even higher in associated coastal waters. Rich crops are found in regions where upwelling leads to high primary production. In any particular area there is a considerable tendency for the zooplankton to be distributed patchily, both horizontally and vertically. Raymont (1966) has summarized zooplankton biomass

lata for various oceanic and coastal waters. In addition to areal variations the zooplankton biomass shows considerable change with depth.
Frequently there is a rich layer lying close to the surface and another
at a depth of several hundred metres. However, the distribution may be
much influenced by the vertical diurnal or seasonal migration of the
animals.

Although the zooplankton biomass in tropical and subtropical waters
usually appears to change only slightly throughout the year, very
considerable seasonal variations are found at mid and high latitudes.
These are closely related to the pattern of primary production. Thus, in
the English Channel maxima occur in the spring and early summer, and
sometimes in the autumn, after which there is a rapid decline to a
minimum in January (Harvey et al., 1935).

It should be remembered that in almost all instances the biomass will
be composed of a considerable number of different species, the types
and proportions of which will vary greatly from place to place and often
seasonally. The variety will be greatest in tropical waters and least in
polar waters, where one species, such as *Euphausia superba* (Antarctic)
or *Calanus finmarchicus* (Arctic), may be dominant. Similarly, neritic
and coastal waters will often be characterized by zooplankton populations quite different from those of the adjoining oceanic waters.
Although some zooplankton species may be quite widespread over a
large area, others may be confined to particular water masses. These
indicator species can be of considerable value in studying water movements. Thus, *Sagitta setosa, S. elegans* and *S. serratodentata* (Fig 9.14) have
been extensively used in investigating the water flow pattern around the
British Isles (Russell, 1935; Fraser, 1939, 1952; Southward, 1961, 1962).
Although water temperature and salinity may obviously be significant
factors in determining the domain of any particular zooplankton
indicator species, other more indefinable chemical and biological factors
are probably of greater importance (cf. "good water", p. 191).

D. GRAZING BY ZOOPLANKTON

Zooplankton grazing plays an extremely important part in controlling
the size of the standing crop of phytoplankton in the sea and in the
regeneration of micronutrients. In the fertile regions of the open ocean
the algal standing crop is quite small and relatively constant throughout
the year, and distinct spring blooms only occur in a few regions (e.g. in
some areas of the North Atlantic). In mid and higher latitudes the
removal of phytoplankton cells from the euphotic zone by turbulence
is obviously a major factor in maintaining the standing crop at a low
level, particularly during the winter. However, sufficient zooplankton

survive the winter to keep the numbers of phytoplankton cells in check when light conditions and the stability of the water column become temporarily favourable for their rapid multiplication. During much of the year in these regions a steady state will exist in which primary production in the euphotic zone is balanced by zooplankton grazing in the entire water column.* Strickland (1965) has pointed out that this implies that the standing crop of zooplankton beneath unit area of sea surface is several times larger than that of phytoplankton. In the unproductive oligotrophic waters of the tropical and sub-tropical oceans the standing crop of phytoplankton is small, and since the various species of zooplankton breed at different times of the year, there is a fairly constant grazing population which practically inhibits blooming. Strickland (1965) has likened the productivity of these waters to that of an under-fertilized and spent plot of land, in contrast to the oceans of mid and high latitudes which he has compared with fertile but over-grazed meadow.

In coastal and high-latitude ocean waters there is generally a dearth of herbivores in the upper layers by the end of the winter. Hence, in spring when the hydrographic and light conditions become favourable for the proliferation of phytoplankton, there are insufficient grazers to hold the plant population in check, with the result that a plankton bloom ensues. The magnitude of the bloom is limited by the availability of micro-nutrients and by grazing by the zooplankton crop which increases rapidly after the onset of the bloom. When exhaustion of micro-nutrients begins to limit production, grazing becomes the main factor determining the size of the phytoplankton standing crop, and usually soon causes it to decline drastically. The zooplankton frequently destroy far more plant cells than they need (see, e.g. Cushing, 1964 and p. 212), but these cells when excreted may be used efficiently by other members of the food chain. Mineralization of the excretory products of the zooplankton and detritus proceeds fairly rapidly, and to a limited extent replenishes the euphotic zone with micro-nutrients. It should be remembered that a significant proportion of the particulate faecal material will fall below the euphotic zone before mineralization can occur and may serve as food sources in the intermediate waters. If carnivores, or natural mortality have by this time reduced the population of herbivores sufficiently, conditions will then be favourable for the occurrence of a further phytoplankton bloom. Zooplankton grazing is probably at least in part responsible for the marked patchiness of the phytoplankton crop (see p. 260 and Bainbridge, 1957; Colebrook, 1969).

---

* Although some plant cells will also be lost through death and sinking their number relative to the total production are not likely to be significant.

E. QUANTITATIVE ASPECTS OF ZOOPLANKTON GRAZING

Zooplankton are mainly filter feeders, trapping the suspended particulate matter which they use as their food with the setules of their maxillae. Some species are also able to feed by sweeping their maxillae through the water like nets (Anraku and Omori, 1963). It has been suggested by several workers that herbivores are not just automatic filterers but actively hunt their food. Cushing (1964) considers that the animals eat the cells which they encounter as they swim through the water. Probably all these processes play some part in the feeding of zooplankton in the sea.

It is believed that active reproduction of herbivorous copepods such as *Calanus* is restricted to times when food is abundant and superfluous feeding occurs (Cushing, 1964). The production of eggs is a function of food concentration (Marshall and Orr, 1964) and it ceases when food is scarce. During the period of reproduction the diet of many species of zooplankton must be augmented with supplies of animal food. Only a few measurements of the filtration rates of zooplankton have been made. Raymont (1966) suggests a mean rate of 70 ml/day for *Calanus*. It seems likely that most coastal waters at the time of the spring bloom contain a great excess of plant cells over that which is actually needed by the zooplankton population for its metabolic purposes. However, it is uncertain whether enough would be present in winter to satisfy its needs, and Cushing's (1959a) results suggest that *Calanus* is undernourished during the winter.

The determination of the rate of zooplankton production in the sea is fraught with many difficulties. Mullin (1969) has reviewed several of the procedures which have been used for this purpose and has pointed out the many uncertainties involved in them. In general the number of unverifiable assumptions and over-simplifications which it is necessary to make, added to our lack of knowledge of the biology and physiology of the zooplankton population, makes these estimates of somewhat dubious value. The published values for zooplankton production have been tabulated by Mullin (1969). They range from ca 50–200 mg C/m² per day in the water column in productive regions, such as George's Bank, Gulf of Maine and the North Sea, to < 10 mg C/m² per day in unproductive water.

REFERENCES

Anderson, G. C. (1964). *Limnol. Oceanogr.* 9, 284.
Anraku, M. and Omori, M. (1963). *Limnol. Oceanogr.* 8, 116.
Ansell, A. D., Raymont, J. E. G., Lander, K. F., Crowley, E. and Shackley, P. (1963). *Limnol. Oceanogr.* 8, 184.

Ansell, A. D., Coughlan, J., Lander, K. F. and Loosmore, F. A. (1964). *Limnol Oceanogr.* **9**, 184.

Antia, N. J., McAllister, C. D., Parsons, T. R., Stephens, K. and Strickland, J. D. H. (1963). *Limnol. Oceanogr.* **8**, 166.

Bainbridge, R. (1957). *Biol. Rev.* **32**, 91.

Belser, W. L. (1963). *In* "The Sea" (M. N. Hill, ed.), Vol. 2. Interscience, New York.

Bernard, F. (1953). *Deep Sea Res.* **1**, 34.

Bernard, F. (1958a). *C.R. Acad. Sci., Paris.* **247**, 2045.

Bernard, F. (1958b). *Rapp. Proc.-Verb. Cons. Int. Explor. Mer.* **144**, 103.

Bernard, F. (1961). *In* "Marine Microbiology" (C. H. Oppenheimer, ed.). Thomas, Springfield, Illinois.

Blinks, L. R. (1954). *In* "Autotrophic Micro-organisms", (B. A. Fry and J. L. Peel, eds.). Cambridge University Press, London.

Braarud, T. (1935). *Hval. Skr. Nr.* **10**, 1.

Braarud, T. (1961). *In* "Oceanography" (M. Sears, ed.), Publ. No. 67, AAAS Washington D.C. p. 271.

Braarud, T. (1962). *J. Oceanogr. Soc. Japan, 20th Anniv. Vol.* 628.

Braarud, T., Gaarder, K. R., and Grøntved, J. (1953). *Rapp. Cons. Perm. Int. Explor. Mer.* **133**, 1.

Brown, A. H. (1953). *Amer. J. Bot.* **40**, 719.

Brown, A. H., and Weis, D. (1959). *Plant Physiol.* **34**, 224.

Calvin, M., and Bassham, J. A. (1962). "The Photosynthesis of Carbon Compounds." Benjamin, New York.

Christensen, T. (1964). *In* "Algae and Man" (D. F. Jackson, ed.). Plenum Press, New York.

Colebrook, J. M. (1969). *Progr. Oceanogr.* **5**, 115.

Conover, S. A. M. (1956). *Bull. Bingham. oceanogr. Coll.* **15**, 62.

Conover, R. J. (1968). *Am. Zool.* **8**, 107.

Cooper, L. H. N. (1933). *J. mar. biol. Ass. U.K.* **18**, 729.

Cooper, L. H. N. (1957). Paper presented at Symposium of the International Council for the Exploration of the Sea, Bergen 1957. Preprint B/No. 24.

Cowey, C. B. (1956). *J. mar. biol. Ass. U.K.* **35**, 609.

Curl, H. and McLeod, G. C. (1961). *J. mar. Res.* **19**, 70.

Currie, R. (1962). *Proc. roy. Soc.* A **265**, 341.

Cushing, D. H. (1959a). *Fish. Invest. London, Series II.* **22**, 1.

Cushing, D. H. (1959b). *J. Cons. int. Explor. Mer.* **24**, 455.

Cushing, D. H. (1963). *J. mar. biol. Ass. U.K.* **43**, 387.

Cushing, D. H. (1964). *In* "Grazing in Terrestrial and Marine Environments" (D. J. Crisp, ed.). Blackwell, Oxford.

Davis, H. C. (1963). *Arch. Hydrobiol. (Plankt.).* **59**, 145.

Digby, P. S. B. (1953). *J. Anim. Ecol.* **22**, 289.

Droop, M. R. (1954). *Nature, Lond.* **174**, 520.

Droop, M. R. (1961). *J. mar. biol. Ass. U.K.* **41**, 69.

Edmunds, L. N. (1965). *J. cell. comp. Physiol.* **66**, 159.

Eppley, R. W. and Strickland, J. D. H. (1968). *Adv. Microbiol. Sea,* **1**, 23.

Fogg, G. E. (1957). "The Metabolism of Algae". Methuen, London.

Fogg, G. E. (1963). *Brit. Phycol. Bull.* **2**, 195.

Foxton, P. (1956). *Discovery Repts.* **28**, 195.

Fraser, J. H. (1939). *J. Cons. int. Explor. Mer.,* **14**, 25.

Fraser, J. H. (1952). *Mar. Res. Scot.*, 1952, No. 2, 1–52.
Fraser, J. H. (1955). *Mar. Res. Scot.*, 1955, No. 1, 1–12.
Fraser, J. H. (1962). "Nature Adrift." Foulis, London.
Fritsch, F. E. (1935) (1945). "The Structure and Reproduction of the Algae."
    Vol. I, 1935; Vol. II, 1945. Cambridge University Press, London.
Gibor, A. (1956). *Biol. Bull. Woods Hole.* 111, 230.
Goldman, C. R. (1965). "Primary Productivity in Aquatic Environments" (C. R.
    Goldman, ed.) *Mem. Ist. Ital. Idrobiol.* Vol. 18 Suppl.
Grice, G. D. and Hart, A. D. (1962). *Ecol. Monogr.* 32, 287.
Gross, F., Marshall, S. M., Orr, A. P. and Raymont, J. E. G. (1947). *Proc. Roy.
    Soc. Edin.* 63,(B), 1.
Gross, F., Nutman, S. R., Gauld, D. T. and Raymont, J. E. G. (1950). *Proc. Roy.
    Soc. Edin.* 64(B), 1.
Guillard, R. R. L. and Cassie, V. (1963). *Limnol. Oceanogr.* 8, 161.
Harris, E. (1959). *Bull. Bingham. Oceanogr. Coll.* 17, 31.
Hart, T. J. (1934). Discovery Repts 8, 1.
Hart, T. J. (1942). Discovery Repts 21, 263.
Harvey, H. W. (1955). "The Chemistry and Fertility of Sea Water", 224 pp.
    Cambridge University Press, London.
Harvey, H. W., Cooper, L. H. N., Lebour, M. V., Russell, F. S. (1935). *J. mar.
    biol. Ass. U.K.* 20, 407.
Haxo, F. T. (1960). *In* "Comparative Biochemistry of Photoreactive Systems"
    (M. B. Allen, ed.), p. 339. Academic Press, New York.
Haxo, F. T. and Blinks, L. R. (1950). *J. gen. Physiol.* 33, 389.
Hendey, N. I. (1964). *Fish. Invest. Lond.* Ser. 4, 317 pp.
Hentschel, E. (1933–36). *Wiss. Ergebn Deut. Atlant. Exped.* "Meteor" 11, 1.
Hinshelwood, C. N. (1946). "The Chemical Kinetics of the Bacterial Cell." Oxford
    University Press, London.
Hinshelwood, C. N. (1952). *J. chem. Soc.* 745.
Holmes, R. W. (1956). *Bull. Bingham. Oceanogr. Coll.* 16, 1.
Holmes, R. W. (1957). *In* "Treatise on Marine Ecology and Paleoecology" (J.
    W. Hedgpeth, ed.), *Geol. Soc. Amer.*, Memoir 67, Vol. 1, pp. 109–128.
Hulburt, E. M. (1962). *Limnol. Oceanogr.* 7, 305.
Hulburt, E. M. and Rodman, J. (1963). *Limnol. Oceanogr.* 8, 263.
Jerlov, N. G. (1968). "Optical Oceanography." Elsevier, Amsterdam.
Jitts, H. R. and Scott, B. D. (1961). *Limnol. Oceanogr.* 6, 116.
Johnson, M. W. (1963). *Limnol. Oceanogr.* 8, 89.
Johnston, R. (1963). *Limnol. Oceanogr.* 8, 270.
Johnston, R. (1964). *J. mar. biol. Ass. U.K.* 44, 87.
Jones, E. C. (1962). *J. Cons. int. Explor. Mer.* 27, 223.
Jorgensen, E. G. and Steeman Nielsen, E. (1965). *Mem. Ist. Ital. Idrobiol.* 18,
    (Suppl), 37.
Kain, J. M. and Fogg, G. E. (1958). *J. mar. biol. Ass. U.K.* 37, 781.
Ketchum, B. H. (1939). *Amer. J. Bot.* 26, 399.
Ketchum, B. H. (1962). *Rapp. Proc. Verb. Cons. Perm. Int. Explor. Mer.* 153, 142.
Kimball, J. F., Corcoran, E. F. and Wood, E. J. F. (1963). *Bull. mar. Sci. Gulf
    Caribb.* 13, 574.
Lasker, R. (1960). *Science, N.Y.* 131, 1098.
Lewin, R. A. (1962). "Physiology and Biochemistry of Algae" (R. A. Lewin, ed.),
    Academic Press, New York.

Lewin, J. C., Lewin, R. A. and Philpott, D. E. (1958). *J. gen. Microbiol.* **17**, 373.
Lillick, L. C. (1940). *Trans. Amer. Phil. Soc.* **31**, 193.
Lorenzen, C. J. (1963). *Limnol. Oceanogr.* **8**, 56.
Lovegrove, T. (1970). *In* "Some Contemporary Studies in Marine Science" (H. Barnes, ed.). George Allen and Unwin, London.
McAllister, C. D., Parsons, T. R. and Strickland, J. D. H. (1960). *J. Cons. perm. int. Explor. Mer.* **25**, 240.
McAllister, C. D., Parsons, T. R., Stephens, K. and Strickland, J. D. H. (1961). *Limnol. Oceanogr.* **6**, 237.
McLeod, G. C. and Kanwisher, J. (1962). *Physiol. Plant.* **15**, 581.
Maddux, W. S. and Jones, R. F. (1964). *Limnol. Oceanogr.* **9**, 79.
Mallams, A. K., Waight, E. S., Weedon, B. C. L., Chapman, D. J., Haxo, F. T., Goodwin, T. W. and Thomas, D. M. (1967). *Chem. Commun.* 301.
Margalef, R. (1958). *In* "Perspectives in Marine Biology" (E. Buzzati-Traverso, ed.). University of California Press, Berkeley, Calif.
Margalef, R. (1961). *Invest. Pesq.* **18**, 1.
Marshall, S. M. and Orr, A. P. (1952). *J. mar. biol. Ass. U.K.* **30**, 527.
Marshall, S. M. and Orr, A. P. (1955a). "The biology of a Marine Copepod *Calanus finmarchicus* (Gunnerus)." Oliver and Boyd, Edinburgh.
Marshall, S. M. and Orr, A. P. (1955b). *Deep Sea Res.* **3** (Suppl) 110.
Marshall, S. M. and Orr, A. P. (1958). UNESCO Conference on radio-isotopes in scientific research. UNESCO/NS/RIC 137. 6 pp.
Marshall, S. M. and Orr, A. P. (1964). *In* "Grazing in Terrestrial and Marine Environments" (D. J. Crisp, ed.). Blackwell, Oxford.
Menzel, D. W. and Ryther, J. H. (1960). *Deep Sea Res.* **6**, 351.
Menzel, D. W. and Ryther, J. H. (1961a). *Deep Sea Res.* **7**, 282.
Menzel, D. W. and Ryther, J. H. (1961b). *Deep Sea Res.* **7**, 279.
Menzel, D. W. and Ryther, J. H. (1961c). *J. Cons. int. Explor. Mer.* **26**, 250.
Menzel, D. W. and Spaeth, J. P. (1962). *Limnol. Oceanogr.* **7**, 151.
Menzel, D. W., Hulburt, E. M. and Ryther, J. H. (1962). *Deep Sea Res.* **10**, 209.
Monakov, A. V. and Sorokin, Yu I. (1961). *Trud. Inst. Biol. Vodoahranilish.* **4**, 251.
Mullin, M. M. (1969). *Oceanogr. Mar. Biol. Ann. Rev.* **7**, 293.
Newell, G. E. and Newell, R. C. (1963). "Marine Plankton." Hutchinson, London.
Oh'Eocha, C. (1965). *In* "Chemistry and Biochemistry of Plant Pigments" (T. W. Goodwin, ed.). Academic Press, London.
Parke, M. and Dixon, P. S. (1968). *J. mar. biol. Ass. U.K.* **48**, 783.
Petipa, T. S. (1966). "Physiology of Marine Animals". Scientific Publishing House, Moscow.
Pinter, I. J. and Provasoli, L. (1963). *In* "Marine Microbiology" (C. H. Oppenheimer, ed.). Thomas, Springfield, Ill.
Prescott, G. W. (1969). "The Algae." Nelson and Sons, London.
Pringsheim, E. G. and Wiessner, W. (1961). *Arch. Mikrobiol.* **40**, 231.
Provasoli, L. (1963). *In* "The Sea", Vol. 2 (M. N. Hill, ed.). Interscience, New York.
Rabinowitch, E. I. and Thomas, G. J. B. (1960). *Science, N.Y.* **132**, 422.
Raymont, J. E. G. (1963). "Plankton and Productivity in the Oceans." 660 pp. McMillan Co., New York.
Raymont, J. E. G. (1966). *Adv. Ecol. Res.* **3**, 117.
Reid, J. L. (1962). *Limnol. Oceanogr.* **7**, 287.
Riley, G. A. (1943). *Bull. Bingham Oceanogr. Coll.* **8(4)**.

Riley, G. A. (1944). *Amer. Scient.* **32**, 129.

Riley, G. A. (1946). *J. mar. Res.* **6**, 54.

Riley, G. A. (1956). *Deep Sea Res.* **3**, (Suppl) 224.

Riley, G. A. (1959). *Bull. Bingham Oceanogr. Coll.* **17**, 83.

Riley, G. A. (1963). *In* "The Sea", Vol. 2 (M. N. Hill, ed.). Interscience, New York.

Riley, G. A., Stommel, H. and Bumpus, D. F. (1949). *Bull. Bingham Oceanogr. Coll.* **12**, 169.

Riley, J. P. and Segar, D. A. (1969). *J. mar. biol. Ass. U.K.* **49**, 1047.

Round, F. E. (1965). "The Biology of the Algae." Arnold, London.

Russell, F. S. (1935). *J. mar. biol. Ass. U.K.* **20**, 309.

Ryther, J. H. (1954). *Biol. Bull. Woods Hole.* **106**, 198.

Ryther, J. H. (1956a). *Limnol. Oceanogr.* **1**, 72.

Ryther, J. H. (1956b). *Limnol. Oceanogr.* **1**, 61.

Ryther, J. H. (1959). *Science, N.Y.* **130**, 602.

Ryther, J. H. (1969). *Science, N.Y.* **166**, 72.

Ryther, J. H. and Guillard, R. R. L. (1959). *Deep Sea Res.* **6**, 65.

Ryther, J. H. and Menzel, D. W. (1959). *Limnol. Oceanogr.* **4**, 492.

Ryther, J. H., Yentsch, C. S., Hulburt, E. M. and Vaccaro, R. F. (1958). *Biol. Bull.* **115**, 257.

Saruhashi, K. (1953). *Pap. Met. Geophys. Tokyo.* **3**, 202.

Sheldon, R. W. and Parsons, T. R. (1967). "A Practical Manual on the Use of the Coulter Counter in Marine Science." Coulter Electronic Sales Co. of Canada.

Shiriaishi, K. and Provasoli, L. (1959). *Tohuku J. agric. Res.* **10**, 89.

Silva, P. C. (1962). *In* "Physiology and Biochemistry of Algae" (R. A. Lewin, ed.). Academic Press, New York.

Southward, A. J. (1961). *J. mar. biol. Ass. U.K.* **41**, 17.

Southward, A. J. (1962). *J. mar. biol. Ass. U.K.* **42**, 275.

Smayda, T. J. (1963). *In* "Marine Microbiology" (C. H. Oppenheimer, ed.). Thomas, Springfield, Ill.

Smayda, T. J. (1970). *Oceanogr. mar. biol. Ann. Rev.* **8**, 353.

Smith, G. M. (1951). *In* "Manual of Phycology" (G. M. Smith, ed.), p. 13. Chronica Botanica Co., Waltham, Mass.

Sorokin, C. and Krauss, R. W. (1958). *Plant. Physiol.* **33**, 109.

Spencer, C. P. (1957). *J. mar. biol. Ass. U.K.* **33**, 265.

Steele, J. H. (1956). *J. mar. biol. Ass. U.K.* **35**, 1.

Steele, J. H. (1958). *Rapp. Proc. Verb. Cons. Perm. Int. Explor. Mer.* **144**, 79.

Steele, J. H. (1965). *Spec. Publs. int. Commn Northwest Atlantic Fish* No. 6, 463.

Steele, J. H. and Baird, I. E. (1962). *Limnol. Oceanogr.* **7**, 42.

Steele, J. H. and Yentsch, C. S. (1960). *J. mar. biol. Ass. U.K.* **39**, 217.

Steeman Nielsen, E. (1952). *J. Cons. int. Explor. Mer.* **18**, 117.

Steeman Nielsen, E. (1954). *J. Cons. int. Explor. Mer.* **19**, 309.

Steeman Nielsen, E. (1964). *J. Ecol.* **52** (Suppl). 119.

Steeman Nielsen, E. and Hansen, V. Kr. (1959a). *Deep Sea Res.* **5**, 222.

Steeman Nielsen, E. and Hansen, V. Kr. (1959b). *Physiol. Plant,* **12**, 353.

Steeman Nielsen, E. and Jensen, A. (1957). *Galathea Rept.* **1**, 49.

Steeman Nielsen, E. and Park, T. S. (1964). *J. Cons. int. Explor. Mer.* **29**, 19.

Strickland, J. D. H. (1958). *J. Fish. Res. Bd. Can.* **15**, 453.

Strickland, J. D. H. (1960). *Fish Res. Bd. Can. Bull.* **122**.

Strickland, J. D. H. (1965). *In* "Chemical Oceanography" (J. P. Riley and G. Skirrow, eds), Vol. 1, p. 477. Academic Press, London.

Strickland, J. D. H. and Parsons, T. R. (1968). "A Practical Handbook of Sea Water Analysis." *Fish Res. Bd. Can. Bull.* **167**, 311 pp.

Sutcliffe, W. H. (1965). *Limnol. Oceanogr.* **10** (Suppl), R253.

Sverdrup, H. U. (1953). *J. Cons. int. explor. Mer.* **18**, 287.

Syrett, P. J. (1956). *Arch. Biochem. Biophys.* **75**, 117.

Tanada, T. (1951). *Amer. J. Bot.* **38**, 276.

Teixeira, C. (1963). *Bol. Inst. Oceanogr. Universidade Sao Paulo.* **13**, 53.

Thomas, W. H. (1958). *In* "Quarterly Progress Report No. 4, Scripps Institute of Oceanography, Tuna Research Program".

Tranter, D. J. and Newell, B. S. (1963). *Deep-Sea Res.* **10**, 1.

UNESCO (1966). Monographs on Oceanographic Methodology, No. 1, Determination of photosynthetic pigments in sea water.

UNESCO (1967). Report of Working Group 20 on radiocarbon estimation of primary production. UNESCO, Paris.

Vaccaro, R. F. (1963). *J. mar. Res.* **21**, 284.

Van Norman, R. W. and Brown, A. H. (1952). *Plant. Physiol.* **27**, 691.

Verduin, J. (1960). *Limnol. Oceanogr.* **5**, 372.

Vlymen, W. J. (1970). *Limnol. Oceanogr,* **15**, 348.

Vollenweider, R. A. (1950). *Schweiz. Z. Hydrol.* **12**, 193.

Vollenweider, R. A. (1969). "Measuring Primary Production in Aquatic Environments." Blackwell, Oxford.

Walne, P. R. (1956). *Fish Invest. Ser. Lond. Ser. II.* **20** : (9).

Walne, P. R. (1963). *J. mar. biol. Ass. U.K.* **43**, 767.

Westlake, D. F. (1963). *Biol. Rev.* **38**, 385.

Wimpenny, R. S. (1936). *Fish. Invest. Great Britain* Ser. II. 15(3), 1.

Wimpenny, R. S. (1938). *J. Cons. int. Explor. Mer.* **13**, 323.

Wimpenny, R. S. (1958). *Rapp. Proc. Verb. Cons. Perm. Int. Explor. Mer.* **144**, 70.

Wood, E. J. F. (1959). *J. mar. biol. Ass. India.* **1**, 26.

Wood, E. J. F. (1963a). *In* "Marine Microbiology" (C. H. Oppenheimer, ed.). Thomas, Springfield, Illinois.

Wood, E. J. F. (1963b). *Oceanogr. mar. biol. Ann. Rev.* **1**, 197.

Wood, E. J. F. (1964). *Nova Hedwigia.* **8**, 5 and 453.

Wood, E. J. F. (1967). "Microbiology of Oceans and Estuaries." Elsevier, Amsterdam.

Zenkevitch, L. A. (1963). "Biology of the Seas of the U.S.S.R." 955 pp. Allen and Unwin, London.

Chapter 10

Marine Sediments

I. INTRODUCTION

The ocean floor has a wide diversity of topographical features (see Chapter 3), most of which are covered with a blanket of unconsolidated sediment. This sedimentary layer has an average thickness of ca 500 m, but varies considerably from place to place. It is thinnest, or absent completely, on the tops of hills and seamounts, and thickest in some deep-sea basins where it can exceed 1,000 m. The oldest sediments found in the deep-sea areas are Jurassic in age, i.e. not much more than ca 100 million years old. This represents only a few percent of geological time and raises fundamental questions of earth history. Several theories have been advanced to explain the absence of soft unconsolidated deep-sea sediments older than the Jurassic: (i) the ocean floors are geologically young; (ii) rates of sedimentation were much slower in past geological times, with the result that pre-Jurassic sediments have too small a thickness to be identified; (iii) pre-Jurassic sediments were deposited and have subsequently been consolidated, or even transformed by metamorphism from material below, into compact rocks; (iv) ocean-floor spreading has resulted in the transfer of deep-sea sediments under the continental masses.

Each of these various theories has its own supporters but in the present state of knowledge it is not possible to decide which, if indeed any, offers the true explanation for the absence of geologically old deep-sea sediments. These are problems of geophysics, a science which is now entering into one of its most exciting phases, and are closely related to the development of the earth as a planetary body. Such fundamental questions in earth history, however, are not concerned only with the physics of the planet but are closely bound up with equally fundamental chemical problems.

The sediments found on the deep-sea floor are unique, and represent a type completely different from those of the continental and near-shore environments; their origin is closely linked with the geochemical history of the earth. Near-shore sediments are composed of solid material brought to the sea mainly by the action of rivers; calcium carbonate of a biogenous origin; and evaporite deposits. The sediments are deposited on the seaward extensions of the land areas, i.e. up to and including the continental slope. Environments of deposition include: beaches, deltas, estuaries, tidal flats, lagoons, fiords, off-shore basins, submarine canyons and the shelf itself. These sediments have been extensively sampled and studied, and many comprehensive treatises are available dealing with their conditions of deposition, mineralogy and geochemistry. However, it is only in the last two or three decades that deep-sea sediments have been studied in detail. As a result of these studies oceanographers, geologists and geochemists are beginning to build up a picture of the conditions on the deep-sea floor. The picture is yet by no means complete, but many major advances have been made in recent years which contribute to our understanding of these unique sediments. Because of this, it is now possible to describe the mineralogy and geochemistry of deep-sea sediments in some detail, and this is the overall purpose of the following chapters. A general framework is presented within which the mineralogy and chemistry of deep-sea sediments can be described in terms of fundamental earth science, and some of the great advances made in marine sedimentary geochemistry in recent years are fitted into it. Continental and near-shore sediments are mentioned in the text only in so far as they relate directly to the processes of deep-sea sedimentation.

II. The Marine Sedimentary Environment

Marine sediments may be divided into two broad categories according to whether they are deposited in the *deep-sea* or the *near-shore* environment. This is a genetic classification since the conditions of sedimentation, such as rate of deposition, influence of land-derived material, of local biological conditions, and the physico-chemical state of the overlying water and the sediment-water interface, differ in the two marine environments. Deep-sea sediments may also be further divided into *pelagic* and *non-pelagic* types.

This is a fundamental distinction which has been used for many years in marine geology. In general, the term pelagic refers to those deep-sea sediments in which the land-derived mineral components* have been deposited from a dilute suspension which has a long residence time in sea water, and which has settled slowly from the upper layers of overlying water. Once they have settled, these components may be redistributed by sea-bottom processes (see also p. 310), but the important

---

* These may constitute between ca 5 and 90% of the sediments.

distinction is that they originated in surface, or near-surface, waters. In contrast, non-pelagic deep-sea sediments contain land-derived components which have not been in suspension for long periods in the upper layers of the water column, and which have usually been deposited by bottom processes such as turbidity currents etc. However, the terms pelagic and non-pelagic are not well defined. For example, the majority of pelagic sediments are deposited in the abyssal depths, i.e. in depths of water exceeding 1,000 m. However, the depth of water itself is not a definitive parameter of this type of sediment since pelagic deposits may be found close to coastal areas where the supply of land–derived material is low, or where topographical conditions prevent such material from being deposited. Pelagic deposits may also occur in mid-ocean areas, far from the land masses, in depths of water less than 1,000 m. It is evident, therefore, that although depth of water and distance from land do exert a control on sedimentation, other parameters must be introduced to enable a comprehensive classification of marine sediments to be made.

The components of marine sediments are derived from a variety of sources (see Chapter 12), but most of the sediments contain at least some land-derived material, and attempts have been made to classify them according to the nature of this material, particularly its particle size and rate of deposition.

The particle size distribution in the land-derived (mainly clay) fraction of marine sediments is now known in some detail (see for example, Revelle, 1944; Goldberg and Griffin, 1964; Griffin et al., 1968). Beltagy and Chester (1970) determined the size distribution of particles in the carbonate-free surface samples of a number of North Atlantic deep-sea sediments. The results of this study (excluding deep-sea sands) are listed in Table 10,1.

TABLE 10.1

The average particle size distribution in the carbonate-free surface samples of 31 Atlantic Ocean sediments. Percentage by weight of each size fraction (from Beltagy and Chester, 1970)

>27·5 μm	16–30 μm	8–16 μm	4–8 μm	2–4 μm	<2 μm
1·8	2·0	4·5	9·4	7·7	75·1

It can be seen from Table 10.1 that quantitatively the $<2~\mu m$*

---

* In most grade size analysis schemes the "clay" size is taken as $<4~\mu m$. The $<2~\mu m$ fraction is a division commonly used by clay mineralogists. It is particularly convenient for the study of deep-sea sediments because >60% of the land-derived components of the sediments are in this size class.

fraction is by far the most important single size component of the land-derived fraction of deep-sea sediments. The distribution of this clay size fraction in sediments from all the major oceans has been tabulated by Griffin *et al.* (1968)—see Table 10.2—who have shown that on average it makes up ca 60% of the land-derived material. In most particle size analyses of deep-sea sediments the clay material is simply reported as a <2 μm fraction, but Sackett and Arrhenius (1962) have shown that in a Pacific clay ca 45% of the particles are in fact <1 μm in size.

TABLE 10.2

The average content (wt.%) of the <2 μm fraction in sediments from the major oceans (from Griffin et al., 1968)

Type	Location	Number of samples	Percentage by weight of <2 μm fraction
Pelagic sediment	Atlantic	23	58
	Pacific	22	61
	Indian	52	64
Continental Shelf sediment	Atlantic Coast of U.S.	12	2
	Gulf of Mexico	11	27
	Gulf of California	15	19
	Sahul Shelf, N.W. Australia	9	72
Suspended river sediment	33 U.S. rivers	1026	37

Clearly, therefore, the land-derived mineral material in most deep-sea sediments has been deposited from a suspension composed largely of colloidal size particles. There are few estimates of the concentration of this suspended mineral matter in open ocean waters, since most determinations are of total detritus, which usually contains a high proportion of organic matter (see for example, Barash, 1962). However, those estimates that are available show that this material is present in a very dilute suspension. For example, Armstrong (1958) found the concentration in deep water from the eastern Atlantic to range between 0·05 and 1·00 mg/l. Arrhenius (1963) used the concentration of particulate aluminium in Pacific ocean waters as an indicator of silicate material, and estimated its concentration to be ca 23 μg/l.

This suspended mineral matter in deep-sea areas is not evenly distributed throughout the water column, and may be concentrated into specific layers. For example, Ewing and Thorndike (1965) reported ca 2·5 µg/l in a sample of north-west Atlantic water from a depth of 4,030 m, but found no measurable suspended matter in water from a depth of 3 m at the same location. An extensive study of suspended particulate matter in the World Ocean has been made by Jacobs and Ewing (1969). They found that the total amount of suspended matter decreased in the order South Atlantic >North Atlantic >North Pacific >South Pacific. For deep-sea waters the majority of samples were in the range 0 to 49·5 µg/l, and the next largest number in the range 50 to 99·5 µg/l. Although these values include organic material, they do serve to illustrate the fact that the concentration of suspended matter in pelagic areas of the oceans is low. This contrasts with near-shore localities where concentrations several orders of magnitude higher have been reported (Emery, 1960).

In addition to general patterns of the areal distribution of suspended matter in deep-sea areas, Jacobs and Ewing (1969) also reported the presence of nepheloid layers which contained higher concentrations of particles than normal ocean waters. Such layers, which were found above areas where thick deposits of pelagic sediments occur, extend upwards from the sea floor for several hundred metres, and led Jacobs and Ewing to postulate the occurrence of lutite-flows,* dilute equivalents of turbidity currents. They originate when the small-size material at the sediment-water interface is disturbed and reactivated into a flow. These authors further suggested that such lutite flows may be responsible for the distribution of certain pelagic sediments, because the suspended material which they contain will be transported by deep currents, and will follow bottom topography until the particles eventually settle out.

Relatively little is known of the particle size distribution of the land-derived mineral material in deep-sea waters. According to an estimate made by Sackett and Arrhenius (1962), which was based on the distribution of particulate aluminium species, there are two size fractions in Pacific pelagic waters; a coarse fraction with a particle size >0·5 µm, and a fine fraction with particle sizes in the range 0·01 to 0·5 µm. The average ratio of the coarse to fine material was found to be 1·6, which is slightly higher than that of 1·3 obtained from the particle size analysis of a North Pacific pelagic sediment. This difference may be caused by

---

* The term lutite is applied to sediments, or sediment suspensions, which are poorly sorted and in which an important fraction consists of particles < 4 µm in size—see also p. 296.

the settling out of the particles which will result in a different distribution in the water column and sediment surface.

The settling history of suspended particles in deep-sea areas is influenced by a number of factors, all of which tend to modify simple gravitational settling. Some of these factors, such as salinity, temperature and turbulence are a direct result of the sea water environment. Other factors depend on the nature of the particles themselves, e.g. irregular particles will have settling rates different from those predicted by Stokes' Law. For larger particles gravitational settling is also modified by aggregation in the gut of filter feeding organisms (Rex and Goldberg, 1958; Arrhenius, 1963). Delany *et al.* (1967) found that particle size concentration gradients existed in Equatorial Atlantic deep-sea sediments for particles >4 μm. These gradients were thought to be due largely to the transport of particles from the Euro-African mainland by the wind, and in order to explain the lack of significant lateral spread by currents it was suggested that particles >4 μm in the surface layers are aggregated by filter feeding organisms into larger faecal pellets which then sink rapidly upon excretion. From theoretical calculations Arrhenius (1963) has shown that small particles (< 0.5 μm in diameter) sink at a faster rate than Stokes' Law predicts for simple gravitational settling. In contrast, settling times estimated from coagulation rates are much smaller than those actually observed, a discrepancy which may be caused by the organic matter present in deep-sea waters.

In deep-sea areas, therefore, the land-derived mineral material in the waters is present in the form of a dilute colloid which has a slow rate of coagulation. This is reflected in deep-sea sediments which have a large proportion of material <2 μm in size. Another characteristic feature of this colloidal suspension is its slow rate of deposition (of the order of millimetres per thousand years), which results in a wide areal distribution of particles <2 μm in size. Near-shore waters, however, have a more concentrated mineral suspension in which the particles have a larger median diameter—according to figures quoted by Griffin *et al.* (1968) the suspended sediment in thirty three American rivers has a < 2 μm fraction which averages only ca 37% by weight (on a carbonate-free basis)—and a relatively high rate of coagulation. The result is that this concentrated, rapidly flocculating mineral suspension varies locally in composition, has a small areal distribution, and may be deposited at rates exceeding 100 mm per thousand years.

In view of these inherent differences between the deep-sea and the near-shore environments, Bramlette (1961) and Arrhenius (1963) have suggested that pelagic deep-sea sediments should be defined on the basis of a maximum value for the rate of deposition of the land-derived mineral component, and that this limiting value should be in the range of millimetres per thousand years. Some examples of sedimentation rates of pelagic sediments are given in Table 10.3. The values are either quoted on a carbonate-free basis, or are for samples containing less than 5% total carbonate, and are thus an estimate of the sedimentation rate of the land-derived mineral matter (plus any authigenic minerals, such as montmorillonite or the zeolites, and biogenous opal).

TABLE 10.3

*Rates of sedimentation of pelagic sediments**

Location	Rate (expressed in mm/10^3 yr)	
	Mean	Range
South Pacific	0·45	0·3–0·6
North Pacific	1·5	0·4–6·0
South Atlantic	1·9	0·2–7·5
North Atlantic	1·8	0·5–6·2

* The sedimentation rates are expressed on a carbonate-free basis, or are for cores which contained <5% total carbonate. The rates were determined by the Io/Th method and are modified from Goldberg and Koide (1961), Goldberg et al. (1964) and Goldberg and Griffin (1964).

TABLE 10.4

A general scheme for the classification of deep-sea sediments†

I. INORGANIC DEEP-SEA SEDIMENTS

These are defined as consisting of <30% organic skeletal remains.

A. NON-PELAGIC INORGANIC DEEP-SEA SEDIMENTS

These sediments have some, or all, of the following characteristics: (i) they are usually deposited beyond the continental slope, often adjacent to the land masses (some turbidity current deposits are an exception to this and may be found in abyssal plains far from land areas). (ii) an important fraction of the land-derived mineral matter is of silt and sand size, i.e. >4 μm. There is no general agreement on how much material >4 μm in size must be present before the sediment is classified as non-pelagic. Revelle (1944) suggested a limit of 20% and Shepard (1963) increased this to 30%. Griffin et al. (1968) have shown that the <2 μm fraction makes up on average about 60% by weight of pelagic sediments from the major oceans. This <2 μm fraction was taken by the authors as the upper limit for the "clay" fraction. However, in the system used by

† In this classification the terminology used in naming the major groups is based on that suggested by Goldberg (1954) and the general scheme is modified from Revelle (1944) and Shepard (1967).

most sedimentologists the upper size limit for clay is <4 μm, and silt is in the range 4 to 62·5 μm (Shepard, 1954). Beltagy and Chester (1970) have shown that in twenty six North Atlantic deep-sea sediments the <4 μm fraction makes up ca 83% by weight. However, it is unlikely that an absolute limit to the amount of silt-sized material in any particular sediment can be set, and indeed, such a limit is undesirable since particle size distribution alone is not a reliable distinguishing parameter. (iii) the land-derived mineral matter has not been suspended for long periods in the upper water layers, and has a rate of deposition in excess of a few mm per thousand years. It has been transported mainly by bottom processes, e.g. turbidity currents. (iv) the sediments may exhibit fine sedimentary structures, e.g. thin laminations in graded beds. (v) the biogenous fraction may be composed of pelagic or non-pelagic organisms. (vi) the sediments may contain more than ca 1% organic carbon.

The non-pelagic deep-sea sediments may be sub-divided into the following classes.

(i) Non-Pelagic Lithogenous Sediments ("Muds")

These sediments are sub-divided according to particle size and colour. In previous classifications the term *mud* has been confined to non-pelagic sediments, and *clay* has been used only for those having a pelagic origin. For example, Revelle (1944) distinguished four types of non-pelagic deep-sea muds; clayey muds, silty muds, sandy muds and sands. However, the term mud is ill defined and more recently, the system of nomenclature suggested by Shepard (1954), based on sand-silt-clay ratios, has been applied to these sediments. In this system, clay has a median particle size <4 μm, silt a median particle size between 4 and 62·5 μm and sand a median particle size $>62·5$ μm. Sediments with varying amounts of each size class are then described by means of a triangular diagram. Thus, a sediment may be designated a deep-sea, non-pelagic sandy silt, or clayey silt, etc. Colour is also used to describe these sediments. According to Kuenen (1950) blue "mud" is the most common type, and red and yellow "mud" are varieties of this which owe their colour to the weathering conditions prevailing on the adjacent land areas. Other varieties include green "mud", a type rich in glauconite, and grey "mud" ("mud" is used here as applied by Kuenen (1950) and Revelle (1944) to indicate a deep-sea, non-pelagic sediment containing <30% organic skeletal material).

(ii) Glacial Marine Sediments

These are often greenish or grey in colour and usually contain material which is largely of silt size. They differ from other deep-sea sediments in that they consist mainly of rock flour produced by the action of ice on land, and are usually transported by ice-rafting. They occur particularly around Antarctica, where they extend to the margin of the pack ice, but have also been found further north in the South Atlantic and Indian Oceans, and also in the North Atlantic (Shepard, 1967). According to Ericson *et al.* (1961) glacial marine sediments have a complete absence of particle size sorting, a feature which readily distinguishes them from turbidity current deposited sediments.

(iii) Turbidite and Slide Deposits

Turbidite sequences consist of coarse layers of sand with considerable amounts of silt, clay and even gravel sized material, and often contain considerable amounts of detrital shell material. The actual material making up the sequences may have a near-shore origin, but the sequences themselves are found in both pelagic and non-pelagic deep-sea sediments. Heezen (1963) has listed the following properties as being generally characteristic of turbidite deposits: (i) they have sharply defined bases; (ii) they have a particle size grading from coarse at the bottom to fine at the top; (iii) the contact of the beds may be blurred at the top due to burrowing by organisms; (iv) the turbidite layers have physical, chemical, and biological characters, very different from those which would be expected. of sediments directly deposited in the deep-sea environment, e.g. the occurrence of shallow water organisms in deep-sea areas; (v) they are confined to submarine canyons, the continental rise and the abyssal plains.

Material which has resulted from sliding (or slumping) movements is a distinct deposit characteristically different from turbidite sequences. Slide deposits commonly contain a high proportion of clay sized material, which is less stable to movement than sand and gravels, and do not show the grading typical of turbidites. The essential differences between turbidite and slide deposits have been tabulated by Kuenen (1956).

iv) Non-Pelagic Inorganic Mineral Sands (excluding quartz sands)

In addition to sands of a turbidite origin, which often consist largely of quartz, some deep-sea sands are composed of varying proportions of inorganic mineral grains of a type not usually transported to mid-ocean areas from the continents. Usually these minerals have a local origin, either from volcanic activity or submarine weathering. Often the proportion of these grains is small and they are mixed with normal deep-sea sediment components, e.g. volcanic foraminiferal sand. Such mineral sands can, however, form discrete sediment types. An example of this is the samples of the mineral sands described by Fox and Heezen (1965). These were collected along the Mid-Atlantic Ridge between 57° S. and 38° N., and consisted predominantly of volcanic minerals of local origin, either as relatively pure sediments, or mixed with foraminiferal material.

3. PELAGIC INORGANIC DEEP-SEA SEDIMENTS

Pelagic sediments are defined as having some, or all, of the following characteristics: (i) the land-derived mineral matter (lithogenous component) has been in suspension for long periods in the upper water layers and has accumulated slowly—in the order of millimetres per thousand years—in the deep water oxygenated areas usually far from the land masses; (ii) a high proportion (ca 70%) of the land-derived material, is of clay dimensions (<4 μm). A large proportion of this is, in fact, <2 μm in size and has been deposited from a dilute mineral suspension; (iii) the iron compounds of the land derived minerals in the surface sediments are in an oxidized state and are usually brown or red in colour; (iv) when the sediments exhibit stratified layers these often have a mottled appearance due to reworking by organisms; (v) the biogenous fraction is composed of the skeletal remains of pelagic organisms; (vi) the sediments often contain authigenic minerals, e.g. ferro-manganese nodules and zeolites, and consequently have an enhanced concentration of trace elements such as Cu, Pb, Ni, Co, etc.; (vii) the sediments usually contain less than 0.5% organic carbon.

The pelagic deep-sea sediments are sub-divided into the following classes:

(i) Pelagic Lithogenous Clays

These sediments contain $<30\%$ organic skeletal remains, and have ar inorganic fraction composed predominately of lithogenous minerals, i.e chiefly land-derived clays and quartz. They are often brown or red in colou and have frequently been termed *red clay*, although there is little justificatior for this term. Kuenen (1965) has suggested the name *deep-sea clay*, but thi is rejected in the present classification since it would also include hydro genous clays. Lithogenous clay may be diluted with other material, ane when such material exceeds ca 5% of the sediment a qualifying name may b used e.g. *calcareous, siliceous* or *volcanic lithogenous clay*.

(ii) Pelagic Hydrogenous Clays

In order to avoid confusion with previous terminologies the term "clay" i used to describe pelagic hydrogenous deposits (excluding ferro-manganes nodules) regardless of particle size. Hydrogenous clays contain $<30\%$ organic skeletal remains and have an inorganic fraction composed predomin antly of hydrogenous minerals, i.e. chiefly authigenic montmorillonite ane phillipsite, after which the clay is named. Some difficulty arises in placing ferro-manganese nodules in this classification. These nodules cover vas areas of the deep-ocean floor, and are an important form of pelagic sedimenta tion. However, they appear to be more common on the sediment surface although this may simply reflect the paucity of sub-surface surveys. Therefore until more detailed data is available the term "ferro-manganese nodul deposits" is useful to describe the surface distribution of marine sediments but it is best used if the sediment on which the nodules lie is also describec e.g. ferro-manganese nodules on lithogenous clay.

II. BIOGENOUS DEEP-SEA DEPOSITS

These are defined as consisting of $>30\%$ skeletal organic remains. Som deep-sea sediments consist of non-pelagic organisms which have bee transported from near-shore areas by, for example, turbidity currents However, these form discrete types, and usually the organisms in deep-se sediments are pelagic in origin. In the past, these sediments have been calle *pelagic oozes* although in some recent classification schemes the term "ooze" ha been abandoned in favour of a more detailed lithological description—see fo example, Ericson *et al.* (1961). To avoid confusion with the most widely use schemes the term ooze is retained in the present classification. Deep-sea ooze are sub-divided according to whether the predominant organism is calcareou or siliceous.

A. CALCAREOUS OOZES

These sediments contain $>30\%$ organically derived $CaCO_3$ and are classi fied according to the predominant type of organism present, e.g. *globigerin* (or the general term *foraminiferal*), *pteropod* or *coccolith* ooze. In some class fication schemes the particle size of the total sediment is used as a descriptiv parameter, e.g. foraminiferal sand, silt etc. If the shell fragments cannot b identified the general term calcareous ooze may be applied. If the nor calcareous fraction exceeds ca 20% qualifying adjectives may be used, e.g *lithogenous foraminiferal ooze, volcanic pteropod ooze*, etc.

B. SILICEOUS OOZES

These sediments contain $<30\%$ organically derived $CaCO_3$ and $>30\%$ organically derived SiO_2 (opal). They are classified according to the principal organism into radiolarian and diatom oozes. If appreciable amounts of non-siliceous material are present, qualifying adjectives, such as those given above for the calcareous oozes may be used.

C. CORAL REEF DEBRIS

In his recent classification of deep-sea deposits Shepard (1967) distinguishes the deposits formed of coral reef debris from other pelagic sediments. These coral reef debris deposits include coral sands and coral muds.

Dating of marine sediments is usually carried out by a number of techniques based on the decay of the natural radio-nuclides present in them, or by magnetic reversal procedures. Radio-nuclide dating techniques have been reviewed by Turekian (1965), and Koczy (1963). The two most widely used methods are those based on radio-carbon and those utilizing members of the uranium, thorium and actino-uranium decay series. Unfortunately, the results given by the various methods are often not in agreement. For example, dates of North Atlantic sediments given by the Io/Th (i.e. $^{230}Th/^{232}Th$) method (Goldberg et al., 1964) differ from those obtained by radiocarbon techniques (Ericson et al., 1961) by a factor of ten. The rates of deposition quoted in Table 10.3 were all determined by the Io/Th method, and although this does not imply that they are absolute rates of deposition it does give a comparison of relative rates in the sediments from various oceans. Rates of sedimentation have not been constant throughout geological history. In the Atlantic Ocean, for example, major changes associated with glaciation occurred ca 11,000 and ca 75,000 years ago (Broecker et al., 1960), and ca 115,000 years ago (Goldberg and Griffin, 1964). The values given in Table 10.3 are for the upper portions of sediment cores, and may be taken to represent post-glacial sedimentation rates.

Sedimentation rates of land-derived material depend on a number of complex factors, e.g. local topography, water currents, catchment areas and particularly on the transport paths (atmospheric, hydrospheric or glacial) of the detrital solids. The effects of these factors have been discussed by Griffin et al. (1968)—see also Chapter 11. However, from the average values given in Table 10.3 it is evident that pelagic sedimentation rates in the World Ocean are of the order of millimetres per thousand years, in accordance with the definition of pelagic sediments suggested by Arrhenius (1963). From the measurements that are available for nearshore areas, it would appear that sedimentation rates are probably several orders of magnitude higher in these areas.

The characteristics discussed above, i.e. sedimentation rates and particle size of the land-derived inorganic detritus, are used to define the differences between near-shore and deep-sea pelagic sediments. However, there are other sediments deposited in the deep-sea areas which are not pelagic in character. These have been variously described as hemi-pelagic, deep-sea terrigenous and non-pelagic sediments. The terminology is confused, and criteria to be used to distinguish pelagic from non-pelagic deep-sea sediments have not been generally agreed. There is no doubt, however, that some of the sediments deposited in deep-sea areas are sufficiently different in character from the pelagic types to warrant a separate classification. For example, sediments deposited on the abyssal plains are often different from those of the adjacent hills and rises (Heezen and Laughton, 1963). On the plains themselves, the normal pelagic sediments may be interbedded with silt, sand and gravel layers typical of near-shore areas, and sequences such as these have been reported from all known abyssal plains. Some, but not all, of these sequences are formed by the action of turbidity currents which bring about the smoothing out of the unprotected deep-sea areas close to the continental shelf slopes into flat abyssal plains (Kuenen, 1967). Other deposits can result from the slide or slump of continental slope material without the action of turbidity currents. Some of the sediments associated with the mid-ocean ridge systems are also discrete types, such as the sands described by Fox and Heezen (1965), which are composed of locally derived grains, and ponded ridge flank sediments which also contain coarse local material (Siever and Kastner, 1967).

Another sediment deposited in deep-sea areas which is obviously not pelagic is that referred to as glacial marine, a type which contains a high proportion of silt-sized ice-rafted material. Both glacial marine and turbidite sediments are relatively easy to differentiate from pelagic sediments because of their predominantly coarse particle size. There are other deep-sea sediments, however, which cannot easily be classified as pelagic or non-pelagic. These are the so-called terrigenous or hemi-pelagic muds which are often deposited in deep-sea areas around the borders of the continents.

Parameters which have been widely used in the past to distinguish between pelagic and non-pelagic deep-sea sediments include, in addition to particle size distribution and rate of deposition, the trace element content and the state of oxidation of the component minerals. In pelagic areas the overlying waters are oxygenated, and oxidizing conditions prevail over much of the ocean floor. In contrast, under certain special conditions of restricted circulation, water in near-shore areas may become oxygen deficient or even totally anoxic. Various intermediate stages of oxidation can exist in marine sediments depending mainly on the length of time the sediment is in contact with oxygenated bottom waters, i.e. on the rate of sedimentation. For the slowly deposited pelagic sediments this time will be long, and Bramlette (1961) has suggested that a high state of oxidation is a criterion of the upper layers of these sediments. A rough guide to the state of oxidation of a sediment can often be obtained from its colour, since this is largely controlled by the state of oxidation of the iron compounds present, and by the nature and state of any organic material which it contains. Revelle (1944) has summarized this relationship as follows: a red-brown colour indicates the presence of ferric hydroxide; a blue-grey colour indicates the presence of ferrous sulphide and probably finely divided organic matter; a dark grey or black colour indicates a sediment rich in organic matter. Characterizing a sediment by means of its colour is not a particularly satisfactory classification system. This has been shown in the past by the use of the term "red clay" since many sediments described by this term are not red, and are sometimes not even clay. Nonetheless, the colour relationships suggested by Revelle (1944) are reflected to some extent in the various marine environments, and pelagic clays are usually brown or red in colour, whereas non-pelagic muds are often blue, grey, black or green below the sediment surface.

The particle size distribution of non-pelagic deep-sea sediments differs from that of the pelagic types in that a larger proportion of the land-derived material is of silt size, i.e. $> 4 \mu m$, although the actual amount is not defined precisely. The larger median particle size of these non-pelagic sediments results in the individual particles having a rate of deposition in excess of that found in pelagic sediments, i.e. greater than a few millimetres per thousand years.

One particularly important difference between the pelagic and non-pelagic environments is the concentration of certain trace elements in sediments deposited under the former conditions. These trace elements include Cu, Pb, Zn, Ni and Co, and their concentration in Pacific pelagic sediments results largely from the presence of micro-manganese nodules which are often dispersed throughout pelagic sediments (see also Chapter 12). Another feature which is characteristic of pelagic sediments is the evidence they show of reworking by organisms, often evidenced by a mottled appearance. Such reworking is a function of their slow rate of deposition, and non-pelagic and near-shore sediments are deposited too quickly to allow reworking to occur to any extent. For this reason Ericson et al. (1961) found that mottling due to burrowing by mud-feeding animals is a reliable parameter by which slowly deposited sediments may be distinguished from layers of catastrophic deposition, e.g. those formed by turbidity currents in deep-sea areas.

There can be little doubt, therefore, that pelagic sediments, which are deposited in deep-sea areas by the slow deposition of small particles from the overlying waters, are a discrete class of sediment. However, the terms pelagic and non-pelagic are not defined in ways which are acceptable to all marine geologists. For this reason it is not possible to make a rigorous classification of deep-sea sediments, and no attempt to do so will be made here. Instead, a number of criteria will be given, which, if taken together, may be used to characterize pelagic from non-pelagic deep-sea sediments. Of the various possible criteria given in the literature the one suggested by Bramlette (1961) and Arrhenius (1963) (i.e. rate of sedimentation) is probably the best single differentiating factor, since basically the major gross differences between pelagic and all other marine sediments are a result of low rate compared to high rate sedimentation.

III. The Classification of Deep-sea Sediments

In the present context deep-sea sediments are defined as those deposited beyond the continental slope, i.e. beyond the seaward extension of the land areas. Some of the difficulties inherent in classifying these deep-sea sediments have been discussed in the preceeding section. Criteria which have been used in the past in classification schemes include: (i) gross lithological characters, particularly particle size distribution and rate of sedimentation; (ii) chemical composition; (iii) the nature and origin of the component minerals; (iv) geographical location, i.e. environment of deposition.

(I) The usefulness of particle size distribution and rate of deposition as definitive parameters of marine sediments have been discussed above.

Ericson et al. (1961) made a classical study of a series of Atlantic deep-sea cores, and classified them according to a number of lithological characteristics. These included colour, mottling effects caused by burrowing animals, texture, particle size (including grading effects) and shell content. In this classification the term *clay* was not used to describe a particle size class, but was replaced by *lutite*. The objection to the use of the term clay was that it implied a definite mineralogical composition. The name lutite was applied to those sediments in which an important fraction consisted of particles below 4 μm in size. Such sediments may contain considerable amounts of silt-sized material, and the major distinction between a lutite and a silt is that the latter is a well sorted sediment, and the former is poorly sorted. The distinction between these two types of deep-sea sediments was made within the following particle size classification: silt, 4-62·5 μm; fine sand 62·5-250 μm; medium sand, 250-500 μm; coarse sand, 500-2,000 μm; granules 2,000-4,000 μm. Silts and sands were further classified as quartz or calcareous types depending on the dominant mineral present. In calcareous sands the carbonate material is usually of a biogenous origin, i.e. shell debris, and the sediment may be named after the predominant organism, e.g. foraminiferal sand.

(II) An example of a classification of deep-sea sediments based on purely chemical composition is that given by El Wakeel and Riley (1961). In this scheme the sediments were divided into four chemical types: those containing >60% total carbonate were termed *calcareous oozes* and were sub-divided according to the principal organism present, e.g. *globigerina ooze*; those having a total carbonate content of 5–60% were termed *calcareous clays*; those with <5% total carbonate and with between 52 and 58% SiO_2 were termed *argillaceous* or *deep-sea clays*; those with <5% total carbonate and >58% SiO_2 were termed *siliceous oozes*.

This classification is somewhat unsatisfactory since both deep-sea clays and siliceous oozes are defined in terms of total SiO_2, and the term siliceous is restricted to those sediments with >58% SiO_2. This is an attempt to chemically define "excess SiO_2", i.e. SiO_2 over that present in the clay minerals and quartz. However, difficulties arise for those sediments which are diluted with between 5 and 60% total carbonate and which would be classified according to the chemical scheme as calcareous clays regardless of silica content. In an attempt to overcome this difficulty El Wakeel and Riley (1961) recalculated their analyses on a water, organic carbon and carbonate-free basis and classified any sediment which then contained >65% SiO_2 as siliceous. Thus, a sediment containing, for example, 30% total carbonate and >65% SiO_2 on a carbonate-free basis would be termed a calcareous-siliceous clay.

(III) Classification of deep-sea sediments according to the origin of their major components is attractive since it produces a genetic interpretation of the sediment. This genetic approach is extremely useful both as a basic framework within which to describe each of the component phases—see Chapter 12—and also as a basis to describe the total sediment. There have been a number of classifications based on the origin of the sedimentary components, and those suggested by Goldberg (1954) and Arrhenius (1963) which are basically similar in content, although not in terminology, are particularly useful. Of the various components in deep-sea sediments only three are present in such amounts that they can dominate the sediments and so be useful in a classification

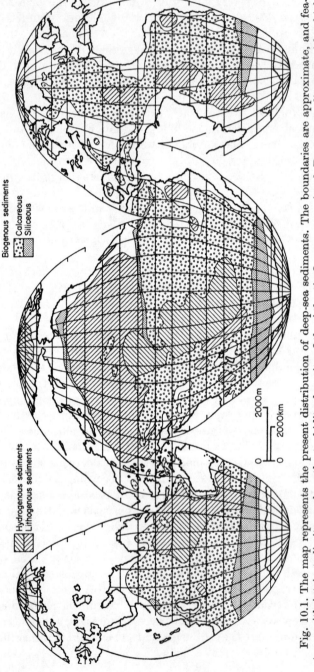

Fig. 10.1. The map represents the present distribution of deep-sea sediments. The boundaries are approximate, and features with intricate limits, such as the turbidite deposits of the Atlantic Ocean are omitted. Deep-sea sediments with a high proportion of land-derived material are not sub-divided into pelagic and non-pelagic types (see text), but are both referred to as lithogenous sediments. (Modified from Arrhenius, 1963).

scheme. These components are: (i) those produced by weathering of the lithosphere, which are termed *lithogenous* by Goldberg and *terrigenous* by Arrhenius; (ii) those which are totally or partially precipitated from solution in sea water and are termed *hydrogenous* by Goldberg and *halmeic* by Arrhenius; (iii) those of biological origin, termed *biogenous* (Goldberg) and *biotic* (Arrhenius).

(IV) The environmental approach to the classification of deep-sea sediments has been advocated by Kuenen (1950, 1965) whose classification scheme includes all marine sediments. The sediments are grouped according to environment, and deep-sea deposits are sub-divided into hemi-pelagic and pelagic according to their distance from the land areas and their median particle size—see also Revelle (1944).

A classification of deep-sea sediments is attempted here only in order that the subsequent descriptions of the chemistry and mineralogy of the sediments can be made within a general framework, but it must be realized that, at present, no single classification of these deep-sea sediments has universal approval. It is also likely that in any classification scheme there will be a number of deep-sea sediments which are difficult to place in any particular category. Examples of such sediments include coral reefs, the deposits in the so-called ponded basins on the flanks and central parts of the Mid-Atlantic Ridge, and the lithified carbonates recently reported in deep-sea areas (see for example, Fischer, 1967, Thompson *et al.*, 1968). The classification of deep-sea sediments is given in Table 10.4, p. 289.

IV. THE DISTRIBUTION OF DEEP-SEA SEDIMENTS

The distribution of the major types of deep-sea sediments is shown Fig. 10.1. This map is based on that compiled by Arrhenius (1963) and represents conditions at the surface of the ocean floor at the present day. The boundaries are only approximate, but nonetheless certain general trends in the distribution of the sediments can be established; (i) lithogenous clays and calcareous oozes are the predominant types of deposit; (ii) hydrogenous clays cover large areas in the South Pacific and Indian Oceans but not in the Atlantic; (iii) siliceous sediments are largely confined to the Antarctic region, to a band in the North Pacific (both diatom ooze) and to the Equatorial Pacific (radiolarian ooze).

REFERENCES

Armstrong, F. A. J. (1958). *J. mar. Res.* **17**, 23.
Arrhenius, G. O. S. (1963). *In* "The Sea" (M. N. Hill, ed.), Vol. 3. Interscience, New York, p. 655.

Barash, M. S. (1962). Abstract in *Deep Sea Res.* **10** (1963), 532.

Beltagy, A. I., and Chester, R. (1970). In preparation.

Bramlette, M. M. (1961). *In* "Oceanography." *Am. Ass. Adv. Sci. Publ.* **67.**

Broecker, W. S., Ewing, M., and Heezen, B. C. (1960). *Am. J. Sci.* **258**, 429.

Delany, A. C., Delany Audrey, C., Parkin, D. W., Griffin, J. J., Goldberg, E. D. and Reinmann, B. E. F. (1967). *Geochim. cosmochim. Acta,* **31**, 885.

El Wakeel, S. K., and Riley, J. P. (1961). *Geochim. cosmochim Acta,* **25**, 110.

Emery, K. O. (1960). "Geology of the Sea Floor off Southern California." John Wiley, New York.

Ericson, D. B., Ewing, M., and Heezen, B. C. (1961). *Bull. geol. Soc. Am.* **72**, 193.

Ewing, M. and Thorndike, E. M. (1965). *Science N.Y.,* **147**, 1291.

Fischer, A. G., (1967). *Jour. Geology,* **75**, 488.

Fox, P. J., and Heezen, B. C. (1965). *Science N.Y.,* **149**, 1367.

Goldberg, E. D. (1954). *J. Geol.* **62**, 249.

Goldberg, E. D. and Griffin, J. J. (1964). *J. geophys. Res.* **69**, 4293.

Goldberg, E. D. and Koide, M. (1961). *Geochim. cosmochim Acta,* **26**, 417.

Goldberg, E. D., Koide, M., Griffins, J. J. and Peterson, M. N. A. (1964). *In* "Isotopic and Cosmic Chemistry." North Holland, Amsterdam, p. 211.

Griffin, J. J., Windom, H. and Goldberg, E. D. (1968). *Deep-Sea Res.* **15**, 433.

Heezen, B. C. (1963). *In* "The Sea" (M. N. Hill, ed.), Vol. 3. Interscience, New York, p. 742.

Heezen, B. C. and Laughton, A. S. (1963). *In* "The Sea" (M. N. Hill, ed.), Vol. 3. Interscience, New York, p. 312.

Jacobs, M. B. and Ewing, M. (1969). *Science N.Y.,* **163**, 380.

Koczy, F. F. (1963). *In* "The Sea" (M. N. Hill, ed.), Vol. 3. Interscience, New York, p. 816.

Kuenen, Ph. H. (1950). "Marine Geology." John Wiley, New York.

Kuenen, Ph. H. (1956). *Deep Sea Res.* **3**, 134.

Kuenen, Ph. H. (1965). *In* "Chemical Oceanography" (J. P. Riley and G. Skirrow, eds), Vol. II. Academic Press, London and New York, p. 1.

Kuenen, Ph. H. (1967). *Sedimentology,* **9**, 203.

Revelle, R. R. (1944). Carnegie Institute, Washington, Publication 556.

Rex, R. W. and Goldberg, E. D. (1958). *Tellus,* **10**, 153.

Sackett, W. M., and Arrhenius, G. O. S. (1962). *Geochim. cosmochim. Acta,* **26**, 955.

Shepard, F. P. (1954). *J. sedim. Petrol.,* **24**, 151.

Shepard, F. P. (1963). *In* "The Sea" (M. N. Hill, ed.), Vol. 3. Interscience, New York, p. 480.

Shepard, F. P. (1967). "Submarine Geology." Harper, New York.

Siever, R. and Kastna, M. (1967). *J. mar. Res.,* **25**, 263.

Thompson, G., Bowen, V. T., Melson, W. G. and Cifelli, R. (1968). *J. sedim. Petrol.,* **38**, 1305.

Turekian, K. K. (1965). *In* "Chemical Oceanography" (J. P. Riley and G. Skirrow, eds), Vol. 1. Academic Press, London and New York, p. 81.

Chapter 11

Processes Involved in the Formation of Deep-Sea Sediments

I. Introduction

The various individual components making up deep-sea sediments are described in detail in Chapter 12. However, in order to appreciate the importance of these components it is necessary to understand something of the fundamental nature of the processes which act on the earth's crust (the lithosphere) to produce the material which eventually forms marine sediments.

There are several stages involved in the formation of a sediment. The processes which release the material from exposed land surfaces are known collectively as weathering and erosion, the latter usually implying the simultaneous removal of the weathered material. The next stage is one of transportation and deposition, and this is followed by diagenesis, i.e. changes which affect the sediment subsequent to its deposition. Each of these processes is described briefly in the following sections.

II. Weathering

Weathering is a combination of processes involving the interaction of the hydrosphere, atmosphere and biosphere on the earth's crust. There are two kinds of weathering; these involve fundamentally different processes. *Mechanical weathering* involves the fragmentation of the original rock material into smaller particles (which range down to sub-micro-metre sizes), but the end product still has the same chemical composition as the original material. *Chemical weathering* involves a chemical attack on the rock material, and the weathering residue has a different chemical composition from the starting material; this often involves the formation of new minerals.

The main agencies active in mechanical weathering are temperature changes, such as those in desert areas of high day and low night temperatures, or those due to the action of frost, the action of organisms, and the physical action of water. The latter includes hydraulic action, e.g. pounding by waves, and also the effect of material suspended in the water acting on itself to produce smaller particles, and on other surfaces, such as a cliff face, to free material which is eventually taken into suspension.

Chemical weathering involves reactions between natural waters (and their dissolved solids and gases) and rock material. The two kinds of weathering seldom act in isolation; they frequently assist each other, although they may act in competition. For example, mechanical weathering may reduce rock particles to such a degree that transportation occurs before chemical weathering can become effective.

Water, the main agent in chemical weathering, is an excellent solvent, and all minerals are soluble in it to some extent. However, under normal surface conditions rock-forming minerals have varying degrees of susceptibility to the processes involved in chemical weathering. For convenience these processes may be sub-divided into a number of types—these include hydration, hydrolysis, reduction, carbonation and solution.

In the production of the material incorporated into marine sediments hydration and hydrolysis are the most important weathering processes. These are closely related phenomena, and Nicholls (1963) has suggested that the first stage in chemical weathering is one of simple hydration, or adsorption of hydrogen and hydroxyl ions from water onto sites on the surfaces of rock-forming minerals where unsatisfied charges exist. The reactions taking place at these sites may be of two kinds, depending on the nature of the site: (i) unsatisfied negative surface charges will attract positively charged particles from the surrounding water, i.e. either hydrogen ions or the positive pole of the water molecule; (ii) unsatisfied positive charges will attract hydroxyl ions or the negative pole of the water dipole. These two reactions may be represented thus,

$$\text{(I) (Lattice } X^-) + H_2O \rightarrow \text{(Lattice } XH) + OH^-.$$

$$\text{(II) (Lattice } X^+) + H_2O \rightarrow \text{(Lattice } XOH) + H^+.$$

where X = unsatisfied surface charge at a site on the mineral lattice.

According to Nicholls (1963) the second stage involves the breaking of structural bonds in the crystal lattices, and the introduction of H^+, OH^- or H_3O^+ into the lattice. Almost all common rock-forming minerals contain element-oxygen bonds, and when a cation—O bond breaks, one hydroxyl ion is attracted to the resulting unsatisfied charge on the cation to give a cation—OH bond. When sufficient cation

—O bonds are broken the cation is liberated into solution. It is apparent, therefore, that the relative strength of the element—O bond will exert a control on the extent to which chemical weathering can affect a particular mineral lattice. Nicholls (1963) has listed the relative strengths of the bonds between oxygen and the common rock forming metallic elements, and these are given in Table 11.1. From this table it can be seen that Na—O bonds are broken more readily than Ca—O bonds etc., and that Si—O bonds are the most stable of the common element—O bonds. For this reason the mineral quartz (SiO_2) is extremely resistant to chemical weathering because of the strength of the Si—O bonds and because, unlike some of the more complex minerals which contain a variety of element—O bonds, all the bonds in the structure are of the same strength and the lattice contains no weak points.

TABLE 11.1

Relative strengths of various element-oxygen bonds in common rock-forming minerals (from Nicholls, 1963)

bond	approximate relative strength
Si—O	2·4
Ti—O	1·8
Al—O	1·65
$Fe^{(+++)}$—O	1·4
Mg—O	0·9
$Fe^{(++)}$—O	0·85
Mn—O	0·8
Ca—O	0·7
Na—O	0·35
K—O	0·25

Goldrich (1938) first established that the orders of stability of the minerals of igneous rocks to chemical weathering was the opposite order to that in which they crystallized during the fractionation of a cooling magma. The reason for this is that the higher the temperature of crystallization the more the mineral is out of equilibrium with conditions at the earth's surface where chemical weathering occurs. The order of stability of igneous minerals towards chemical weathering is quartz > muscovite > potash   feldspar > biotite > hornblende > pyroxene > olivine.

Chemical weathering of rocks is affected by a number of complex factors. These include, in addition to the chemical composition of the individual minerals, climate, drainage and the physico-chemical characteristics of the waters in which the reactions take place. The two physico-chemical parameters which most affect the weathering potential of natural waters are pH and redox potential (E_h). The importance of these has been discussed by Krumbein and Garrels (1952). Nicholls (1963) regards weathering hydrolysis as a form of cation exchange, and expresses the leaching of an element (E) from a lattice as,

$$\text{Lattice}-\text{E}+n\text{H}^+\rightarrow\text{Lattice}-n\text{H}+\text{E}^{n+},$$

where n is a whole number. In a closed thermodynamic system at constant temperature and pressure, an equilibrium should be established such that,

$$\frac{[\text{E}^{n+}][\text{Lattice}-n\text{H}]}{[\text{H}^+]^n[\text{Lattice}-\text{E}]} = k$$

Since Lattice$-n$H and Lattice$-$E are both solids this reduces to

$$\frac{[\text{E}^{n+}]}{[\text{H}^+]^n} = k,$$

i.e. in a closed system, weathering hydrolysis is dependent on the hydrogen ion concentration of the water in contact with the mineral surfaces. In fact, weathering does not take place in a closed thermodynamic system since the water is continually being renewed and equilibrium is rarely attained for most elements. Nevertheless, the pH of the weathering solution plays a large part in determining the extent to which elements are leached from a mineral lattice.

It was shown above that the relative bond strength of an element in a mineral lattice largely determines the degree to which it is leached into solution. The relative bond strength is a function of the factor, Z/r, where Z is the charge on the ion in the bond and r is the radius of the ion. This factor is termed the ionic potential of an element, and it determines the subsequent fate of the elements which are released into the weathering solutions because it is a measure of the surface charge on their ions and so affects their degree of hydration.

Goldschmidt (1934) recognized the importance of the ionic potential of an element in relation to its behaviour in the sedimentary cycle, and he divided the elements into three major groups on this basis (see Table 11.2). The elements of Group 1 have low ionic potentials and remain in ionic solution during weathering and transportation. Those of Group 2 have intermediate ionic potentials and tend to be precipitated as hydroxides. The elements in Group 3 have high ionic potentials and tend to form complex anions with oxygen, and usually remain in solution. These groupings have subsequently been modified by Wickmann (1944), who attempted to find a less empirical basis for the classification.

The solid products which result from the weathering of crustal material are not usually chemically inert during transportation since they are out of equilibrium with the successive environments that they encounter. Nicholls (1963) has classified the detrital material in sedimentary rocks into a number of types on the basis of the chemical

TABLE 11.2

The ionic potentials of some geochemically important elements (ionic potential values from Mason, 1952)

element	ionic potential	
Cs(I)	0·60	⎫
Rb(I)	0·68	⎪
K(I)	0·75	⎪
Na(I)	1·0	⎬ Group 1
Li(I)	1·5	⎪
Ba(II)	1·5	⎪
Sr(II)	1·8	⎭
Fe(III)	4·7	⎫
Al(III)	5·9	⎪
Ti(IV)	5·9	⎬ Group 2
Mn(IV)	6·7	⎪
Si(IV)	9·5	⎭
P(V)	14	⎫
S(VI)	20	⎪
C(IV)	25	⎬ Group 3
N(V)	38	⎭

changes which occur subsequent to weathering. These types include: (i) material which is resistant to chemical weathering; (ii) material which has escaped chemical weathering at its source; (iii) material which represents the solid products of chemical weathering (weathering residues).

(i) Material which is resistant to chemical weathering at the source area will, in general, remain resistant in succeeding environments. The most important example of material in this class is quartz (SiO_2) which is widespread throughout marine sediments. Other resistant minerals include rutile (TiO_2) and zircon ($ZrSiO_4$).

(ii) Material may escape chemical weathering at its source due, for example, to a rapid reduction in grain size which allows it to be removed before the chemical weathering processes can progress to any extent. The subsequent changes affecting this material will depend on its nature, and it is convenient to distinguish two major kinds of material. (*A*) Primary rocks which have not been subjected to chemical weathering in their past history, e.g. igneous rock debris. The changes affecting this material will be similar to those occurring during normal chemical weathering, and also those affecting (iii) below. (*B*) Sedimentary rocks whose components have suffered some chemical weathering prior to deposition. If the degree of weathering has been extreme, the changes occurring during transportation will tend to be the reverse of those occurring during normal chemical weathering. However, if the changes were only slight, normal chemical weathering may proceed in subsequent environments.

(iii) Mineral lattices can suffer varying degrees of chemical weathering. Most of the minerals of igneous, metamorphic and sedimentary rocks are affected to some extent by chemical weathering when they are released from their parent material. However, not all form new mineral phases and some are simply released as *primary* detrital minerals. The changes affecting these minerals will, in general, be a continuation of normal chemical weathering. In the formation of new (*secondary*) minerals, e.g. the clay minerals, the effects of weathering hydrolysis can vary. If the degree of hydrolysis was extreme, degraded clay mineral lattices may be formed, i.e. those from which inter-sheet ions have been stripped. Subsequently, these lattices may be subjected to some degree of reconstitution. An example of this in the marine environment is the reconstitution of degraded illite by potassium ions in sea water. If the weathering hydrolysis was only slight, normal chemical weathering may occur in subsequent environments. The exact nature of the changes which will occur depends on the physico-chemical conditions of each environment, and in particular on how different these environments are.

The products of chemical weathering may, therefore, be divided into four major types: (i) material which has been leached into solution, or which is in colloidal form; (ii) solid material which is resistant to chemical weathering; (iii) solid material which has escaped chemical weathering; (iv) solid material which represents the weathering residues, i.e. the products resulting from weathering hydrolysis.

The products of weathering are functions of the intensity of the weathering processes—depending largely upon local climatic and drainage conditions—and of the geological nature of the source area—depending upon the mineralogy and petrology of the source rocks (Goldberg, 1964). For this reason it is not possible to list specific weathering products from specific source minerals. For example, a granitic rock consisting essentially of quartz, feldspar, muscovite and biotite may have a solid weathering product under temperate conditions in which the mineral proportions are illite >vermiculite >kaolinite > chlorite >quartz. However, under tropical conditions the same rock may yield a weathering residue in which these minerals have entirely different proportions and in which other minerals may also be present, e.g. a laterite consisting essentially of kaolinite, gibbsite, quartz and iron oxides. Because of this dependence on climatic conditions the most important solid weathering residues in marine sediments, i.e. the clay minerals, can be classified according to the type of weathering under which they were produced. For example, kaolinite is the clay mineral characteristic of tropical weathering conditions. This is a broad classification and does not imply that kaolinite cannot be formed under other conditions, but simply that tropical weathering favours this mineral over other clays. The classification will be examined in more detail in Chapter 12.

In general, most land-derived minerals are little affected by weathering processes occurring in sea water. However, volcanic lava, ash and pumice effused onto the sea floor can suffer extensive changes during submarine weathering. These changes include the leaching of elements from hot lava as it comes into contact with cold sea water, and the formation of authigenic minerals such as montmorillonite and the zeolites from volcanic debris. The nature of these changes is discussed in Chapters 12 and 13.

III. Transportation

The proportion of the total amount of material weathered from the continents which is introduced into the oceanic environment, and its subsequent fate in this environment, are controlled by two kinds of processes; the transport paths bringing the weathered material to the oceans, and the transport paths redistributing it within the oceans themselves. The introduction of dissolved material from the land masses is largely confined to stream run-off, but solid material may be transported by a variety of agencies, including wind, water, ice and organisms. Once the material has reached the oceans the most important processes controlling its redistribution are continental margin processes, current action, ice-rafting and movement within organisms. Each of the major transport paths is considered below.

A. WATER TRANSPORT

Kuenen (1965) regards rivers as by far the most important suppliers of sedimentary material to the oceans, and a large fraction of the weathered lithosphere is transported by river agencies. The areas of the major oceans and of the complementary land masses draining into them are given in Table 11.3. However, the grain size and kind of inorganic detritus contributed to the oceans by rivers and streams is not known in detail. In many near-shore areas the influence of river transported detritus often controls the sedimentation processes. For example Emery (1960) has estimated that in the coastal area off southern California streams contribute about ten times as much sediment to the ocean as do wind and local cliff erosion combined, and Kaplan and Rittenberg (1963) have concluded that recent detrital sediments produced by weathering and erosion of the lithosphere, are the most important type of sediment being deposited in basin areas. The grain size of the suspended material in river and estuarine waters varies with local meteorological and tidal conditions, and median diameters of between 1 and 50 μm have been recorded (Revelle and Shepard, 1939 Emery, 1960).

TABLE 11.3

Oceanic areas and complementary land areas draining into them. Units of thousands of square kilometres (from Goldberg, 1967)

Ocean	Area	Land area drained	Percentage (land-ocean)
Atlantic	98,000	67,000	68·5
Indian	65,500	17,000	26·0
Antarctic	32,000	14,000	44·0
Pacific	165,000	18,000	11·0

The total amount of sediment introduced annually into the oceans by rivers has been estimated at ca $3\cdot6 \times 10^{15}$g/year (Turekian, 1969), but very little of this material above ca 4 μm in size reaches the deep-sea areas directly from river transport. A proportion of the finer grained material will escape gravitational settling, although much of it will be deposited in near-shore areas partly after size modification by flocculation. The processes involved in the flocculation of clay minerals have been extensively studied by several authors, e.g. Whitehouse and McCarter (1958), Whitehouse et al. (1960), Welder (1959), and have been summarized by Postma (1967). The results of these studies indicate that river waters carry mainly unflocculated clay material, but that in saline waters flocculation occurs, and clay particles do not settle as single solid

rains, or in irreversible solid states of aggregation. According to Postma
1967) clay particles of colloidal and sub-colloidal dimensions have a
harge which is usually negative. This charge is balanced by a "double
ayer" of hydrated cations, and the thickness of this layer depends
n various factors, including the total ionic concentration of the
urrounding liquid phase. An increase in the total ionic concentration
f the surrounding solution tends to decrease the thickness of the double
ayer. If the thickness decreases below a critical value flocculation
ccurs and small particles conglomerate into larger units and tend to
ettle. This occurs at fresh water—saline water boundaries and causes
unflocculated river borne clay material to flocculate as it enters the
narine environment. In saline waters, the settling velocities of clay
articles are controlled by ion-fixation mechanisms among other
rocesses. Whitehouse *et al.* (1960) have given the name "coacer-
ates" to clay mineral settling units in saline waters; these are
lefined as thermodynamically reversible assemblages of solid clay
articles which settle as a solid-rich liquid phase. As the water becomes
nore saline, the coacervates become denser by expelling water. The
novement of clay mineral mixtures across fresh water-saline water
oundaries at high current velocities usually results in the selective
ransport of different clay species (Whitehouse and McCarter, 1960).
'or example, illite and kaolinite flocculate more quickly at low
hlorinities than montmorillonite. At high current velocities this can
esult in the deposition of illite and kaolinite with the retention of
nontmorillonite in suspension. The settling rate of sediment sus-
ensions from flowing water is thus dependent on the concentration
nd nature of the suspended particles, the kinetics of flocculation, and
he current velocity.

  The influence of river-transported detritus on sedimentation in marine
reas varies locally and depends on a large number of factors which
nclude the size, geological nature and climate of the land area drained,
he topography of the sea floor, distance from land and the influx of
naterial from other sources, such as wind, current and ice transport,
nd biological precipitation. In some oceanic areas the deposition of
iver-transported material dominates the pattern of sedimentation. For
xample in parts of the Equatorial Atlantic, particularly south of ca 10°
N., the land-derived component of the sediments is rich in tropical
veathering products derived from the African and South American
ontinents. The major rivers draining the surrounding area are the
Niger, the Congo and the Amazon, and according to Griffin *et al.* (1968)
he sediments in the low latitude equatorial regions are mainly com-
osed of material supplied by these rivers which has been subsequently

distributed by Equatorial current systems. In the area between c 10°N. and 35° N. eolian transport is also an important input mechanisr (see also p. 334).

Other examples of the influence of river transport on deep-se sediments are seen off the coast of South America where tongues of hig illite content extend seawards from the mouths of the Magdalena Orinoco and Amazon rivers (Griffin *et al.*, 1968). The St. Lawrenc River in the North Atlantic has a similar run-off area, dominated by a association of illite and chlorite in the surrounding deep-sea sediments In the Indian Ocean significant amounts of sedimentary materials ar contributed to the adjacent sediments by the Ganges and Indu Rivers.

In areas where river transport is, or has previously been, importan much of the material introduced from the continents is deposited o the continental shelf and slope regions, which act as an intermediat sediment trap. These shelf and slope materials are an important sedi ment source and are distributed into deep-sea areas by continenta margin processes, such as sliding (or slumping) and turbidity cur rents.

There are, therefore, two major kinds of processes which redistribut solid material introduced into the oceans. (i) Continental margin pro cesses act on material dumped on the shelf and slope areas. This materia will tend to have a larger particle size and a more varied minera composition than the detritus carried directly into deep-sea areas fron the land surfaces. The shelf and slope material will be transported b processes such as sliding, slumping and turbidity currents, movement which are restricted to the ocean bottom. For this reason the distribu tion of continental margin material will be controlled by bottom topo graphy, and such material will tend to be deposited as flat areas such a abyssal plains, and is important in the formation of non-pelagic deep-se sediments (see also p. 285). (ii) Normal oceanic circulation processe act mainly on material which escapes deposition on the continenta margins. This material has a small particle size and a long residence tim in sea water. The particles settle out from the upper layer of the wate column, and sea floor topography need not act as a transport barrier The currents distributing this small-sized material include surface intermediate and deep-water currents. Material from these processes i particularly important in areas remote from the continents, e.g. the flank of the Mid-Oceanic Ridge system and topographic highs, which ar isolated from bottom transport processes (see Turekian, 1965), and i important in the formation of pelagic deep-sea sediments (see also p 284).

B. WIND TRANSPORT

For hundreds of years mariners have been aware that atmospheric dust is carried into deep-sea areas. Darwin (1846) was perhaps the first natural scientist to realize that this dust was an important transport pathway for sedimentary material. The presence of such dust in marine sediments was established by Murray and Renard (1891), who reported the occurrence of wind transported particles in deep-sea sediments off the coasts of Africa and Australia. One of the earliest detailed studies of wind transported (eolian) particles in marine sediments was made by Radczewski (1939). He found that the sediments off the West Coast of Africa in the vicinity of the Cape Verde islands are rich in rounded quartz grains coated with red-brown iron oxide. The grains are similar to those found in the adjacent Sahara Desert, and Radczewski concluded that they had been brought to the sediments by the wind. Dust from the Asian deserts, consisting mainly of quartz and feldspar, is carried by winds into the North Pacific (Futi, 1939), and its presence in the deep-sea sediments of this area was detected in a classical study made by Rex and Goldberg (1958) who were the first to show the oceanwide importance of the wind as a major transporting agent. Brown (1952) has shown that volcanic dust from the Cape Verde Islands is carried across the Atlantic and that its fall-out zone covers the north equatorial and north temperate regions of the Atlantic. An analysis of wind-borne dust collected about 700 km southwest of the Canary Islands revealed that it was composed of uncoated quartz grains, mica and clay minerals (Game, 1964). Folger et al. (1967) have established that fresh water diatom frustules are a characteristic component of wind blown dust carried westwards across the Atlantic from the African mainland. Detailed studies evaluating the contribution of wind-borne detritus to marine sediments have been made by various authors, and have been reviewed by Delany et al. (1967).

Most of the dust lifted from the continental areas by wind action will be transported within the troposphere, but volcanic eruptions may eject dust into the stratosphere, from which it will eventually settle into the troposphere and behave similarly to tropospheric fall-out (Rex and Goldberg, 1962). In the troposphere, dust transport on a large scale takes place through the movements of the main air masses, i.e. the Equatorial Easterlies (the Trades), the Temperate Westerlies (the Jets) and the Polar Easterlies.

The distribution, quantitative importance, particle size and mineralogy of eolian material in marine sediments will depend on a large number of factors. According to Goldberg (personal communication) these factors include: (i) strength and circulation patterns of the wind system; (ii) weathering characteristics of the continental source rocks; (iii) the time elapsing between the introduction of the solids to the sea surface and their settling out to the sea bottom; (iv) the relative importance of other transport mechanisms such as ice and current transport; (v) the mechanism by which the particles are removed from the atmosphere, i.e. rain-scrubbing or gravitational settling (see also p. 313).

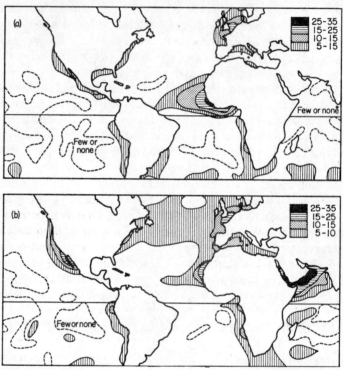

Fig. 11.1. Average frequency of haze (dry aerosol) in the troposphere. (a) Northern winter. (b) Northern summer. Frequencies are given in per cent of total number of observations (from Arrhenius, 1959, after McDonald 1938).

The geological nature of the source area is important since in areas having loose surface deposits, e.g. sandy deserts, there will be a readily available supply of detritus in a form susceptible to wind transport. Evidence of the effect of deserts, or arid regions, on wind-transported material is shown by the correlation between the wind circulation over arid areas and the occurrence of dry atmospheric haze over adjacent seas—see Fig. 11.1. Atmospheric haze is defined as haze generally due to microscopic particles in the atmosphere, e.g. dust, salt, etc., as distinct from mist and fog which are caused by droplets of water (Arrhenius, 1959). Figure 11.1 also illustrates other features which are characteristic of wind transport. One of the most important of these is the striking variation in the amount of wind-transported dust between the summer and winter months in the North Atlantic. In summer a band of haze extends across the entire area from Europe to North America in the path of the Westerlies. In winter, however, the haze is virtually absent from the same winds in this area. This has been confirmed by shipboard collections made across the North Atlantic in winter. At every station the dust collected over a 24-hour period amounted to only ca 1 mg of solids. (D. W. Parkin, personal communication.) This compares with maximum daily collections at Barbados in winter of gram quantities from approximately the same volume of air. The probable reason for this is that Barbados lies on the Trade Wind tracks which include a large part of Africa in their collection areas. In contrast, the Atlantic Westerlies collect largely from North America, and as most of this area will be frozen for a major part of the winter the amount of dust raised into the atmosphere will be much smaller than in the summer months. However, the dust collected at Barbados, which had a year round average of ca 2.5 $\mu g/m^3$ air, showed a pronounced seasonal variation. In the summer months, the amount of dust collected was about an order of magnitude higher than that in the winter months. According to Prospero (1968) this is probably due to a southerly movement of the African dust "belt" in winter which results in the dust being transported to the south of Barbados in this season. Another feature illustrated by the haze distribution map is the importance of wind circulation patterns on the transport of dust. This is shown by the frequency of heavy haze in the summer months in the northern Indian Ocean resulting from dust carried by the southwest monsoon. The influence of atmospheric circulation on the transport of aerosols is also shown in the distribution of wind-transported components in marine areas, e.g. the Trade Winds carry dust from the Euro-African mainland across the Atlantic; and the Westerlies transport material from Australia almost all the way across the South Pacific.

The fall-out of dust from the troposphere is controlled largely by gravitational settling, and by co-precipitation and scavenging during rain precipitation. Delany et al. (1967) have made a detailed study of air-borne dust collected at Barbados and found that there is a marked difference between the particle size distribution of the dust and that of the Atlantic sediments, particularly in the size range 2-4 μm (see Table 11.4). According to Delany et al. (1967) if rain-scavenging processes were the dominant fall-out mechanism for dust particles <2 μm, then the deposited dusts would have the same size distribution as those in the air. Because of this, the differences between the size distribution of the transported and deposited dusts were interpreted as resulting from gravitational settling which will result in a different distribution within the dust column and the settled sediments. However, it is difficult to interpret the differences in particle sizes between the collected dusts and deep-sea sediments because the collection efficiency of the meshes used by Delany *et al.* (1967) falls off rapidly for particles of a few micrometres in size.

Atlantic deep-sea sediments adjacent to the African coast have between 50 and 60% of their total land-derived solids <2 μm in size. Since Delany *et al.* (1967) concluded that much of this material has an eolian origin they suggested that aggregation of the small-sized material must have occurred within the dust column in order for it to fall out close to the African coast. However, Beltagy and Chester (1970) have considered the transport of small sized particles in the Trade Winds

TABLE 11.4

Size analyses of dust samples and Atlantic sediments (wt. %). (From Delany et al., *1967)*

Size fraction (μm)	Barbados dust samples		Sediment Zep 23, Abyssal Hills east of Mid-Atlantic Ridge	Sediment Zep 25, off African coast
<2	43·5	47·3	54·8	62·8
2–4	27·7	34·9	8·5	7·7
4–8	23·3	15·6	10·7	6·5
8–16	5·2	1·9	11·3	7·2
16–44	0·3	0·4	13·0	9·1
44–74			0·7	3·0
>74			0·9	3·7

and have shown that, in theory, a particle with a diameter of ca 0·5 μm could travel around the earth before falling out due to gravity, and that particles with a diameter of ca 0·1 μm can travel around the earth about three times. This led to the concept of first, second, third, etc., generation particles, a generation being equivalent to one complete orbit of the earth. The implication of this is that at all points in space in the Trade Wind system there should be a background of small-sized particles which will be falling out over the entire wind path. Thus, a dust assemblage which falls out close to the African coast will contain, in addition to large particles, a small-sized fraction which may be second or third generation. However, because of its relatively long residence time in the air* this small-sized dust will be susceptible to rainfall

* Of the order of ca 30 days.

2M– MICA
MASS CONCENTRATION

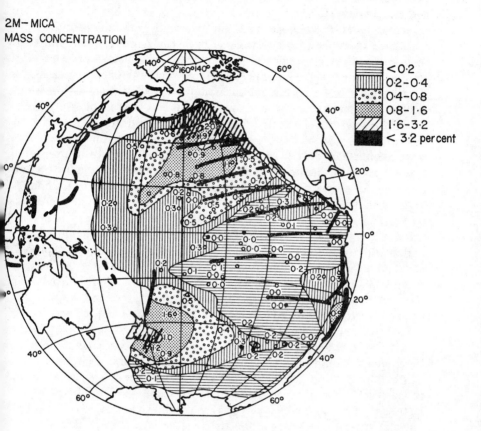

Fig. 11.2. The distribution of mica at the sediment surface of the Pacific Ocean. Features of particular interest are the zones of wind-transported mica from the Australian continent and from the Atacama Desert, and the apparent absence of a similar transport from the Asiatic continent in intermediate northern latitudes —see text. The distribution of mica off the North American continent is complex. South of the Murray Fracture Zone eolian transport prevails; north of the zone turbidite transport appears to be responsible for much of the sediment distribution. Those geomorphological features, which are effective in protecting the oceanic basins against the inflow of turbidites (fracture zones, trenches and troughs), are indicated on the map; extensive accumulation of turbidites in the north east Pacific and off New Zealand are indicated by black and white striping (From Arrhenius, 1965).

precipitation, and much of it may fall out over areas of high rainfall, e.g. the Amazon Basin.

Arrhenius (1965) has stressed the importance of rain precipitation within dust carrying wind masses in both the northern and southern hemispheres. As an example he cites the distribution of $2M$ mica (muscovite) in the Pacific, and points out that several features in its distribution can be correlated with rainfall precipitation within wind circulation patterns. For example, there is no area of high mica concentration in a latitudinal zone extending off the east coast of Japan in the path of the strong Westerlies which are known to carry large amounts of dust from the Asian continent. Arrhenius interprets this as a result of the failure of the wind-transported material to extend far over the ocean because of precipitation in the meridional zone of high rainfall over the east of Japan—see Fig. 11.2. In contrast, material originating in the Australian desert areas is carried half way across the Pacific Ocean through a region of relatively low rainfall.

Rex and Goldberg (1962) have sub-divided the wind-transported material in marine sediments into three genetic classes: (i) extra-terrestrial material; (ii) biological material originating from the continents; (iii) inorganic solids, including volcanic debris, originating from the continents. To these may be added a fourth category; man-made contaminants.

(i) The extra-terrestrial material which is introduced into marine sediments via the atmosphere is considered in detail in Chapter 12.

(ii) Biological material has been identified in atmospheric dusts collected over many oceanic areas (see for example, Ehrenberg, 1847). Detailed studies by Kolbe (1955, 1957) have shown that dust transported diatom populations may contain fresh-water as well as marine species. Delany et al. (1967) have reported that the organic components of dust collected in Barbados contained various marine organisms, fungus hyphae and fresh-water diatoms, one species of which inhabits cold running waters and may originate from a mountain region in an area such as Spain. Other biological material included spores, pollen and insects.

(iii) The inorganic continental weathering products supplied to sediments by wind transport consist chiefly of the clay minerals, quartz and feldspars. Other minerals present in minor amounts include micas, calcite, dolomite, amphiboles, pyroxenes, anatase, rutile, haematite, goethite, talc and serpentine. The importance of this wind transported inorganic material in Atlantic sediments has been stressed by Delany et al. (1967), who concluded from mineralogical evidence that the land-derived fraction of the sediments deposited east of and on the Atlantic Ridge is derived wholly from wind-transported material originating in the arid regions of the Euro-African mainland. Bricker and Prospero (1969) collected atmospheric dust on the island of Bermuda and concluded that the dust, which averaged ca 6·5 $\mu g/m^3$ air in August 1968, was transported from the Euro-African mainland. Windom (1969) has investigated the dust content of a number of permanent snow fields, and by comparing the mineralogy, size distribution, and sedimentation rate of the dusts with those of adjacent deep-sea sediments, he concluded that the North and South Pacific and Central Atlantic areas may receive between 25 and 75% of their detrital phases from the fall-out of atmospheric dust. Prospero and Bonatti (1969) have used a shipboard mesh technique to collect eolian dust in the Eastern Equatorial Pacific. These authors found two distinct dust populations, which differ in quantity and mineralogical composition, in this area. To the north of the equator the dust collected averaged ca 0·68 $\mu g/m^3$ air, and consisted predominantly of plagioclase with subordinate amounts of quartz and <5% of each of the major clay species. Prospero and Bonatti (1969) concluded that this northern dust originated in the arid regions of western and southern Mexico. South of the equator the collected dust averaged ca 0·14 $\mu g/m^3$ air. In this region quartz was the dominant mineral in the dust and significant amounts of plagioclase, micas, montmorillonite, kaolinite and chlorite were also present. According to Prospero and Bonatti (1969) this dust was transported from the arid region of Peru and the Atacama Desert of northern Chile. From this study the authors deduced that eolian dust is an important sedimentary component in deep-sea sediments of the Pacific west of Central America and the south-western United States. Shipboard atmospheric dust collections have also been made in the Bay of Bengal during the summer monsoon season, and the results have been reported by Goldberg et al. (1969). The mineral assemblages in the dust are similar to those of the Indian Ocean equatorial sediments, but are different from those introduced by rivers draining into the Bay of Bengal.

Volcanic ejection products make a significant contribution to marine sediments in some areas, but only the finer material (volcanic ash) is transported over large areas by wind action*. Ash horizons are found in many localities, e.g. the Mediterranean (Mellis, 1954), Gulf of Alaska (Nayudu, 1964), North Atlantic (Bramlette and Brady, 1940), Equatorial Pacific (Worzel, 1959). According to Nayudu (1964) the distribution of ash over wide areas is common after large-scale eruptions, and the pattern of distribution is largely controlled by the dominant wind direction, particularly high altitude winds. Volcanic ash consists of

---

* Dust directly raised by the wind from the land areas has its highest concentration at low latitudes. In contrast, Kuenen (personal communication) has pointed out that the distribution of volcanic dust is independent of latitude.

glass shards together with much smaller amounts of mineral and crystal fragments. Selective transport is operative in the relative distribution of these two ash components, and glass particles are carried over greater distances than mineral grains.

(iv) Man-made contaminants include industrial waste products expelled into the atmosphere, e.g. carbon and metal flakes, etc.; nuclear bomb debris; pesticides and other agricultural chemicals; and products of social activities, e.g. particulate lead from petrol fumes.

C. ICE TRANSPORT

Sedimentary material is both introduced into the oceans and re-distributed within them by ice transport mechanisms. Ice in the marine environment can originate either as *land ice*, which is formed on land areas mainly as ice caps or glaciers, or as *sea ice*, which is formed by the freezing of sea water. Shelf ice which extends seawards from the land may consist of land ice, or sea ice or a mixture of the two. In the sea there are two principal forms of free-floating ice; icebergs which are land ice that has broken away from its parent formation, and ice floes, which originate as sea ice.

On land, ice in the form of sheets and glaciers is an effective agent of erosion and transportation. The mechanisms of erosion include both tearing (or plucking) material from underlying rock surfaces, and the action of embedded material. Powdered rock, or rock flour, is often produced by these processes. Thus, the icebergs formed from land ice by calving can contribute a variety of continental material to the oceans. Sea ice which drifts in the form of ice floes can also incorporate continental material by freezing around beach and river deposits, or by having material dropped on its surface by river or wind action.

Icebergs are transported from polar regions towards the equator and reach much lower latitudes than ice floes. They are far more common in the southern hemisphere which has ca 93% of the total mass of the world's icebergs. At present, the iceberg limit is ca 55° S. and ca 40° N., but this can be exceeded in areas which have strong currents directed towards the equator. In the Pleistocene period the limits extended as far as 35° S. in the southern hemisphere and 35° N. in the northern hemisphere (Fairbridge, 1966).

The presence of large-sized ice-rafted material in marine sediments has been reported by many investigators. For example, Needham (1962) identified a large number of rocks and fragments of ice-rafted origin in the Atlantic off the Cape of Good Hope. In the Pacific, ice-transported pebbles have been found in the vicinity of sea mounts (Kimo *et al.*,

1956; Pratt, 1961). Clearly, ice-rafted material will make its highest contribution to sediments deposited in polar areas—see for example, Lisitsin (1960)—but the influence of ice weathering is also apparent in the distribution of the clay sized material over large areas of the deep-sea floor. This is shown in the distribution of chlorite, a mineral which appears to be characteristic of glacial weathering. This mineral dominates the clay assemblages in zones extending from the polar regions for about 20° in the northern hemisphere and about 10° in the southern hemisphere (Griffin et al., 1968).

D. TRANSPORT BY ORGANISMS

In deep-sea areas transport by organisms is difficult to evaluate, but is probably not very important. Sea mammals, birds and fishes can carry sedimentary material in their stomachs, and driftwood can enclose even large boulders. However, the total contribution of the combined action of these agencies to deep-sea areas compared to that of other forms of transport is probably negligible, except in high latitudes where transport by sea mammals may be significant. The processes of organic transportation have been reviewed by Emery (1963).

There are, however, two important ways in which marine organisms can affect deep-sea sediments; by disturbing deposited sediments by burrowing, and by ingesting and aggregating suspended particles which are subsequently excreted as faecal pellets and sink to the bottom of the water column. Both of these aspects have been discussed in Chapter 10.

E. SUMMARY OF TRANSPORTATION PROCESSES

In a study of the distribution of clay minerals in oceanic areas Griffin et al. (1968) have defined the major pathways by which land-derived solids are introduced into particular oceanic provinces, and have related these pathways to sedimentation rates. Their findings are listed in Table 11.5. From this table it is evident that the transport paths have a geographical dependence and are dominated by wind, river and ice mechanisms. There are two striking features in the distribution of sedimentation rates which reflect the natures of the transport paths and the surrounding continental areas. (i) The lowest average sedimentation rates are found in the South Pacific sediments, which are about four times lower than those in the North Pacific. In both regions the continentally derived solids are supplied largely by wind transport (with the exception of the Northeast Pacific where turbidite deposits are common), but since there is more land area per unit of ocean area in the North Pacific the input of land-derived material is greater and is the dominant supply mechanism. In contrast, in the South Pacific,

TABLE 11.5

Rates of accumulation and sediment transport paths of the land-derived solids in various oceanic areas (from Griffin et al., 1968)

Ocean	Province	Transport paths	Rates of accumulation (mm/10³ years)
Atlantic	50° N	Glacial from Arctic Polar easterlies Jet stream	
	15° N.–50° N.	Jet stream	0·3–0·5 (Mid-Atlantic Ridge Valleys)
		St. Lawrence River	2–7 (From continents to flanks of Ridge)
	15° N.–15° S.	Trades	0·7–0·8 (Mid-Atlantic Ridge Valleys)
		South American and African rivers	1–8 (From continents to flanks of Ridge)
	15° S.–50° S.	Jet Stream	0·2–0·4 (Mid-Atlantic Ridge Valleys)
			2–6 (From continents to flanks of Ridge)
	50° S.	Glacial from Antarctic Polar easterlies Jet stream	2
Pacific	50° N.	Glacial from Arctic Polar easterlies Jet stream Alaskan rivers	0·8
	15° N.–50° N.	Jet stream	0·4–2·0
	15° N.–15° S.	Trades	0·3–6·0
	15° S.–50° S.	Jet stream	0·3–0·8
	50° S.	Glacial from Antarctic Polar easterlies Jet stream	0·5–2·3
Indian	15° S.–20° N.	Trades	1
	20° S.–50° S.	Jet stream	0·5–0·8
	50° S.	Glacial from Antarctic Polar easterlies Jet stream	

most of the minerals are of a volcanic origin, e.g. montmorillonite. (ii) In the Atlantic Ocean the rates of sedimentation of the Mid-Atlantic Ridge deposits are much lower than those of sediments nearer to the continents. According to Griffin *et al.* (1968) this is so because the ridge deposits are remote from land and are prevented by bottom topography from receiving the river-transported detritus which is important in other Atlantic areas. These ridge deposits will, therefore, receive mainly eolian material, and for this reason will have a lower sedimentation rate. In the Atlantic, the equatorial regions as a whole receive a smaller input of land-derived solids than the mid-latitude regions. This is reflected in clay sedimentation rates which are higher in mid-latitude areas (Turekian, 1964), and in the distribution of detrital quartz which shows a two-fold increase in the mid-latitude compared to the equatorial region in the northern hemisphere (Beltagy and Chester, 1970).

Much of the discussion in subsequent chapters is related to deep-sea sedimentation in the Atlantic and Pacific Oceans. The transportation processes which supply sedimentary material differ in importance in the two oceans, and it is useful at this stage to examine some of the principal differences between the Atlantic and the Pacific. These are summarized below: (i) exclusive of adjacent seas the Atlantic has an area of ca 98 million square kilometres, and the Pacific ca 165 million square kilometres; (ii) there is an almost continuous belt of trenches surrounding the Pacific Ocean; the only break being in the northeastern region along the west coast of North America. This belt of trenches protects most of the Pacific from the action of turbidity currents, and turbidite deposits are confined to the northeastern region and to an area off New Zealand—see Fig. 11.2. The central regions of the Atlantic are not protected by a trench belt, and turbidite deposits are common in all regions, excluding those of the Mid-Ocean Ridge system— see Fig. 11.3; (iii) there are fewer rivers draining into the Pacific than into the Atlantic (see Table 11.3), and in the Pacific much of the river-borne material is deposited in the marginal trenches which act as sediment traps; (iv) hilly terrain is much more common in the Pacific than in the Atlantic because in the Pacific the irregularities are not smoothed out by processes such as turbidity current deposition; (v) there is more extensive vulcanism and seismic activity in the Pacific than in the Atlantic Ocean, and seamounts are more common in the former ocean; (vi) in the Atlantic Ocean catchment area there is a great desert (the Sahara) in the path of the Trade Winds. Because of this, Atlantic deep-sea sediments contain a higher proportion of eolian material than those of the Pacific; (vii) there are more areas remote from land in the

Pacific than in the Atlantic Ocean; areas >1,500 km from land are virtually absent from the Atlantic, but amount to ca 15% of the Pacific (Bramlette *et al.*, 1955).

IV. DIAGENESIS

Diagenesis refers to the processes which affect a sediment subsequent to its deposition. In the present context the processes are limited to those which occur in the unconsolidated deep-sea sediment layers, and are thus less dramatic than where the deep-burial processes of compaction and cementation result in the formation of consolidated sediments.

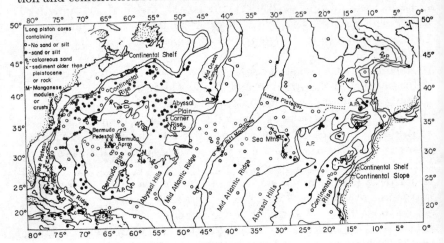

Fig. 11.3. The distribution of deep-sea turbidite sands in the North Atlantic. No cores obtained from abyssal plains are devoid of silt or sand layers. In contrast, sediments from the Bermuda Rise and the Mid-Atlantic Ridge system are pelagic in character, except in certain rare areas near the crests of the Mid-Atlantic Ridge, or on the flanks of seamounts, where local winnowing has altered this pattern (from Heezen and Laughton, 1963; after Heezen *et al.*, 1969).

The components of deep-sea sediments may be classified in various ways—see Chapter 12—but it is convenient at this stage to make a basic distinction between solid and dissolved material in the sediment—interstitial water complex. Much of this solid material is not completely stable, and may undergo considerable changes. The nature of some of these diagenetic changes will be discussed when the mineralogy and chemistry of the components of deep-sea sediments are described—see Chapters 12 and 13. These changes, which can have a profound effect on the composition of deep-sea sediments, include: (i) the formation of new minerals, e.g. quartz (see also p. 357) and certain zeolites (see also p. 369), both of which are formed from silica dissolved in interstitial waters; (ii) the modification of pre-existing minerals. The clay minerals are particularly susceptible to these kinds of changes which result from an exchange between those ions dissolved in interfacial and interstitial waters, and those held in surface and inter-sheet positions in the clay structures. This may result in some exchangeable cations in the clays being rendered non-exchangeable; (iii) the complete or partial dissolution of minerals. Biogenous skeletal carbonates and opaline silica are particularly prone to post-depositional dissolution; (iv) the post depositional mobilization and migration of elements (see also p. 405); (v) The physical effects resulting from depth of burial, or overburden, in the sediment column. Although lithified carbonates and chert have been found in deep-sea areas, most of the sediments are unconsolidated. The burial of continental and near-shore sediments results in compaction, i.e. a decrease in volume (porosity) and a reduction in water content. Recent work carried out on deep-sea sediments of the experimental Mohole (Rittenberg et al., 1963) indicates that there is apparently no decrease in the water content to depths of ca 100–150 m. However, there was an increase in mechanical strength with depth which Rittenberg et al. (1963) have interpreted as resulting from weak cementation, or the establishment of some kind of bonding between the clay minerals; (vi) the effects resulting from the depth of overlying waters, e.g. temperature and hydrostatic pressure differences. These can be particularly important in dissolution equilibria (see also p. 372); (vii) other effects resulting from time spent on the ocean floor. Included here are such processes as burrowing by organisms, and the partial or complete removal of organic matter in oxygenated deep-sea areas.

It may be concluded that diagenetic changes are extremely important in modifying the composition of deep-sea sediments. It should be stressed that these phenomena, and others which have not been mentioned, do not act in isolation, but tend rather to form a part of the complex series of modifications occurring at the sediment—water interface and at depth in deep-sea sediments.

REFERENCES

Arrhenius, G. O. S. (1959). In "Researches in Geochemistry" (P. H. Abelson, ed.). John Wiley, New York.
Arrhenius, G. O. S. (1965). In "Advances in Earth Science" (P. M. Hurley, ed.). M.I.T. Press, Cambridge, Mass. p. 155.
Beltagy, A. I. and Chester, R. (1970). In preparation.
Bramlette, M. N. and Brady, W. H. (1940). U.S. Geol. Surv. Prof. Paper, **196A**, 34 pp.

Bramlette, R. R., Arrhenius, G. O. S. and Goldberg, E. D. (1955). *Geol. Soc. Am. Sp. Paper* **62**, p. 221.

Bricker, O. W. and Prospero, J. M. (1969). Abs. no. M99, *Trans. Am. Geophys. Un.* **50**, 176

Brown, W. F. (1952). *Mon. Weath. Rev. U.S. Dep. Agric.* **80**, 59.

Chester, R. and Hughes, M. J. (1969). *Deep-sea Res.* **16**, 639.

Darwin, C. (1846). *Q. J. Geol. Soc. Lond.* **2**, 26.

Delany, A. C., Delany Audrey, C., Parkin, D. W., Griffin, J. J., Goldberg, E. D. and Reinmann, B. E. F. (1967). *Geochim. cosmochim. Acta*, **31**, 885.

Ehrenberg, C. (1847). *Abh. K. Akad. Wiss.* **269**, 460.

Emery, K. O. (1960). "Geology of the Sea Floor off California." John Wiley, New York.

Emery, K. O. (1963). *In* "The Sea" (M. N. Hill, ed.), Vol. 3. Interscience, New York, p. 776.

Fairbridge, R. W. (1966). *In* "The Encyclopedia of Oceanography" (R. W. Fairbridge, ed.). Reinhold, New York, p. 781.

Folger, D. W., Burckle, L. H. and Heezen, B. C. (1967). *Science N.Y.*, **155**, 1243.

Futi, H. (1939). *J. met. Soc. Japan, Ser.* 2, **17**, 473.

Game, P. M. (1964). *J. sedim. Petrol.* **34**, 355.

Goldberg, E. D. (1964). *Trans. N.Y. Acad. Sci.* **27**, 7.

Goldberg, E. D. (1967). *In* "Submarine Geology" (F. P. Shepard, ed.). Harper, New York.

Goldberg, E. D., Griffin, J. J. and Fisher, F. (1969). Abs. no. 081, *Trans. Am. Geophys. Un.* **50**, 196.

Goldrich S. S. (1938). *J. Geol.* **46**, 17.

Goldschmidt, V. M. (1934). *Geol. För. Stockh. Förh.* **56**, 385.

Griffin, J. J., Windom, H. and Goldberg, E. D. (1968). *Deep-Sea Res.* **15**, 433.

Heezen, B. C. and Laughton, A. S. (1963). *In* "The Sea" (M. N. Hill, ed.), Vol. 3. Interscience, New York, p. 312.

Heezen, B. C., Thorp, M. and Ewing, M. (1969). *Geol. Soc. Am. Spec. Paper* **65**, 126 pp.

Kaplan, I. R. and Rittenberg, S. C. (1963). *In* "The Sea" (M. N. Hill, ed.), Vol. 3. Interscience, New York, p. 583.

Kimo, H., Fisher, R. L. and Nasu, N. (1956). *Deep-Sea Res.* **3**, 126.

Kolbe, R. W. (1957). *Science N.Y.*, **126**, 1053.

Kolbe, R. W. (1965). *Rep. Swed. deep-sea Exped.* **6**.

Krumbein, W. C. and Garrels, R. M. (1952). *J. Geol.* **60**, 26.

Kuenen, Ph. H. (1950). "Marine Geology." John Wiley, New York.

Kuenen, Ph. H. (1965). *In* "Chemical Oceanography" (J. P. Riley and G. Skirrow, eds), Vol. 2. Academic Press, London and New York, p. 1.

Lisitsin, A. P. (1960). *Deep-Sea Res.* **1**, 89.

McDonald, W. F. (1938). *U.S. Dept. Agr. Weather Bureau*, No. 1247.

Mason, B. (1952). "Principles of Geochemistry." John Wiley, New York.

Mellis, O. (1954). *Deep-Sea Res.* **2**, 89.

Murray, J. and Renard, A. F. (1891). *Scient. Rep. Challenger Exped.* **3**.

Nayudu, Y. R. (1964). *Marine Geol.* **1**, 194.

Needham, H. D. (1962). *Deep-Sea Res.* **9**, 475.

Nicholls, G. D. (1963). *Science Prog. LI(201)*, **12**.

Postma, H. (1967). *In* "Estuaries" (G. H. Lauff, ed.). American Association for the Advancement of Science, Washington, D.C.

Pratt, R. M. (1961). *Deep-Sea Res.* **8**, 152.

Prospero, J. M. (1968). *Bull. Am. met. Soc.* **49**, 645.

Prospero, J. M., and Bonatti, E. (1969). *J. Geophys. Res.* **74**, 3362.

Radczewski, O. E. (1939). *In* "Recent Marine Sediments" (P. D. Trask, ed.). American Association Petroleum Geologists, Tulsa, p. 496.

Revelle, R., and Shepard, F. P. (1939). *In* "Recent Marine Sediments" (P. D. Trask, ed.). American Association Petroleum Geologists, Tulsa, p. 245.

Revelle, R., Bramlette, M., Arrhenius, G. O. S. and Goldberg, E. D. (1955). *Geol. Soc. Am. Spec. Paper.* **62**, 221.

Rex, R. W. and Goldberg, E. D. (1958). *Tellus*, **10**, 153.

Rex, R. W. and Goldberg, E. D. (1962). *In* "The Sea" (M. N. Hill, ed.), Vol. I. Interscience, New York, p. 295.

Rex, R. W., Syers, J. K., Makeson, M. L. and Clayton, R. H. (1969). *Science N.Y.*, **163**, 277.

Rittenberg, S. C., Emery, K. O., Hulsemann, J., Degens, E. T., Fay, R. C., Reuter, J. H., Grady, J. R., Richardson, S. H. and Bray, E. E. (1963). *J. sedim. Petrol.* **33**, 140.

Turekian, K. K. (1964). *Trans. N.Y. Acad. Sci.* **26**, 312.

Turekian, K. K. (1965). *In* "Symposium on Marine Geochemistry." Occ. Publ. No. 3, Narragansett Marine Lab., Univ. of Rhode Island, p. 41.

Turekian, K. K. (1969). *In* "Handbook of Geochemistry" (H. Wedepohl, ed.). Springer-Verlag, Berlin, p. 197.

Welder, F. A. (1959). *Tech. Rept. 12, Coastal Studies Inst.*, Baton Rouge, Louisiana.

Whitehouse, U. G., and McCarter, R. S. (1958). *Proc. 5th. Natl. Conf. Clays, Clay Minerals*, **81**.

Whitehouse, U. G., Jeffrey, L. M. and Debrecht, J. D. (1960). *Proc. 7th. Natl. Conf. Clays, Clay Minerals*, **1**.

Wickmann, F. E. (1944). *Arkiv. Kemi Mineral Geol.* **19B**, No. 2.

Windom, H. L. (1969). *Bull. geol. Soc. Am.* **80**, 761.

Worzel, J. L. (1959). *Proc. Nat. Acad. Sci.* **45**, 349.

Chapter 12

The Components of Deep-sea Sediments

I. Introduction

Sediments consist of a number of individual components which may be classified in various ways. A fundamental distinction can be made between those components introduced into the waters overlying the sediment as solid phases, and those introduced as dissolved material. This concept leads to a simple two-fold classification of sedimentary components into a detrital fraction—transported as solid material—and a non-detrital, or authigenic fraction—transported as dissolved material (Krynine, 1948). There have been various attempts to enlarge this simple two-fold classification, particularly for the components of deep-sea sediments (see for example: Goldberg, 1954; Arrhenius, 1963). Any classification of the multi-source components of deep-sea sediments must be an arbitrary one, and from purely theoretical considerations no single classification is entirely satisfactory. However, the scheme suggested by Goldberg (1954) is probably the most meaningful in terms of our present understanding of the processes involved in marine sedimentation. In this classification the components of the sediments are divided into five broad categories: lithogenous, cosmogenous, hydrogenous, biogenous and pore solutions. The lithogenous components, which result from the weathering of the earth's surface, and the cosmogenous components, which are solids of extra-terrestrial origin, are introduced into ocean waters as particulate matter. They comprise the detrital faction of the sediments. The hydrogenous material, formed *in* sea water by inorganic reactions, and the biogenous material, produced by organisms, are both formed in the oceanic environment itself and are therefore non-detrital components. In his classification Goldberg (1954) includes pore solutions—i.e. water entrapped in the sediments—as part of the sedimentary complex. In the simple two-fold system these solutions would be classed as non-detrital. In addition to the components listed above a further phase, termed atmogenous, is included in some classifications.

The components of deep-sea sediments are classified according to the geosphere in which they originate, i.e. lithosphere, hydrosphere, biosphere, atmosphere and outer space. The processes involved in the production of solid material in each of these environments are often characteristically different, and Goldberg (1964) has suggested a number of criteria which can be used to distinguish between the various classes. These criteria include chemical composition, size distribution, isotopic composition, geographic occurrence, association with other minerals, and form or habit. Arrhenius (1963) has pointed out that not enough is known of the ultimate origin and mode of accretion of some of the minerals in marine sediments to permit a rigid classification scheme to be made. Nevertheless, from a descriptive standpoint Goldberg's classification provides a useful framework within which it is convenient

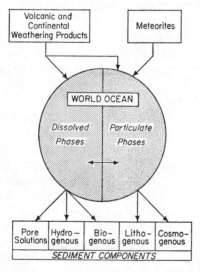

Fig. 12.1. The classification of the components of deep-sea sediments. (Goldberg, 1964).

to describe the components of deep-sea sediments. This classification is illustrated schematically in Figure 12.1, and the individual components are described in subsequent sections. The discussion is mainly confined to those mineral components which are important in deep-sea sediments. Those which are restricted to near-shore sediments are not described unless they can affect the overall geochemistry of the oceans, e.g. near-shore phosphates are included, but glauconite is not.

II. LITHOGENOUS COMPONENTS

In Goldberg's classification, lithogenous components are defined as those arising from land erosion, submarine volcanoes or from underwater weathering where the solid phase undergoes no major change during its residence in sea water. Included in this category in the present context is material thrown directly into the sea from terrestrial volcanoes, e.g. wind-transported volcanic ash. Some of the lithogenous material in deep-sea sediments may have an organic origin, but the vast majority of lithogenous components are inorganic and are incorporated into the sediments either as rock fragments or, more usually, as discrete single mineral particles.

A. ROCK FRAGMENTS

Rock fragments in marine sediments may be divided into two general classes:

(i) products of weathering. These are found mainly in near-shore areas, and result from weathering of the adjacent land masses. They occur more rarely in deep-sea areas, and are usually deposited there by the action of ice rafting, or result from the weathering of local rocks, e.g. on the Mid-Ocean Ridge system.

(ii) products of submarine and terrestrial volcanic activity. These are found extensively in deep-sea sediments, and are often concentrated in regions of active vulcanism. The fragments may be composed of a variety of materials and are collectively called pyroclasts. According to Ninkovich et al. (1964), the transportation and deposition of volcanic material depends on a number of factors which include the initial force of the eruption, the angle of ejection, the height attained by the particles (particularly fine ash), the velocity and turbulence of the winds transporting the ash, and the composition, size, shape and density of the ejected particles. The largest material can be transported for only relatively short distances through the atmosphere (sub-aerial volcanoes), or through the hydrosphere (submarine volcanoes). However, pumice—a general term for highly vesicular solidified lava of any chemical or mineralogical composition—may be transported over longer distances because of its porous nature. Pumice is often classified on the basis of the parent lava, e.g. trachytic, andesitic or basaltic pumice, and is frequently abundant in areas of once-active vulcanism, such as the Azores, although small amounts have been identified in many marine sediments (Murray and Renard, 1891).

Volcanic glass is another form of pyroclastic material which is commonly found in marine sediments. Bonatti (1965) has divided this

volcanic glass into two genetic types: (i) glass formed by the direct interaction of lava with sea water during submarine volcanic activity. This
glass is extremely hydrated and is very susceptible to decomposition; (ii) glass formed either in sub-aerial eruptions and subsequently
transported to marine areas, or in submarine eruptions where it
is prevented from reacting with sea water. This type of glass will
only become hydrated and altered slowly. Volcanic glass, like pumice,
is classified according to the nature of the parent lava; it often makes
up a large proportion of volcanic ash deposits. For example, Ninkovitch
et al. (1964) found that up to 70% of the ash deposits resulting from
volcanic activity in the South Sandwich Islands consisted of glass.

Peterson and Goldberg (1962) found two types of volcanic glass in
South Pacific pelagic sediments. One, a clear colourless glass, was
associated with alkali felspars and quartz. The other, a dark reddish
brown highly altered glass, was associated with calcic felspars. In many
localities basaltic volcanic glass (sideromelane) is often highly altered;
the original vitreous matter being transformed into palagonite. The
process is brought about by submarine weathering, and although the
details are not fully understood, it is probably initiated when hot lava
interacts with sea water, and proceeds inwards from the surface of glass
fragments. This general process was suggested by Nayudu (1962), who
also found that palagonite itself may undergo alteration.

B. MINERAL GRAINS

The main contribution made by lithogenous material to deep-sea
sediments is in the form of individual mineral grains. Virtually all the
minerals found in the terrestrial environment are represented in deep-sea
sediments, but only those which are quantitatively important will be
discussed here.

(i) Quartz

Silica in the form of quartz (SiO_2) remains virtually unaltered during
weathering and transportation processes and is found in many marine
sediments, often as a major component. Almost all of this quartz has a
lithogenous detrital origin, from terrestrial rocks or submarine (volcanic)
sources and has been transported to the sediments by wind, water,
ice, or organic agencies. Detailed studies on the origin of the quartz in
marine sediments have been made by various authors. Murray and Renard
(1891) suggested a wind-transported origin for the rounded quartz
grains often found in deep-sea sediments far from the land areas.
Radczeweski (1939) found that Sahara Desert iron oxide-coated quartz
grains were present in sediments off the Cape Verde region and assigned

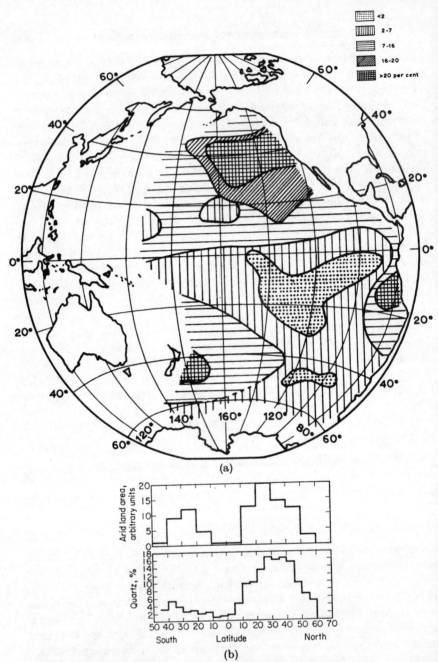

Fig. 12.2. The distribution of quartz in Pacific deep-sea sediments. (a) Areal distribution of quartz in the surface sediments (redrawn from Arrhenius, personal communication; original data after Rex and Goldberg, 1958). (b) Latitudinal dependence of the arid land areas surrounding the Pacific Ocean, and of the quartz contents of Pacific deep-sea sediments (Goldberg, 1961).

them an eolian origin, a finding which was later confirmed by Griffin and Goldberg (1963). Arrhenius (1963) has postulated that much of the fine-grained quartz in pelagic sediments has been water transported, and Goldberg and Griffin (1964) have suggested that the high quartz content of Atlantic equatorial deep-sea sediments south of ca 10° N. results from transport by rivers draining adjacent continental areas. In high latitudes of the Atlantic and Pacific Oceans drifting ice carries quartz to deep-sea areas. The importance of volcanically derived quartz in deep-sea sediments has been noted by Peterson and Goldberg (1962) for the South Pacific, and by Goldberg (1964) for the Antilles Arc region of the Atlantic.

An extensive study of the distribution of quartz in the sediments of the eastern Pacific Ocean has been made by Rex and Goldberg (1958). In these sediments the quartz grains are well sorted with a maximum concentration between 1 and 20 μm. The areal distribution of quartz showed a marked latitudinal dependence with maximum concentrations at ca 30° N. and, less well defined, at ca 35° S.—see Figure 12.2. The highest quartz concentrations in these mid-latitudes were in sediments deposited far from land (in areas inaccessible to ice-rafting or turbidity current transport), and could not be correlated with Pacific Ocean water circulation patterns. However, the distribution of quartz did show a relationship with exposed arid land areas around the Pacific. From mineralogical evidence Rex and Goldberg (1958) rejected volcanic activity as a quantitatively important source of the quartz, and concluded that the latitudinal dependence could be explained only by evoking atmospheric circulation as the major transporting agent.

In addition to latitudinal variation in the quartz contents of surface sediments in the eastern Pacific, Rex and Goldberg (1958) found a distinct decrease in quartz with depth in several sediment cores. An abrupt decrease in quartz values was not found in all the cores studied, and when present it was usually accompanied by a change in gross sediment type. Chester and Hughes (1969) investigated the detailed distribution of quartz in a North Pacific lithogenous clay core, and found that although there was no abrupt change in the distribution of quartz with depth there was a change in its particle size distribution at a depth of ca 240 cm. In the lower portion of the core there was about twice as much quartz in the <2 μm fraction of the sediments, and the change corresponded to variations in certain chemical parameters. Rex and Goldberg (1958) have suggested that the changes in the distribution of quartz with depth were the result of climatic changes which accompanied the Tertiary/Quaternary transition. The decrease in the amount of quartz, and the change in its particle size distribution in Tertiary sediments, could have resulted from several factors. These include: a decrease in the arid land source area; weaker transporting wind systems; an increase in the importance of other forms of transport, e.g. water currents; a change in the predominant atmospheric fall-out mechanism, e.g. rain precipitation or gravitational settling, which could result in a different particle size distribution in the sediments (Arrhenius, 1963; Chester and Hughes, 1969; see also p. 313).

In contrast to the sediments of the North Pacific, which contain up to ca 25% quartz, those of the South Pacific have a relatively low quartz content—rarely exceeding ca 6%, even in mid-ocean areas. Much of the quartz in South Pacific sediments is derived from acid volcanic eruptions (Peterson and Goldberg, 1962). This agrees well with the overall sedimentary supply system in the South Pacific, an area which, according to Griffin and Goldberg (1963), receives only small amounts of continentally-derived detrital minerals and large amounts of pyroclastic material (see also p. 348).

TABLE 12.1

The average particle size distribution of quartz (as a percentage of the total quartz) in North Atlantic deep-sea sediments (from Beltagy and Chester, 1970)

Equatorial sediments (17 samples)		Mid-latitude sediments (15 samples)	
Size fraction	% Quartz	Size fraction	% Quartz
>8 μm	9·8	>8 μm	44·1
4–8 μm	17·3	4–8 μm	15·0
2–4 μm	15·2	2–4 μm	9·9
<2 μm	57·6	<2 μm	30·9

The distribution of quartz in sediments from the Atlantic Ocean has been studied in detail by Goldberg and Griffin (1964), and Beltagy and Chester (1970). In order to interpret the distribution of quartz in terms of source areas and major transport paths, Beltagy and Chester (1970) divided the north Atlantic into two major regions; the Equatorial Region (ca 10° N. to ca 35° N.), and the Mid-latitude Region (ca 35° N. to ca 50° N.). The authors found that the distribution of quartz is significantly different in these two regions of the North Atlantic. The average total quartz content (on a carbonate-free basis) in the Equatorial Region is ca 11·5%, compared to ca 24% in the Mid-latitude Region—see Figure 12.3. There is also a difference in the particle size

TABLE 12.2

The particle size distribution of quartz (as a percentage of the total quartz) in an Equatorial trans-Atlantic section (from Beltagy and Chester, 1970)

Size fraction	Sediment*						
	A	B	C	D	E	F	G
>8 μm	29·8	6·1	0·0	0·0	0·0	0·0	0·0
4–8 μm	18·7	20·8	23.6	0·0	16·7	0·0	11·7
2–4 μm	25·5	27·6	12·3	0·0	16·1	0·0	10·4
<2 μm	26·0	45·4	64·0	100·0	67·1	100·0	77·8

* See Figure 12.3 for sediment positions.

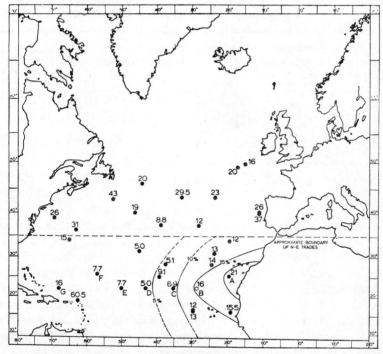

Fig. 12.3. The distribution of quartz in North Atlantic deep-sea sediments (weight %, on a carbonate-free basis). The letters A, B, C, D, E refer to the location of sediments mentioned in Table 12.2 (from Beltagy and Chester, 1970).

distribution of the quartz (see Table 12.1); the mid-latitude sediments containing a much higher proportion of larger sized quartz. In the Equatorial Region the total quartz decreases both in quantity and size away from the African coast (see Figure 12.3) across the Atlantic

(Table 12.2). Clearly, the Mid-Atlantic Ridge did not act as a barrier in this region, and Beltagy and Chester concluded that much of the quartz has an eolian origin from a source area in the African deserts, and that material from the American mainland only begins to influence deep-sea sediments in this region west of ca 65° W. The areal distribution of quartz in the Mid-latitude Region is complex and no simple trends are apparent. It is evident that it has been transported by a variety of mechanisms which include wind, water circulation systems turbidity currents, continental margin processes and ice transport.

TABLE 12.3

The quartz contents of various size fractions from North Atlantic deep-sea sediments (wt.%, on a carbonate-free basis; from Beltagy and Chester, 1970)

Size fraction	Average of 17 equatorial sediments	Average of 15 mid-latitude sediments	Deep-sea sand	Average of two near-shore sediments*
>8 μm	44	49	78	41·5
4–8 μm	28	42	52·5	39
2–4 μm	26	31	57·5	26·5
<2 μm	7·7	10	17	8·2

* Taken from the continental shelf off Norway.

Beltagy and Chester (1970) found that although the contribution made to the total quartz by each sediment size fraction varies geographically (Table 12.2), certain overall trends are apparent in the distribution of quartz as a function of particle size in all deep-sea sediments. These trends are listed in Table 12.3 from which it can be seen that there is an overall decrease in the proportion of quartz in the fractions of decreasing particle size. In the majority of the deep-sea sediments about half of the >8 μm size class consists of quartz, whereas the <2 μm class contains only ca 10% quartz. Deep-sea sands, which consist largely of quartz grains, have a corresponding increase in this mineral in all size fractions.

(ii) Clay minerals

(a) Introduction. Most clay minerals are stable weathering residues produced by the decomposition of silicate minerals. Others, such as some types of chlorite, are metamorphic minerals, and a few result

from hydrothermal processes. Almost all of the clay minerals found in terrestrial environments have also been found in marine sediments, but the only common varieties in deep-sea sediments are members of the

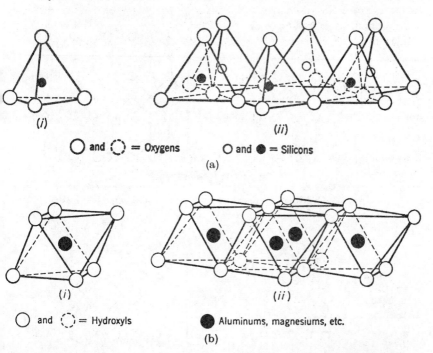

(i)

(ii)

◯ and ⟡ = Oxygens   ◯ and ● = Silicons

(a)

(i)

(ii)

◯ and ⟡ = Hydroxyls   ● Aluminums, magnesiums, etc.

(b)

Fig. 12.4. The structure of the principal clay mineral species. (a) Structure of the clay mineral "tetrahedral" layer, showing (i) a single silica tetrahedron and (ii) the sheet structure of silica tetrahedrons arranged in a hexagonal network (from Grim, 1968). (b) Structure of the clay mineral "octahedral" layer, showing (i) a single octahedral unit and (ii) the sheet structure of octahedral units (from Grim, 1968). (c) Schematic structural diagrams of the principal clay species (from Mason, 1952). See p. 336.

chlorite, illite, kaolinite and *montmorillonite* groups, or are mixed-layer clays. A detailed mineralogical description of the clay minerals is clearly outside the scope of the present work, and the following discussion is limited to a much simplified treatment. For more detailed information the reader is referred to Grim (1968), and Deer *et al.* (1962).

The basic unit of structure in the clay minerals is the SiO_4 tetrahedron in which the silicon atom is situated at the centre, and the four corners are occupied by oxygen atoms. In the clay minerals the tetrahedra are arranged in the form of a sheet, and the structure of the minerals consists of two kinds of layers. (i) A *tetrahedral* layer in which the SiO_4

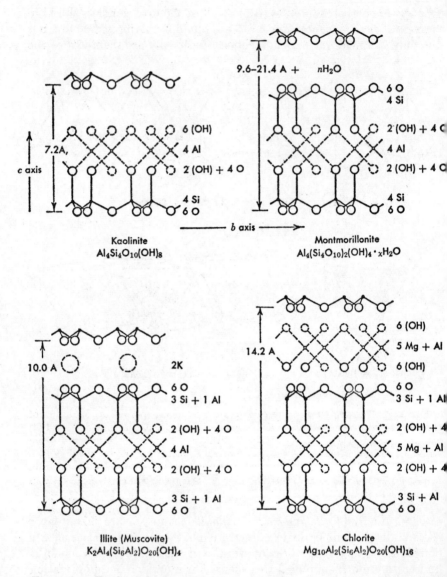

Fig 12.4 (c) For legend see p. 335.

tetrahedra are linked by three of their four corners to form a hexagonal sheet in which their tips all point in the same direction, and their bases are in the same plane—see Figure 12.4a. (ii) An *octahedral* layer which consists of two sheets of closely packed oxygen atoms or hydroxyl ions with cations of aluminium, iron or magnesium between them in octahedral co-ordination—see Figure 12.4b.

Although these two kinds of layers, i.e. tetrahedral and octahedral, may be combined in several ways, there are only two common combinations: *Type I*, in which the basic structure consists of one tetrahedral layer linked to one octahedral layer by shared oxygen atoms common to both layers; *Type II*, in which the basic unit consists of two tetrahedral layers which are linked by a single octahedral layer. The structure of the main groups of clay minerals in deep-sea sediments can be interpreted in terms of these two structural types.

Type I clay minerals

The major clay mineral group having a Type I structure is the *kaolinite* (or kandite) group. The structure of this group consists of a two layer sheet composed of one tetrahedral and one octahedral layer, and these individual two-layer basic sheets are linked together only by weak hydrogen bonds. The general formula of the group may be written $Al_4[Si_4O_{10}](OH)_8$—the square brackets are used to indicate the tetrahedral layer and have no chemical significance.

Type II clay minerals

These include illites, montmorillonites and chlorites. The basic Type II unit is composed of a "sandwich" of two tetrahedral layers (in which there is usually some substitution of Si^{4+} by Al^{3+}) with one octahedral layer between them. The structure of the three major groups of Type II clay minerals can be interpreted in terms of the basic three layer unit; the fundamental differences between them being in the manner in which the units are bonded together.

In the *illite* group the negative charge on the basic unit which results from the substitution of Al^{3+} for some of the Si^{4+} in the tetrahedral layer is balanced by potassium ions. These are introduced into the structure in inter-unit, i.e. inter-sheet, positions, and serve to bind the units together by ionic bonding. The members of the illite group are similar in both structure and composition to the mica muscovite, and are, in fact, largely derived from this mineral. The general formula for the group may be written as $K_2Al_4[Si_6Al_2O_{20}](OH)_4$.

Many illites are essentially clay-sized muscovites, but others are mixed layer muscovite-montmorillonite types, and some are mechanical mixtures of muscovite and montmorillonite (Mason, 1952). For this reason, the term illite is not confined to a specific clay mineral, but is applied to a general group. Although this is a convenient general term,

Yoder (1957) has suggested that a precise definition of the nature of the mica, and of any mixed layer mineral present, should be specified. The fundamental mica structure, i.e. the Type II sheet structure, can exist as several polymorphs, four of which are commonly found in nature. Yoder (1957) has designated these polymorphs 1M, $2M_1$, $2M_2$ and 3T micas, in which the first symbol gives the number of layers in the repeat unit cell; the second the crystal symmetry, where M is monoclinic and T is trigonal; and the subscript indicates that more than one kind of polymorph can exist. The stacking of the layers in the mica types listed above is regular, but random stacking can occur (particularly in the clay minerals), and Yoder refers to these structures as 1Md* micas. Muscovite mica is commonly found as the $2M_1$ polymorph and less commonly as the 1Md, 1M or 3T forms. Biotite mica usually occurs as the 1M polymorph, but is also found as the 1Md, $2M_1$ and 3T forms.

Before Yoder made his study 1M and 1Md muscovite had not been identified in nature, but subsequently they have been found to have a common occurrence among some natural minerals, two of which are illite and glauconite. In fact, according to Yoder, previous studies of illite indicated that it usually consists of mixed layers of 1Md or 1M muscovite and montmorillonite, and may contain $2M_1$ or 3T muscovite as an additional phase.

In the *montmorillonite* (or smectite) group any substitution of silicon in the tetrahedral layers is balanced by the introduction of exchangeable cations, such as sodium and calcium, held between (and around the edges of) the three layer units. In montmorillonites, water molecules lie between the three layer Type II units, and it is the water content which is the characteristic feature of this clay mineral group. The inter-sheet water molecules may be replaced by other molecules and this property is utilized in the identification of montmorillonites by X-ray diffraction methods. In the most commonly used technique the water is replaced by ethylene glycol, which causes the mineral lattice to expand. Many investigators who use this technique report all minerals affected by ethylene glycol as "expanding lattice" clays, and these may include not only true montmorillonites, but also degraded illites with inter-sheet water molecules. The general formula of the group may be written $Al_4[Si_4O_{10}]_2(OH)_4 . nH_2O$, although iron-containing montmorillonites are also known, e.g. nontronite, which is common in South Pacific deep-sea sediments.

In the *chlorite* group of minerals the net negative charge which results from the substitution of Si^{4+} by Al^{3+} in the tetrahedral layers of the Type II units is balanced by the introduction of a "brucite" layer into the structure. The formula of the mineral brucite is $Mg_3(OH)_6$, but in the

---

* d indicates a disordered structure.

brucite layer of the chlorite structure there is replacement of some Mg^{2+} by. Al^{3+}, and this gives the layer a net positive charge. In the chlorite structure the "brucite" layers lie between the basic three-layer Type II units. The general formula of the group may be written $Mg_{10}Al_2[Si_6Al_2O_{20}](OH)_{16}$, but may be modified to an iron-rich form which is common in deep-sea sediments.

The structures of the principal clays are illustrated in Figure 12.4c. In addition to the four major clay groups, several other clay minerals are often important in deep-sea sediments. These include vermiculite, mixed layer clays, and chain structure clays such as attapulgite, sepiolite and palygorskite.

b) Clay minerals in the marine environment. Clay minerals are a major constituent of soils, and the kind of clay mineral introduced into the oceans depends largely on the type of soil in the source area. The amount of clay minerals supplied to the oceans is a function of the local effectiveness of transport mechanisms, and the subsequent fate of the clay is determined by distribution processes within the oceans themselves. However, recent studies have shown that gross trends in clay mineral distribution in deep-sea sediments are controlled largely by the predominant clay composition of the world's soils.

Although the distribution of clay minerals in deep-sea sediments is dependent on the nature of the source area soil, clay minerals do undergo modification in the marine environment. These modifications appear to include the transformation of one clay mineral into another, but at present the data given in the literature is confusing and this topic is not reviewed in detail here. In addition, some less drastic changes occur. These include surface adsorption and the reconstitution of degraded clay mineral lattices; changes which occur as a result of their transfer to sea water with its high salinity. Degraded clay mineral lattices, i.e. those which have had cations stripped from inter-sheet positions, are particularly susceptible to reconstitution in sea water. For example, illite which has lost inter-sheet potassium will tend to absorb this element (or magnesium) from sea water (see also p. 62). Changes which affect the basic clay mineral lattices, i.e. chemical diagenesis, are not well understood. Some clay species are obviously detrital in origin, e.g. kaolinites which are products of tropical weathering and which abound in equatorial deep-sea sediments. Other relationships, such as that between 2M muscovite and disordered 1M illites in deep-sea sediments have not been established, although age determinations have shown that most of the illite in the surface layers of the sediments is detrital in origin (Hurley *et al.*, 1963).

Holland (1965) has suggested that particulate silicate matter intro‑
duced into the oceans by river input can undergo three kinds of reac‑
tion; base exchange, reconstitution of degraded clay mineral lattices
and the production of new clay minerals (e.g. the alteration of mont
morillonite to chlorite and illite; Whitehouse and McCarter, 1958–59)

From these considerations it is possible to classify the major clay
minerals in marine sediments into a number of genetic types: (i) those
which are primary metamorphic minerals and not chemical weathering
products, e.g. some chlorites; (ii) those which are detrital weathering
residues and which have undergone no chemical changes, or in which the
changes are confined to minor adsorption reactions, e.g. tropical
kaolinites; (iii) those which are detrital mineral residues, but which have
undergone more important adsorption reactions, including the recon
stitution of degraded lattices, e.g. the formation of illite from degraded
illite; (iv) those which are formed from other detrital clay minerals by

TABLE 12.4

*The average concentration of the major clay species in the <2 μm clay fraction
of deep-sea sediments (from Griffin et al., 1968)*

Oceanic area	Chlorite	Average % clay Montmorillonite	Kaolinite	Illite
North Atlantic	10	16	20	55
South Atlantic	11	26	17	47
North Pacific	18	35	8	40
South Pacific	13	53	8	26
Indian	12	41	17	33

chemical diagenetic alterations which modify the basic clay mineral
lattice, e.g. illite and chlorite forming from montmorillonite; (v) those
which are formed from non-clay material by oceanic weathering pro‑
cesses, e.g. the formation of montmorillonite from the products of
submarine vulcanism.

(c) *The distribution of clay minerals in deep-sea sediments.* As a result of
several recent detailed studies the areal distribution of the major clay
mineral species in deep-sea sediments is now reasonably well understood.
The most comprehensive of these studies was made by Griffin et al.
(1968) who related the clay mineral distribution in the <2 μm fraction
of deep-sea sediments to the sources and transport paths of solid phases
from the continents, and to the introduction of submarine volcanic

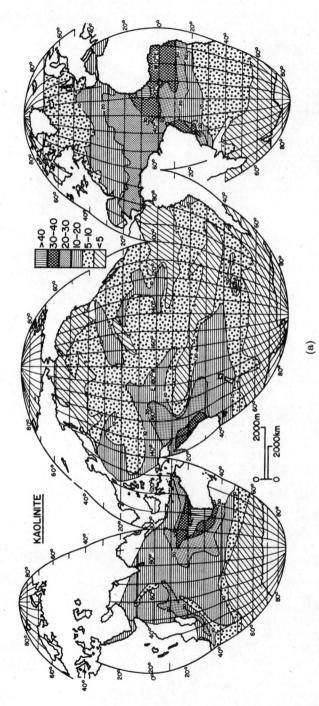

(a)

Fig. 12.5.(a) The distribution of the major clay mineral species (as a percentage of the four major clay minerals) in the <2 μm fractions of deep-sea sediments (Griffin *et al.*, 1968). (a) The distribution of kaolinite. (b) The distribution of chlorite. (c) The distribution of illite. (d) The distribution of montmorillonite.

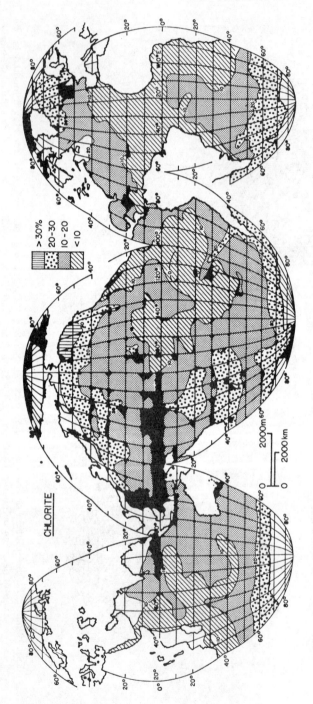

Fig. 12.5(b). For legend see p. 341.

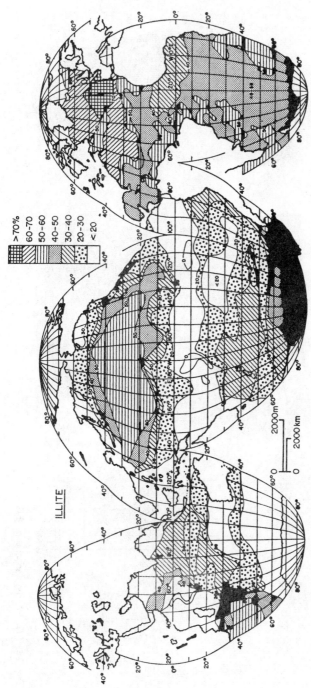

Fig. 12.5(c). For Legend see p. 341.

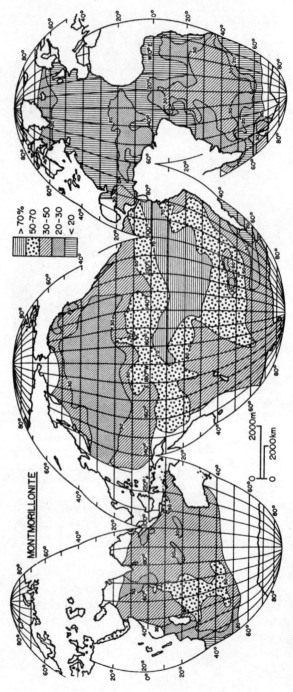

Fig. 12.5(d). For legend see p. 341.

materials into the oceans. The results of this, and other studies, are summarized in the following section in terms of the major clay species described above. The distributions of each species in the <2 μm fraction of the sediments of World Ocean are illustrated in Figure 12.5, a, b, c and d, and average concentrations in the individual oceans are given in Table 12.4, p. 340.

(1) Kaolinite. Although kaolinite can be formed under a variety of geological conditions, it is a characteristic product of intense tropical and desert weathering, and is concentrated in deep-sea sediments in areas between ca 25° N. and ca 25° S.—see Figure 12.5a. Because of its abundance in tropical areas Griffin *et al.* (1968) describe it as the *low latitude* clay mineral.

The high concentration of kaolinite in the <1 μm fraction of Atlantic equatorial sediments was noted by Yeroshchev-Shak (1961), and was later confirmed in the <2 μm fractions by Goldberg *et al.* (1963), Biscaye (1964) and Goldberg and Griffin (1964). All these authors have concluded that the high concentration of kaolinite in equatorial sediments results from tropical weathering products which are transported to the oceans from the African and South American mainlands by river and wind transport, and are confined to low latitudes by the equatorial current systems. The important rivers in the area are the Congo and the Niger in Africa, and the São Francisco in South America. In the area adjacent to the west coast of Africa eolian transport is important where off-shore currents flow towards the equator, i.e. from ca 10° N. to ca 35° N. (Beltagy and Chester, 1970).

On average, Pacific Ocean sediments contain only about half as much kaolinite as those of the Atlantic. The land area adjacent to the East Pacific, i.e. the west coast of the American continent, is mountainous, and the high altitude conditions of soil formation do not favour the production of kaolinite (Griffin *et al.*, 1968). In addition, few rivers drain into this region of the Pacific. Thus, the massive tropical kaolinite reservoir which is so important in the Atlantic catchment area is absent. In addition, the mountain ranges with their high rainfall tend to act as a barrier to any eolian material which may be carried across the Atlantic Ocean and American mainland from the African deserts, since the eolian suspensions are susceptible to washing-out by rainfall precipitation. The western Pacific has a higher content of kaolinite than the eastern area, particularly in the region around Australia. An area of high kaolinite concentration extends from the coast of Australia and branches into two long tongues, one to the southeast and one to the northeast (see Fig. 12.5a). According to Griffin *et al.* (1968), the kaolinite in the sediments off the north-east coast originates from river run-off, whereas that in the south-east branch probably originates in the desert areas of the Australian interior and has been transported by wind action.

The Indian Ocean has two areas of high kaolinite concentration. One extends seawards from Australia, and the decrease in kaolinite concentration with increasing distance from land clearly indicates that the kaolinite originates from the desert soils of Western Australia. The second, which is found around Madagascar, is probably due to river run-off from adjacent tropical land masses.

(2) *Chlorite.* Chlorite is strongly concentrated in the polar regions of the World Ocean and is termed the *high latitude* clay mineral by Griffin, *et al.* (1968)—see Figure 12.5b. The rocks of the polar land masses are abundant in chlorite. This undergoes little chemical weathering in polar conditions, where mechanical particle disintegration is the predominant weathering process, and ice rafting the principal transport mechanism. This chlorite may be regarded as a primary mineral, whereas most of the chlorites in low latitude sediments are probably weathering residues.

The clay mineral distribution maps produced by Griffin *et al.* (1968) indicate the main chlorite features in the Atlantic Ocean. They show concentration in the polar regions, and an area of high chlorite content adjacent to the St. Lawrence River, the latter resulting from run-off over a catchment area which is rich in chlorite originating from glacial till soils. The lowest concentrations of chlorite occur in tropical areas, i.e. it has an inverse relationship to kaolinite (see p. 345).

The distribution of chlorite in the Pacific Ocean is more complicated than the relatively simple pattern found in the Atlantic. In the Pacific, the chlorite distribution is dominated by three major features. (i) The highest chlorite concentration is found in high latitude areas, but in the northern hemisphere the chlorite-rich zone extends along the continental margin to ca 20° S., being particularly apparent off the coast of Alaska. The chlorite in these regions is thought by Griffin, *et al.* (1968) to have a continental detrital origin. (ii) There is a band of high chlorite content off the Hawaiian Islands. This chlorite probably has a secondary authigenic origin and is formed from volcanic material, or possibly from gibbsite. (iii) There are two zones of high chlorite content off the coast of Australia. The zone to the northeast is rich in basic volcanic mineral material and the chlorite is confined to the <2 μm size class. This led Griffin *et al.* (1968) to suggest that it may be associated with volcanic activity. The chlorite in the southern zone occurs in a different kind of mineral assemblage, and is found in both the <2 μm and the silt size fractions. According to Griffin *et al.* (1968) this suggests differences in both the chlorite source and the transport mechanism for the two zones; the southern zone receiving an eolian input of chlorite from the Australian mainland, whereas in the north-eastern zone the chlorite is formed by *in situ* weathering processes.

In the Indian Ocean there are no distinct areas of high chlorite concentration but, in general, values are slightly higher in the eastern parts of the region. This is probably due to the presence of kaolinite which acts as a diluent in the sediments adjacent to the African continent.

(3) *Illite*. Members of the illite group are the most common clay species in deep-sea sediments, and unlike chlorites and kaolinites they are not concentrated in particular latitudes—see Figure 12.5c. The detrital nature of the illites in Atlantic surface sediments has been confirmed by Hurley *et al.* (1963), who demonstrated that they are of the order of hundreds of millions of years old and are obviously derived from ancient terrestrial rocks. Most of the illites in deep-sea sediments originate from muscovite-type micas, and can be formed under a wide variety of geological conditions. For this reason the illite distribution in the World Ocean is controlled by the extent of the adjacent land area, by the efficiency of local transport mechanisms, and by dilution with other clays such as kaolinites. The main feature of the illite distribution is that it has a higher concentration in the northern than in the southern hemisphere. This reflects the larger amount of surrounding land area in the former hemisphere.

In the North Atlantic, which has the highest illite content of the World Ocean, the most striking feature is a broad band off the Spanish coast which consists of >70% illite. According to Griffin *et al.* (1968), this illite originates in the North American continent and is transported by the jet-stream Westerlies. To the west of this band the input from the St. Lawrence river system is evident. However, the most apparent areas of river transport are found in the South Atlantic, where tongues of high illite content extend seaward from the mouths of several rivers. In the equatorial areas the decreasing illite concentrations are attributable to dilution by kaolinite—see Figure 12.5c.

The sediments of the North Pacific Ocean contain much more illite than those of the South Pacific, reflecting the higher continental detrital inputs in the north. The highest illite contents in the North Pacific are in the central ocean area between 20° and 40° N. This corresponds to a zone of high quartz content which, according to Rex and Goldberg (1958), has an eolian origin, and Griffin *et al.* (1968) relate the illite input in this central band to the same jet-stream transport from the arid Euro-Asian land areas. In general, the illite content of the South Pacific sediments does not exceed ca 30%, except in isolated areas.

This is due to the relatively low detrital input in the South Pacific, and the dilution of other clays by montmorillonites. In this area the highest illite values are found in a band extending seaward from eastern Australia, a band in which the illite probably has an eolian origin (Griffin *et al.* 1968).

According to Gorbunova (1966), the distribution of illite in the Indian Ocean is dominated by inputs from rivers draining into the Bay of Bengal and the Arabian Sea in the north, and from the Antarctic and South Africa in the south. There is also a zone of low illite off the west of Australia, a region where other clays are diluted by a large input of kaolinite.

(4) *Montmorillonite.* The montmorillonites (or strictly, the "expanding lattice clays") in deep-sea sediments may be formed in at least three different ways: (i) as "primary" detrital weathering residues; (ii) as "secondary" detrital degraded lattices of the mica or chlorite types which have had their inter-sheet K^+ ions replaced by water molecules; (iii) as *in situ* products of the submarine weathering of volcanic material. Both the "primary" (i) and "secondary" (ii) montmorillonites can undergo modification in the marine environment. For example, the "secondary" montmorillonites formed from degraded lattices may have these lattices reconstituted on contact with sea water. This process may occur through a number of stages which involve the formation of mixed layer clays. Such mixed layer clays may also originate from "primary" lattices; and samples of this type have been identified in sediments off the North American coast by Yeroshchev-Shak (1964), who interpreted them as representing the relics of land-derived montmorillonites undergoing alteration in the marine environment. The distribution of montmorillonite is shown in Figure 12.5d.

In the North Atlantic Ocean montmorillonite is fairly evenly distributed among the sediments, with slightly higher amounts in equatorial regions. The latter may be due to the influence of detritus carried by the Amazon, or to volcanically derived montmorillonite from the Caribbean area. Compared with the North Atlantic, the central area of the South Atlantic has a higher content of montmorillonite which decreases towards the continental margins. According to Griffin *et al.*, (1968) this is the result of one, or both, of the following processes: (i) wind transported montmorillonite originating in the South Pacific being deposited over the South Atlantic, and being diluted in coastal areas by riverborne detritus rich in other clays; (ii) an input of volcanic materials from the Atlantic Ridge. Biscaye (1964), from a study of montmorillonite in the Atlantic, concluded that there was neither evidence of an association between montmorillonite and volcanic material, nor any features which indicated that the montmorillonite had formed from degraded lattices which were being reconstituted by the fixation of K or Mg from sea water (c.f. Yeroshchev-Shak, 1964), and he concluded that the montmorillonites were land-derived clays (see also Berry and Johns, 1966). This tends to support the suggestion made by Griffin *et al.* (1968) that in the South Atlantic the montmorillonites are of an eolian origin. There is some evidence, however, that in isolated parts of the Atlantic montmorillonite is forming from volcanic debris, e.g. in the Antilles Arc and Azores areas (Goldberg, 1964).

It is convenient to divide the Pacific Ocean into two broad regions; the North Pacific and the South Pacific, which have different sedimentary characteristics. The North Pacific has more rivers flowing into it and a much greater area of surrounding land than the South Pacific, features which suggest that the contribution of land-derived detritus will be larger in the north. This concept has been developed by Griffin and Goldberg (1963) who suggested that the North Pacific is an area of largely detrital deposition, whereas in the South Pacific authigenic deposition is more important. Evidence for this comes from several sources: (i) in the North Pacific the clay mineral assemblages are complex and variable, and have illites of great age which are obviously detrital in origin. In contrast, the sediments of the South Pacific tend to have a more homogenous clay mineralogy and are rich in authigenic phases; (ii) sedimentation rates of the non-carbonate fractions of the sediments in the North Pacific vary with locality and range between ca 0·5 and ca 6 mm/10^3 years. In the South Pacific there is a more uniform rate of ca 0·4 mm/10^3 years; (iii) in the North Pacific, detrital quartz can make up ca 25% of the sediments (Rex and Goldberg, 1958; Chester and Hughes, 1969), but in the South Pacific the quartz content of the sediments rarely exceeds ca 6%.

The differences between the North and South Pacific are particularly reflected in the distribution of montmorillonite. In the central North Pacific (which is rich in detrital illite) there is an average of ca 30% montmorillonite. In the central South Pacific, however, montmorillonite averages ca 70% of the total clay minerals. There is a great deal of evidence to suggest that the montmorillonite in the South Pacific deep-sea sediments has an authigenic origin, and is formed *in situ* from volcanic debris, particularly basaltic glass. The montmorillonites (particularly nontronite) often occur in association with zeolites and volcanic glass. In the South Pacific the highest concentrations of montmorillonite are found in mid-ocean areas, whereas in the North Pacific this mineral is concentrated along the continental margins and is introduced mainly by river run-off.

In the Indian Ocean montmorillonite is concentrated in central areas away from the land masses. This is interpreted by Griffin *et al.* (1968) as resulting from the large contributions of volcanic debris received by these areas, from which montmorillonite has been formed by *in situ* processes.

Because those areas which have high contents of montmorillonite are associated with vulcanism, Goldberg (1964) has characterized this mineral as indicative of a *volcanic regime*.

(*d*) *Summary.* From the clay mineral distribution patterns described above it is evident that the inter-species relationships in the <2 μm fraction of deep-sea sediments are complex, and all deep-sea sediments usually contain at least a small amount of each of the four major clays. However, certain trends are apparent in those localities where a particular clay is strongly concentrated. This is the result of specific patterns being imposed upon the distribution of the clay "background" material which, because of its small size, has a relatively long residence time in sea water and consequently tends to be well mixed. These distribution trends are summarized below.

(i) Chlorite is the clay mineral characteristic of high latitude areas.

(ii) Kaolinite is the clay mineral characteristic of low latitude areas.

(iii) Montmorillonite is characteristic of those regions rich in volcanic products.

(iv) Illite appears to be formed under under a wide variety of weathering conditions, and its distribution patterns in deep-sea sediments are controlled largely by the amount of dilution with other clay species.

(*iii*) *Feldspars*

Most deep-sea sediments contain small amounts of the feldspar group of minerals. The feldspars are igneous in origin and are one of the most important of all groups of rock-forming minerals, making up ca 60% of the earth's crust. There are three main groups of feldspars; (i) the potassium (or barium) feldspars, of which the most important is orthoclase ($(K,Na)AlSi_3O_8$) and its high temperature variety sanidine, both of which are monoclinic; (ii) the triclinic feldspars, microcline ($KAlSi_3O_8$) and anorthoclase ($(Na,K)AlSi_3O_8$); (iii) the plagioclase feldspars which are triclinic and which are mixtures of two isomorphous end members, albite ($NaAlSi_3O_8$) and anorthite ($CaAl_2Si_2O_8$), with the following albite (Ab): anorthite (An) ratios;

Mineral.	albite,	oligoclase,	andesine,	labradorite,	bytownite,	anorthite,
Ab-An ratio.	0–10	10–30	30–50	50–70	70–90	90–100

Both microcline and orthoclase are often found intergrown with albite to form perthite.

The type of feldspar found in an igneous rock is governed by the type of melt from which the rock is formed. In general, the alkali feldspars are dominant in acidic rocks, whereas in intermediate and basic rocks plagioclase becomes the most important feldspar, with the proportion of anorthite increasing as the rocks become more basic.

The feldspars usually occur in deep-sea sediments as mixed assemblages, and according to Arrhenius (1963) the dominant species in Pacific sediments range between labradorite and oligoclase. Peterson and Goldberg (1962) have made a detailed study of the distribution of feldspars in pelagic sediments of the South Pacific and concluded that they originated from volcanic activity within the area and could be used to identify volcanic processes. Basic vulcanism is dominant over most of the area, and the feldspars are mainly plagioclases of varying composition. A province of acid vulcanism was found associated with the East Pacific Rise; this was characterized by high temperature alkali feldspars (anorthoclase and sanidine) and volcanic quartz.

The detailed distribution of feldspars in the Atlantic Ocean is not as well known as that in the Pacific. Biscaye (1964) found that plagioclase and microcline feldspars were present in the 2–20 μm fraction of all sediments from the Atlantic area. Goldberg et al. (1963) reported that both alkali feldspars and plagioclase occurred in all samples in a section across the Atlantic from Gibraltar to Martinique. These authors concluded that the alkali feldspars in sediments east of the Mid-Atlantic Ridge had an eolian origin, and Delany et al. (1967) later showed that wind-transported dust collected at Barbados contained traces of feldspars. Goldberg and Griffin (1964) have reported that plagioclase is widespread in Atlantic sediments, and is concentrated to some extent in areas of volcanic activity such as the Antilles Arc region.

(iv) Other lithogenous components

A large number of terrestrial igneous, metamorphic and sedimentary (lithogenous) minerals are found in marine sediments. However, apart from quartz, clay minerals and feldspars, they usually make up only a very minor fraction of the sediments deposited in deep-sea areas.

In a study of the mineralogy of a series of Atlantic deep-sea sediments Goldberg and Griffin (1964) identified the following minerals in some, or all, of the samples; amphiboles, pyroxenes, muscovite, biotite, haematite, goethite, anatase, rutile, zircon, tourmaline, kyanite, sillimanite and pyrophyllite. Murray and Renard (1891) listed the minerals found in 390 sediments from all the major oceans and found that feldspars were present in 95% of all the samples, micas in 32%, pyroxenes in 62%, amphiboles in 51% and olivine in 17%.

The nature of the minor lithogenous minerals in deep-sea sediments depends on local geological conditions and on the dominant transport mechanism operative in a given area. For example, Goldberg and Griffin (1964) have shown that in areas of known vulcanism in the Pacific Ocean the minor mineral assemblages are different from those of Atlantic Ocean sediments which, overall, have no significant volcanic contribution. Latitudinal variations occur in the distribution of minor lithogenous minerals, a good example being the concentration of the minerals haematite, goethite, anatase and rutile in Atlantic equatorial sediments. These minerals are the end products of tropical weathering processes. Gibbsite, a hydrated aluminium oxide, is also common in soils formed by tropical weathering, and has been reported in Atlantic Equatorial deep-sea sediments, particularly off the coasts of Africa and Brazil (Biscaye, 1964).

TABLE 12.5

The hydrogenous components in deep-sea sediments

I. PRIMARY MATERIAL—formed directly from sea water.
Carbonates, e.g. calcite, aragonite, dolomite.
Phosphates, e.g. francolite, phosphate nodules, phosphorite-rock
Silicates, e.g. quartz overgrowths
Sulphates, e.g. barite, celestite
Oxides and hydroxides, e.g. ferro-manganese nodules, goethite
Adsorbed components (including those incorporated into deep-sea sediments by ion-exchange reactions), major elements—mainly alkali metals—trace elements, organic material.

II. SECONDARY MATERIAL—this results from the submarine weathering of pre-existing minerals, which are mainly igneous.
Montmorillonite (including nontronite)
Zeolites

Other lithogenous minerals found in marine sediments include calcite, dolomite, apatite, chromite, garnet, serpentine, tourmaline, zircon, etc. and it is likely that small amounts of all the terrestrial minerals are present in the sediments.

Most of the lithogenous components in deep-sea sediments are either rock fragments or mineral weathering residues. These residues consist either of unaltered, or partially altered, primary minerals such as quartz and feldspars, or secondary minerals incorporating the solid products of weathering, e.g. the clay minerals. In addition, lithogenous components introduced from the continents include material precipitated from weathering solutions. Iron and manganese oxide mineral particles are particularly important members of this category since trace elements adsorbed by them in the terrestrial environment can be carried on then into the oceans. Aluminium hydroxide particles and gels are also important and may possibly act as loci for clay mineral formation.

III. HYDROGENOUS COMPONENTS

Hydrogenous components are defined as those which result from the formation of solid matter in the sea by inorganic reactions, i.e. non-biological processes (Goldberg, 1954). Chester and Hughes (1967) divided the hydrogenous components of deep-sea sediments into two broad groups: (i) primary material, which is formed directly from sea water; (ii) secondary material, which results from the submarine alteration of other pre-existing minerals. A summary of this classification is given in Table 12.5.

A. PRIMARY MATERIAL

(i) *Carbonates*

In deep-sea sediments most of the carbonates are biogenous (or lithogenous), but inorganic carbonates may be produced in the marine environment. These include calcite, aragonite, dolomite, magnesite, ferroan magnesite, siderite and hydromagnesite. However, hydrogenous carbonates are not commonly reported in the deep-sea environment although they can occur in the sediments. For example, Bonatti (1966) has described the precipitation of carbonate minerals from hydrothermal solutions generated by submarine vulcanism. The presence of a manganese carbonate in a tropical core from the east Pacific has been reported by Lynn and Bonatti (1965). Dolomite has been identified in deep-sea sediments, although its occurrence is rare. Most of it is probably detrital—see for example, Delany *et al.* (1967)—but Zen (1959) has suggested that the dolomite in sediments of the Peru-Chile Trench has an authigenic origin. Lithified carbonates such as those described by Thompson *et al.* (1968), represent a specialized deep-sea carbonate, the significance of which is not yet fully understood.

(ii) *Phosphates*

There are four types of hydrogenous phosphate material; phosphate minerals such as francolite (and rare earth phosphates), phosphatic nodules, phosphatic pellets and phosphorite-rock.

Deposits of phosphatic nodules, pellets and phosphorite-rock are confined mainly to near-shore areas in depths of water not exceeding ca 1000 m. They have been reported off the coasts of the following

14

localities; Cape of Good Hope, eastern Japan, Spain, Chile, eastern Australia, California, eastern U.S.A., western South America, the Argentine, the Falkland Islands, New Zealand, North West Africa and the Blake Plateau (Murray and Renard, 1891; Dietz et al., 1942; Gorsline, 1963; Mero, 1965; Pratt and McFarlin, 1966; Summerhayes, 1967; Tooms and Summerhayes, 1968).

Phosphatic nodules, which have a specific gravity of 2·7–2·8, are hard and dense, often with smoothly rounded protuberances and cavities, and range in shape from flat slabs to irregular masses. The surface may have a glazed appearance and usually consists either of a thin discoloured layer of phosphorite or of ferro-manganese oxides, when it is black or brown in colour (Dietz et al., 1942). The interior of the nodules often shows no apparent structure, but may display an irregular non-concentric layered structure. The nodules often contain foreign material such as clay minerals, glauconite, carbonaceous matter and siliceous material. Phosphorite-rock may occur massive, and as coatings and cavity fillings on existing rock surfaces; it may have a primary or a replacement origin. Other types of phosphorite-rock include partially or wholly phosphatized limestones and conglomerates containing phosphorite pebbles.

Modern phosphatic assemblages can occur in at least two kinds of association. (i) In areas of high organic productivity. Here the phosphorite-rock consists of a matrix of authigenic phosphate minerals with phosphatized tests of foraminifera, skeletal apatite, opaline silica, glauconite, organic matter, pyrite and terrigenous minerals, i.e. reducing conditions are prevalent during their formation. (ii) In areas which do not have a high organic productivity, particularly in shallow water tropical regions where calcareous deposits are exposed to relatively warm sea water, e.g. on seamounts and drowned coral reefs. In phosphorites formed under these conditions intergrowths with iron and manganese oxide minerals are common, but glauconite and organic matter are absent, i.e. oxidizing conditions are prevalent in the environment (Arrhenius, 1963).

The principal minerals in hydrogenous phosphate deposits are collophane* and francolite, both of which are carbonate-fluor-apatites, and lesser amounts of hydroxy-carbonate-apatites, e.g. dahllite. Chemically, hydrogenous phosphates approximate to the composition of francolite, i.e. $Ca_5(PO_4,CO_3,OH)_3F$, but variations occur as a result of substitutions in the mineral lattice. These substitutions include: Na, Sr, Mn, Mg and

---

* According to Deer, et al. (1962), collophane is often used as a collective term when the apatite-like phase cannot be identified, but may refer to finely crystalline or amorphous carbonate-fluor-apatite.

the rare earths for Ca; and CO_3^{2-}, SO_4^{2-} and SiO_4^{4-} for PO_4^{3-} (Kramer, 1966). The net positive charge which results from the substitution of CO_3^{2-} for PO_4^{3-} is balanced by the introduction of F^- and/or OH^- into the lattice. The trace element geochemistry of marine phosphates has been reviewed by Tooms et al. (1969).

Although some of the phosphate deposits are undoubtedly detrital, others are hydrogenous, and several theories have been put forward to explain the existence of the latter deposits. The two most probable are: (i) that they are primary chemical precipitates; (ii) that they are an incomplete replacement product formed from pre-existing carbonate material.

(i) The chemical precipitation theory (see Kazakov, 1937 and 1950) may be summarized as follows; in the uppermost 100 m of ocean waters phosphorus is taken up by phytoplankton and is returned to solution as orthophosphate ions on death of the organisms when they sink down the water column (Armstrong, 1965). According to Arrhenius (1963), sea water is undersaturated with respect to apatite (but c.f., Kramer, 1964, below) which is thus being dissolved under the temperature and pH conditions existing on the deep-ocean floor. As a consequence of this, any phosphorus which is brought down by organisms does not precipitate from solution upon hydrolysis of the organic matter, and in areas of upwelling (see p. 43) it is returned to the upper water layers. This upwelling of phosphate-rich deep water is accompanied by a decrease in the partial pressure of CO_2 and an increase in pH in the surface layers where the phosphate promotes high productivity. As a result, phosphate saturation can be reached in some shallow low latitude areas where the increased temperature and pH of the water can result in the precipitation of phosphate minerals such as carbonate-fluor-apatite. In addition to the upwelling of phosphate-rich deep water masses, Kramer (1966) has suggested that river run-off of continental water which has flowed over carbonate rocks, and so has a high pH, may cause the precipitation of marine phosphates.

Various authors have attempted to relate the solubility relationship of the phosphate minerals in sea water to natural phosphate precipitation—see for example Kazakov (1937), Dietz et al. (1942), Sillén (1961), Kramer (1964), McConnell (1965) and Pytkowicz and Kester (1967). However, the solubility relationships are not fully understood, and the results of the various studies are difficult to evaluate because they are related to the formation of a variety of phosphate precipitates. These include calcium phosphate, carbonate-fluor-apatite, and hydroxy-carbonate-apatite. Kramer (1964) has recalculated the available data in terms of the precipitation of carbonate-fluor-apatite—the actual phase in marine phosphates—and has found that Pacific Ocean water appears to be slightly supersaturated with respect to this mineral. Because of this, McConnell (1965) has concluded that in normal seawater, natural inhibiting factors must exist to prevent the precipitation of carbonate-fluor-apatite. McConnel also stressed the importance of carbonate or bicarbonate ions in the precipitation of carbonate-fluor-apatite.

In a recent study, Pytkowitz and Kester (1967) have concluded that it is not yet possible to determine the absolute degree of saturation of sea water with respect to calcium phosphate. However, the authors measured the relative calcium phosphate saturation in two regions of the North Pacific. The sediments of one region contained phosphorites, the sediments of the other did not. Pytkowitz and Kester found that there was a higher degree of relative saturation with respect to calcium phosphate in the waters of the region where phosphorites occurred in the sediments. However, this higher degree of relative saturation did not result from a higher total phosphate concentration, but was due to differences in the pH of the waters. In the region in which phosphorites were present in the sediments, the sea water had a higher pH than it had in the region in which they were absent. The lower acidity in the former region caused a larger fraction of the inorganic phosphate to be present in the form of PO_4^{3-} ions. This led the authors to conclude that the presence of phosphorites in the sediments was due to geochemical or biochemical factors which controlled the pH of the waters, rather than to high concentration of inorganic phosphorus.

There is no doubt than an association exists between upwelling and the formation of hydrogenous marine phosphates. However, it is evident that the factors controlling the precipitation of phosphate minerals from sea water are not fully understood. Many details of the reactions are still to be established, and when these are known it may be found that the model for phosphate precipitation given above, which involves supersaturation resulting from the upwelling of phosphate-rich waters, may be a much over-simplified version of the events actually occurring. For example, Tooms et al. (1969) have pointed out that most of the phosphates found in areas of upwelling are not geologically recent deposits, but range in age from Cretaceous to Miocene. One reason for this may be that the warmer seas of the mid-Tertiary promoted phosphate deposition. Kolodny and Kaplan (1970) have also considered the problem of the age of marine phosphatic nodules. The authors determined $^{234}U/^{238}U$ ratios in a number of nodules from off the coast of California, Chatham Rise, the Agulhas Bank, and the Blake Plateau. Nodules from all these regions were found to be of Miocene-Pliocene age.

The authors concluded that the nodules are being eroded rather than deposited, and that it is doubtful if phosphates are being formed under present day conditions in the areas investigated. However, Kolodny and Kaplan (1970) agree that there is an association between upwelling and the formation of marine phosphates, and consider that a combination of upwelling and higher temperatures in the Miocene-Pliocene caused the phosphate deposition; however, they do not exclude the possibility of phosphate deposition at the present time in some areas of upwelling, e.g. the western margins of Africa, and Central and South America.

(ii) The theory that marine phosphates originate from the inorganic replacement of existing carbonate material is based on the fact that phosphate-rich solutions can convert calcite to carbonate-apatite within a sediment (Tooms $et al.$, 1969). The process occurs by the partial replacement of CO_3^{2-} groups by PO_4^{3-} groups, and may take place at the sediment-water interface, or at depth within a sediment, when phosphate is concentrated in interstitial waters.

Although there is no general agreement on the physico-chemical conditions of authigenic marine phosphate formation, it is probable that in areas of upwelling both direct precipitation and replacement mechanisms are involved in the genesis of phosphates.

(*iii*) *Silicates*

Primary hydrogenous silicates which have been found to occur in deep-sea sediments include overgrowths of quartz and alumino-silicates, and chert laminae precipitated from silica-rich interstitial waters (Arrhenius, 1963). Authigenic orthoclase has also been found in a sediment from the western Atlantic. Other authigenic silicates which have recently been reported in deep-sea sediments include palygorskite and sepiolite. Hathaway and Sachs (1965) found sepiolite in samples dredged from the Mid-Atlantic Ridge, and concluded that it had been formed authigenically by the reaction of magnesium in solution with silica released during the weathering of silicic volcanic glass. Heezen $et al.$ (1965) have reported the presence of palygorskite in sediments from the Gulf of Aden and the Red Sea, and it has also been found in deep-sea sediments from the Atlantic (Bonatti and Joensuu, 1967).

(*iv*) *Sulphates*

Primary hydrogenous sulphates include barite, celestite, anglesite, gypsum and anhydrite. The majority of marine sulphate minerals, like the halides, are evaporites, and are usually only found in some near-shore areas where there is a restricted inflow of sea water and extreme solar heat (see also p. 84).

Barite is the only sulphate mineral which is important in deep-sea sediments, and it is produced by both hydrogenous and bi.genous mechanisms. For convenience, both types are considered together in this section. Barite can be precipitated from sea water through the agency of planktonic productivity, and the barium content of deep-sea sediments is usually highest in areas underlying productive waters (Goldberg and Arrhenius, 1958). In Atlantic deep-sea sediments barium is concentrated on topographic highs such as the flanks of the Mid-Ocean Ridge system. According to Turekian (1968), this barite has been precipitated in association with planktonic organisms, and has subsequently sunk down the water column. The solubility of barite increases with pressure and this results in its preferential preservation on topographic highs. Another effect of the increase in barite solubility with depth is the increase in dissolved barium down the water column. Barium profiles of this type have been reported by Chow and Goldberg (1960), who suggested that the release of barium from organisms at the sediment surface also contributes to the higher barium contents of the deep-waters.

The distribution of barite in Pacific deep-sea sediments has been discussed by Arrhenius and Bonatti (1965). These sediments can be divided into three regions on the basis of their barite contents—see figure 12.6. (i) The highest concentrations of barite are found in the region of the East Pacific Rise, where the sediments contain between 5 and 10% barite (on a carbonate-free basis). The authors concluded that barium was enriched in the overlying waters because of volcanic activity, and that barite has precipitated from sea water inorganically as euhedral crystals.* Further evidence for the volcanic origin of the barium has been given by Böstrom and Peterson (1966).

(ii) The sediments of the Equatorial Zone have barite contents of between ca 1 and 5% (on a carbonate-free basis). In order to explain the genesis of this barite, Arrhenius and Bonatti (1965) have suggested that waters which have crossed the East Pacific Rise have been enriched in barium, which in this area has a volcanic origin. At the equator these barium-rich waters pass through a curtain of marine organisms which thrive in the highly productive surface waters. The barium is concentrated by some of the organisms, such as the Xenophyophora which secrete barite granules. These organisms subsequently fall to the sea floor with the result that barite granules, a few micrometres in size, are concentrated in the sediments.

(iii) Other deep-sea sediments in the Pacific have barite concentrations of <1% (on a carbonate-free basis).

* According to Arrhenius and Bonatti (1965) this type of barite contains ca 5% celestite (SrSO₄) in solid solution, and may be referred to as celestobarite.

Barite nodules have been found in the marine environment, but are restricted to shelf areas, e.g. off the coast of Ceylon, the East Indies and California. The genesis of barite nodules has been discussed by Revelle and Emery (1951), and Goldberg *et al.* (1969).

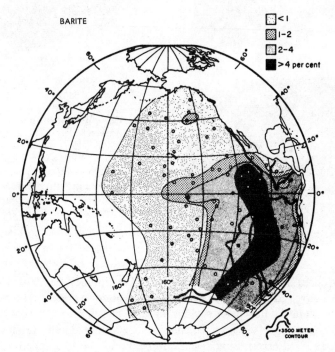

Fig. 12.6. The distribution of barite at the sediment surface of the Pacific Ocean (weight %, on a carbonate-free basis). The map shows the high barite concentrations on, or near, the East Pacific Rise (heavy contours), and the intermediate barite contents in the equatorial zone (Arrhenius, 1965).

(v) Oxides and hydroxides

Iron and manganese oxide minerals are extremely important components of all types of marine sediments. For a description of the iron minerals in near-shore sediments see Borchert (1965). One of the most important occurrences of iron and manganese, particularly in pelagic sediments, is in the form of ferro-manganese nodules, and few single topics in marine sedimentary geochemistry have received as much attention in recent years.

Ferro-manganese nodules were first dredged up during the Challenger Expedition and were described by Murray and Renard (1891). The

(a)I   (a)II

(b)

Fig. 12.7. Manganese nodules from
pelagic environment. (Photographs b
Murphy). (a)I External view of a typ
manganese nodule. (a)II Cross sectio
the nodule showing a concentric serie
light and dark growth zones (see text)
Cross section of a nodule showing ra
bands of foreign material (probably clay

nodules occur on, and in,* a wide variety of sediments, particularly those
having a low sedimentation rate, and are common in the Pacific and
Indian Oceans; it has been estimated that they cover between 20 and
50% of the pelagic sea floor in the south western Pacific (Menard and
Shipek, 1958). Ferro-manganese nodules, which have a specific gravity
of ca 2·4, are porous and often have a bulbous appearance (see Figure
12.7). They have been found in deep-sea sediments from all the major
oceans and in various near-shore and fresh water sediments. The
occurrence of near-shore nodules has been reviewed by Manheim (1965),
and that of pelagic nodules by Mero (1965).

The mineralogy of pelagic ferro-manganese nodules was first investigated in detail by Buser and Grütter, and was described in a number of
papers (Buser and Grütter, 1956; Grütter and Buser, 1957; Buser, 1959).
The main structural elements of the nodules include at least three manganese oxide and hydroxide minerals—δMnO_2, and two forms of manganese
(II) manganite. δMnO_2 forms aggregates of randomly orientated sheets.
The manganites possess a double layer structure, and consist of ordered
sheets of MnO_2 which contain between them disordered layers of metal
ions (chiefly Mn^{2+} and Fe^{3+}) co-ordinated with O^{2-}, OH^- and H_2O. All
nodules contain at least a small amount of iron, and some of the Mn^{2+}

* Present data indicates that macro-nodules are about twice as abundant on the
sediment surfaces as in the first metre of the sediment column.

in the disordered layers of the manganites is replaced by Fe^{3+} to form ferric manganite, and some of the Mn^{4+} in the main layers may also be replaced by Fe^{3+} with the development of a charge which is balanced by the addition of other cations—see below. Two types of manganite were identified on the basis of differences in the disordered $Mn(OH)_2$ layers which can have a basal spacing of 7 or 10 Å. In the 10 Å manganite the water and hydroxyls are probably present as discrete layers, whereas in the 7 Å manganites they form a single layer. According to Arrhenius (1963), the three major manganese minerals in the nodules represent increasing stages of oxidation at formation in the order $\delta MnO_2 > 7$ Å-manganite > 10 Å-manganite. In those nodules which contain more iron than can be held in the manganite phases the iron is present as an amorphous hydroxide, $FeOOH . nH_2O$, some of which may be converted to goethite (α-FeOOH). Trace elements are often concentrated in ferro-manganese nodules, particularly in the disordered manganite layers where Mn^{2+} may be replaced by other polyvalent ions, e.g. Ni^{2+}, Co^{2+}. In addition, trace elements may also be concentrated in the iron phases of the nodules—see later.

In a number of recent studies specific names have been assigned to the mineral phases originally identified by Buser and Grütter (1956). The terminology used is confusing, but in general, 10 Å-manganite is referred to as *todorokite*, and δMnO_2 as *birnessite*. According to some authors, e.g. Cronan and Tooms (1969), 7 Å-manganite and δMnO_2 are mineralogically similar, and differ only in particle size. Because of this they are sometimes both referred to as birnessite.

To summarize, it appears that the manganese minerals in the nodules are dominated by todorokite and birnessite, and the iron minerals by hydrated iron oxide. Croonan and Tooms (1969), and Barnes (1967) have suggested that the mineralogy of the nodules is controlled by their depth of deposition; todorokite occurring at depths of water in excess of ca 3000 m, and δMnO_2 at lesser depths, particularly on topographic highs. However, Price and Calvert (1970) have concluded that the mineralogy of the nodules is dependent on overall oceanic sedimentation patterns, and in particular on the source of the manganese in them (see also p. 367).

In cross section, many ferro-manganese nodules display an alternating concentric series of light and dark bands (see Figure 12.7) which, according to Arrhenius (1963), correspond to growth zones of high and low geothite content. This concentric build-up is often combined with a radial structure in which dendritic growths of manganese dioxide penetrate inwards through several layers (Riley and Sinhaseni, 1958). Often the nodules are built up around a nucleus of foreign material such

TABLE 12.6

Elemental analyses of ferro-manganese nodules (wt. %)

	1	2	3	4	5	6	7	8	9	10
Mn	16·63	19·0	19·18	—	24·2	15·3	16·3	—	18·97	22·06
Fe	13·11	14·0	12·16	11·4	14·0	—	17·5	17·0	11·68	14·58
Ni	0·66	0·46	0·47	0·49	0·99	0·56	0·42	0·68	0·58	0·57
Co	0·42	0·28	0·36	0·39	0·35	0·30	0·31	0·17	0·28	0·34
Cu	0·44	0·55	—	0·39	0·53	0·12	0·20	0·29	0·40	0·33
Pb	0·068	—	—	—	0·090	—	0·10	—	0·10	0·15
Zn	—	—	—	—	0·047	—	—	—	0·04–0·40	0·350
Cr	0·0016	—	—	—	0·001	—	0·002	—	0·001	0·001
V	0·045	—	—	0·053	0·054	0·047	0·070	0·06	0·044	0·059
Bi	—	—	—	—	—	—	—	—	0·001	—
Be	—	—	—	—	—	—	—	—	0·0003	—
Cd	—	—	—	—	0·029	—	0·03	—	0·001	0·001
B	—	—	—	—	0·001	—	—	—	—	0·0295
Ga	—	—	—	—	—	—	—	—	0·001	0·0017
Hg	—	—	—	—	0·003	—	—	—	0·0001–0·001	0·0011
Sn	—	—	—	—	—	—	—	—	—	—
W	—	—	—	—	—	—	—	—	0·038	0·068
Mo	0·034	—	—	0·056	0·052	0·044	0·035	0·054	0·01	0·007
Tl	—	—	—	0·007	—	0·007	—	0·007	0·001	0·0013
Sc	—	—	—	—	0·001	—	0·002	—	—	—
Zr	—	—	—	—	—	—	—	—	—	—
K	—	—	—	—	0·80	—	0·700	—	—	1·00
Na	—	—	—	—	2·6	—	2·3	—	—	2·55

	1	2	3	4	5	6	7	8	9	10
Mg	—	—	1·47	1·60	1·7	1·8	1·7	1·80	—	1·47
Ca	0·227	—	1·81	—	1·9	—	2·7	—	—	1·78
Ba	—	—	—	—	0·18	—	0·170	—	0·15	0·31
Sr	0·797	—	—	—	0·081	—	0·09	—	0·06	0·10
Ti	—	0·810	0·60	0·81	0·670	0·38	0·800	0·37	—	0·61
Al	—	0·730	2·87	—	2·9	—	3·1	—	—	2·36
Si	—	—	8·05	10·0	9·4	7·2	11·0	8·5	—	8·35
P	—	0·540	0·18	—	—	—	—	—	—	0·28
La	—	—	—	—	0·016	—	—	—	0·01–0·1	0·023
Y	—	—	—	—	0·016	—	0·018	—	0·013	0·012
Ce	—	—	—	—	—	—	—	—	0·01–0·5	—
Yb	—	—	—	—	—	—	—	—	0·0025	—
Zr	—	0·0064	—	—	0·063	—	0·054	—	0·040	0·034

1. Average of 139 nodules (Cronan, 1969; in Tooms et al., 1969).
2. Goldberg (1954); average of 33 nodules from the East Pacific.
3. Skornyakova et al., (1962); average of 31 nodules from Pacific Ocean.
4. Willis and Ahrens (1962); average of 7 nodules from Pacific Ocean.
5. Mero (quoted in Arrhenius, 1963); average bulk composition of Pacific Ocean nodules.
6. Willis and Ahrens (1963); average of 8 nodules from Atlantic Ocean.
7. Mero (quoted in Arrhenius, 1963); average bulk composition of Atlantic nodules.
8. Willis and Ahrens (1963); average of 4 nodules from Indian Ocean.
9. Manheim (1965); deep-ocean average nodule composition.
10. Chester (1965); average oceanic nodule composition.

as pumice, consolidated clay, sharks teeth, volcanic glass or igneous rock fragments, and may contain considerable amounts of other minerals such as quartz and clay minerals, etc.

Chemical analyses of ferro-manganese nodules have been made by many authors, and a number of these have been collected together by Mero (1965), and Chester (1965). A summary of some of these analyses is given in Table 12.6. One of the most striking features of the pelagic nodules is their ability to concentrate trace elements from sea water. These trace elements include Cu, Co, and Ni which can attain concentrations of ca 1% in the nodules (Goldberg, 1961). Recent studies have indicated that the trace element content of the nodules is controlled to some extent by their mineralogy. For example, Croonan and Tooms (1969) have shown that Ni and Cu are concentrated in those nodules which are rich in todorokite, and that Co and Pb are concentrated in those nodules in which δMnO_2 predominates.

There have been various theories advanced to explain the origin of ferro-manganese nodules. Basically, there are three major problems which require explanation. These are: (i) the origin of the manganese which is incorporated into the nodules; (ii) the mechanism which brings the manganese to the reaction site; (iii) the mechanism by which the nodules are actually formed. Each of these will now be discussed individually.

(i) The origin of the manganese supplied to the deep-sea areas has been discussed by Petterson (1945), Goldberg (1954), Goldberg and Arrhenius (1958), Lynn and Bonatti (1965), Bonatti and Nayudu (1965), and many other authors. There are three major pathways by which manganese in solution can reach oceanic sediment surfaces; continental weathering, submarine volcanic activity, and migration from sub-surface sediments. Continental run-off and submarine vulcanism have been reviewed by Arrhenius et al., (1964). There is extensive evidence for the introduction of manganese into sea water by hydrothermal solutions locally associated with submarine vulcanism (Bonatti and Nayudu, 1965). The nature of these volcanic submarine phenomena has been discussed by McBirney (1963), and the alteration of solid submarine volcanic material has been described in a number of papers, e.g. Bonatti and Nayudu (1965). However, Arrhenius et al. (1964) have pointed out that manganese is also supplied by weathering processes operative in continental areas, and have concluded that both sources supply manganese to oceanic regions. The importance of each source will vary locally, and Arrhenius et al., have attempted to find a criterion which could be used to characterize nodules which contain manganese dominantly supplied by one of the two processes. They argued that considerable amounts of the transition ions in their lower oxidation states may be released when submarine effusions occur. Since many of these ionic species (including those of iron and manganese) are unstable at the normal oxygen content of sea water, they will be precipitated in areas adjacent to an eruption, or any other kind of hydrothermal activity. The resultant precipitation of disordered solids will take place rapidly from a relatively concentrated solution. In contrast, manganese from continental weathering, and any from volcanic source which is not precipitated locally, will be

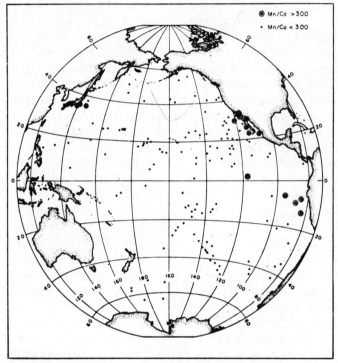

Fig. 12.8. Distribution of the atomic ratio Mn/Co in Pacific ferro-manganese nodules (Arrhenius *et al.*, 1964).

present in dilute solution. Precipitation from such solutions will be slow and will give rise to solids having a high degree of order in their crystal structure. The degree of order itself cannot be used to distinguish between the two types of nodules since the poorly ordered varieties may undergo some degree of recrystallization. For this reason Arrhenius *et al.*, (1964) sought an element which would be closely associated with manganese in the nodules formed from volcanically supplied material. Cobalt was selected as an indicator element, and a study of the Mn : Co ratios of a number of Pacific Ocean nodules showed that the ratios could be split into two broad groups (see Figure 12.8). Nodules from the pelagic areas have Mn : Co ratios of <300 : 1, indicating a relatively high cobalt content and a derivation by rapid precipitation from solutions in which the elements had a dominantly volcanic supply. Nodules from the continental margin regions, in contrast, have Mn : Co ratios of >300 : 1 which Arrhenius *et al.* (1964) interpreted as representing slow precipitation from dilute solutions of dominantly continental origin. However, Manheim (1965) has concluded that volcanic emanations in the Pacific Ocean are totally inadequate to supply the necessary quantities of dissolved manganese, and has concluded that the continents supply the major proportions of this element to Pacific Ocean waters. Price (1967) has also questioned the importance of submarine vulcanism as a source of manganese, and has related the differences between the marginal and deep-sea nodules of the Pacific to a difference in the content of organic matter in the sediments in the two areas. He suggested that the nodules in marginal areas are similar to near-shore nodules, and that the manganese in them has migrated upwards, as a result of reducing conditions, under the influence of active organic constituents in the sediments—see later. It is probable, therefore, that manganese is supplied to the ocean from submarine volcanic activity, continental stream run-off, and post-depositional migration.

(*ii*) Some of the manganese is undoubtedly precipitated directly from sea water, but other explanations for its concentration at the sediment surface have been offered. One of these invokes an organic mechanism for the primary incorporation of manganese into the bottom sediments. For example, Graham (1959) has suggested that stable organic complexes of manganese are formed in marine organisms by metabolic processes, and that these complexes could act as a source of manganese for some nodules. Another explanation involves the upward migration of manganese in the sediments.

Near-shore nodules have several important differences from those deposited in pelagic areas. These differences have been summarized by Manheim (1965) as follows: near-shore (shallow water) nodules have slower accumulation rates, metals in lower oxidation states, a higher content of organic matter, and lower concentrations of trace elements such as Co, Ni, Cu etc., than pelagic, or open-ocean nodules. Manheim has suggested that the major difference between deep-ocean and both near-shore and fresh water nodules is that in the former the metals are supplied from overlying waters, whereas in the latter the metals are supplied from depth within the sediment by post-depositional migration processes. The importance of these migration processes has also been stressed by Price (1967) who has related them to the organic matter content of the sediments. This concept has been elaborated by Price and Calvert (1970) who have concluded that Pacific Ocean nodules have a continuous range of chemical and mineralogical composition between those occurring on the *marginal sea floor*, and those occurring on the tops of *deep-sea seamounts*. These authors consider that the variations in the nodules result from differences in sedimentation patterns in various parts of the Pacific Ocean, and in particular from the differing rates of deposition of the sediments. The rate of deposition effectively controls the amount of organic matter in the sediments, and this, in turn, governs the extent to which post-depositional migration occurs. For Pacific deep-sea sediments the rate of deposition is highest in the marginal sediments which consequently have a relatively high content of organic matter. Price and Calvert (1970) have suggested that because of this, marginal sea floor nodules have acquired their manganese by post-depositional migration from the underlying sediments. Some of this manganese is in a reduced state, and the principal manganese mineral in the nodules is todorokite. In contrast to the marginal sediments, those deposited on top of deep-sea seamounts have a very low rate of sedimentation, and are almost free from organic matter. Diagenetic modification in these sediments is therefore at a minimum, and the manganese in nodules formed under these conditions is derived directly from sea water and is present as δMnO_2(birnessite). Nodules formed on the pelagic sea floor and marginal seamounts are intermediate in composition between the two extreme types described above.

(*iii*) Manganese, present as Mn^{2+}, has been found in some types of ferro-manganese nodules, and since oxidizing conditions normally prevail on the ocean bottom, some process which lowers the redox potential must be operative during the formation of this kind of nodule. Arrhenius (1967) has postulated that the most likely processes causing this are the injection of hydrothermal solutions, or the action of micro-organisms. However, the major part of the manganese in the majority of nodules is in the form of Mn^{4+} (Goldberg and Arrhenius, 1958). Most evidence points to the fact that almost all of the manganese in sea water is present as Mn^{2+} ions, so that for those nodules which have acquired manganese from sea water, oxidation must occur as part of the accretion process on the ocean floor. Goldberg and Arrhenius (1958) have examined this problem and have concluded that ferric oxides provide a surface on which adsorbed manganese could be oxidized catalytically by dissolved oxygen, and that this may be the mechanism occurring on the initiation of nodule formation. Since, in this theory, the oxidation of manganese requires catalysis by and on the iron oxide surface, the amount of manganese deposited will depend on the length of exposure of any surface to sea water, and on the rate of supply of ferric oxides. For this reason, ferro-manganese nodules are found most frequently in areas having low rates of deposition of sedimentary components other than ferric oxides. In areas of rapid deposition, the iron oxides will be buried before nodule formation can proceed to any extent. In general, this theory offers an explanation for the overall distribution of the nodules in oceanic areas, i.e. they have their highest concentration in pelagic areas which have a low rate of clay sedimentation, and in areas which are scoured by bottom currents.

From the various theories discussed above, it will be evident that the genesis of ferro-manganese nodules is not fully understood. The nodules occur in a variety of environments, which include fresh water, near-shore, deep-sea margin and deep-sea localities. There are differences between the nodules from these various environments which appear to depend largely on the following factors: the organic matter content of the sediments (and consequently their redox environment); the depth of the overlying water; the supply and removal mechanisms of trace elements; the rate of accretion of the nodules and the rate of deposition of the sediments; the mechanism by which manganese is introduced into the oceans, and by which it is brought to the sediment surface. The latter includes migration from below, solution of material (e.g. organic aggregates) within the sediment-interstitial water complex, and the removal of manganese directly from sea water into the nodules.

The rate of growth of ferro-manganese nodules has been studied by various authors and the results have been summarized by Arrhenius (1967). No firm conclusion on the rate of accretion can be drawn since the radio-active dating methods employed involve assumptions which may be incorrect. However, data available at present points to the nodules growing at a slower rate than the sediments on, and in, which they occur. This necessitates some mechanism which keeps the majority of nodules permanently at the sediment surface since they are more widespread at the surface than at depth in most oceanic regions. Mechanisms suggested include the action of burrowing organisms, and the assumption that the nodules have a positive buoyancy. Nodules are also common in areas where strong currents are operative, and in these locations the winnowing action may remove sediment particles, leaving the nodules on the surface.

Several authors have shown that the composition of ferro-manganese nodules varies geographically. Mero (1965) has classified the nodules of the eastern Pacific into several "ore provinces" on the basis of variations in their Mn, Ni, Co and Cu content. Willis and Ahrens (1962) have shown that nodules from the Atlantic are also variable in composition, and on average contain less Mn, Co, Cu, Mo and Ti, and more Fe and Ni than those from the Pacific Ocean. Cronan and Tooms (1967) have demonstrated that the composition of ferro-manganese nodules can vary over very short distances.

In addition to macro-nodules which are found largely on the sediment surface, micro-nodules are often found dispersed throughout pelagic sediments. These nodules have a strong influence on the geochemistry of the sediments, and have been discussed in detail by Chester and Hughes (1966, 1969). These authors have also discussed the importance of the iron oxide goethite as a pelagic sedimentary component. Goethite can occur as part of the structure of ferro-manganese nodules (see above), or as a discrete mineral. Goldberg and Arrhenius (1958) have suggested that the non-nodular goethite may have originated as a "free" colloidal iron aquate which was transformed to goethite by ageing. Chester and Hughes (1967) have shown that ca 20% of the hydrogenous iron in a North Pacific core, is associated with goethite. This probably occurs largely as coatings on clay minerals or as individual particles, and has acted as a trace element scavenger.

Iron and manganese oxides, or hydroxides, may also be associated with calcareous pelagic sediments as a coating on calcareous shell debris, or, for manganese, as shell fillings. El Wakeel and Riley (1961) found that a globigerina ooze contained 12.8% Fe_2O_3 (total iron) on a water, organic carbon and carbonate-free basis. This compares with an average Fe_2O_3 content of nine pelagic clays of 8.20%. These authors attributed the high iron content to an oxide or hydroxide of iron which was originally derived from sea water by calcareous organisms and which remained in the sediment after the organisms were dissolved. Iron oxide minerals have also been recently reported in association with hot brines found in the Red Sea (Miller *et al.*, 1966; Craig 1966), but these types are likely to be of restricted importance.

Authigenic oxides of elements other than manganese and iron are not of quantitative importance in pelagic sediments, although their presence in small amounts is occasionally reported. For example, Correns (1954) has suggested that anatase may develop *in situ* in pelagic sediments.

B. SECONDARY MATERIAL

The most important secondary hydrogenous components in pelagic sediments are the clay minerals montmorillonite and nontronite (see Chapter 12–2 (ii)), and the zeolites.

(i) *Zeolites*

These are a group of hydrous sodium calcium alumino-silicates which have a framework-type structure, and usually contain exchangeable cations such as Na^+, K^+, Ca^{2+} and Ba^{2+}. Most of the zeolites found in pelagic sediments are members of the *phillipsite* (($\frac{1}{2}Ca,Na,K)_3Al_3Si_5O_{16}$. $6H_2O)$—*harmatone* ($BaAl_2Si_6O_{16}$. $6H_2O$) series (ideal formula from Deer *et al.*, 1962), of which phillipsite is the most widespread individual mineral, particularly in the South Pacific Ocean where it may constitute >50% of the total sediment (see Figure 12.9).

The mechanism of zeolite formation is not clearly understood, the major problem being whether the silicon and aluminium necessary for the zeolite framework originates from the dissolved silicon and aluminium of sea water, or whether these elements have been inherited from igneous silicates which have undergone weathering on the ocean floor. Bonatti (1963) has studied the zeolites in Pacific Ocean sediments and has concluded that the former mode of origin is unlikely. As an alternative he has suggested that the zeolites were formed from palagonite (see p. 329), and he has observed all stages in the alteration of palagonite to zeolites in some samples. Arrhenius and Bonatti (1965) have also studied the genesis of the zeolites and have concluded that harmatone is formed only in the earlier stages of the interaction between sea water and volcanic glass. Subsequently, the slow growth of phillipsite appears to be the most important process, and this may continue to such an extent that it may engulf the harmatone nucleus and also any particles of nontronite which may be present at the reaction site. In volcanic areas, therefore, it is likely that zeolites (mainly phillipsite) are formed by the modification of pre-existing silicate structures by the incorporation of some cations, such as those of barium and zinc, directly from sea water. The crystals of phillipsite which result from this process may be single or twinned, and are often intergrown with other phases such as nontronite.

Zeolites are found not only in association with volcanic assemblages, e.g. in the South Pacific, but also occur in sediments which do not contain significant amounts of igneous silicates. For example, the dissolving skeletons of opaline organisms may contain microcrystalline zeolites for which the silica has been supplied by dissolution of the organisms (Arrhenius, 1963).

In the Pacific Ocean sediments, phillipsite is the most important zeolite, but in the Atlantic and Western Indian Oceans this mineral is virtually absent from the deep-sea sediments. According to Turekian (1965) the most common zeolite in the latter areas is either *heulandite* ((Ca,Na_2) $Al_2Si_7O_{18}$. $6H_2O$; Deer *et al.*, 1962) or *clinoptilolite*, a high-silica member of the heulandite structural groups of zeolites. However, these zeolites are only present in the sediments in small amounts. Clinoptilolite has also been reported from the Pacific-Antarctic basin where diatomaceous oozes may provide a source of silica (Goodell, 1965).

There are still a number of problems to be resolved before the genesis of the zeolites is fully understood and these have been summarized by several authors, e.g. Sudo and Matsuoka (1959), Mumpton (1960), Coombs *et al.* (1959). In the pelagic sediments of some areas, such as the

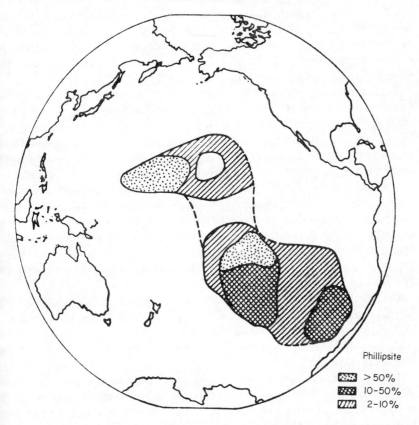

Fig. 12.9. The distribution of Phillipsite at the sediment surface of the Pacific Ocean. (Weight % on a carbonate-free basis); from Bonatti, 1963.

South Pacific, there is an established association between pyroclastic material and the zeolites, of which phillipsite is the most common member. In other areas where there is an "excess" of silica in solution, e.g. from the dissolution of opaline organisms, clinoptilolite is probably the stable zeolite formed (Goodell, 1965).

IV. Biogenous Components

Biogenous components are those produced in the biosphere, and include inorganic shell material and organic matter (Goldberg, 1954).

A. CARBONATES

The importance of skeletal carbonate material as a sedimentary component is shown by the fact that calcareous oozes are the predominant sediment type in many deep-sea areas. These deposits, which contain over 30% of skeletal carbonate (see p. 292) cover almost half of the deep-ocean floor. The most important carbonate-secreting organisms in the upper layers of deep-sea waters are foraminifera, coccolithophorids and pteropods.

Foraminifera are protozoans which have chambered tests, usually ca 0·5–1·00 mm in size (Riedel, 1963). These species live in the upper part of the water column, ca 100–300 m, and secrete calcite tests—see Figure 12.10a. In addition to these forms, benthonic species secreting arenaceous shells of agglutinated foreign grains (such as quartz and sand) are widespread, but only constitute a minor part of pelagic sediments (Riedel, 1963). Quantitatively, the foraminifera are the most important carbonate secreting organisms contributing to Quaternary oceanic sediments. The *globigerina oozes* to which they give rise, cover ca 35% of the deep-ocean floor, and are the most abundant pelagic sediments. In most classification schemes the term globigerina ooze refers, in fact, to a general type of calcareous sediment, rather than one specifically containing only globigerina.

Coccolithopuorids are plants which have a test of jointed or separated circular, elliptical, or angular calcitic plates termed coccoliths, rhabdoliths, etc.—see Figure 12.10b. After the death of the organism the tests usually disintegrate into these plates which may be preserved in the sediments.

Pteropods are pelagic molluscs with tests composed of aragonite (see Figure 12.10c). They occur within the upper few hundred metres of the water column, although some bathypelagic forms are known. Pteropod oozes are confined to areas having a depth of water not exceeding ca 3500 m (Turekian, 1965), and occur in significant amounts only in the Atlantic Ocean.

Calcite and aragonite are both significantly soluble in sea water, and tests composed of these minerals will undergo solution both during descent through the water column and at the sediment-water interface. The distribution of skeletal carbonate material in pelagic sediments depends on the number of organisms in the overlying waters, water circulation patterns and the rate at which dissolution occurs. Factors

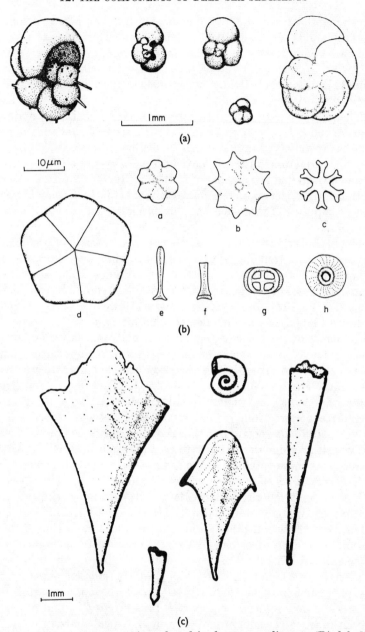

Fig. 12.10. Calcareous organisms found in deep-sea sediments (Riedel, 1963).
(a) Some recent planktonic Foraminfera. (b) Coccolithophorids and related forms
a-c, discoasters; d, pentalith of *Braarudosphaera*; e, f, rhabdoliths; g, h, coccoliths.
(c) Pteropods from a recent sediment.

influencing the dissolution rate include depth of water, rate of deposition of total solids (both carbonate and non-carbonate), the physico-chemical conditions (e.g. partial pressure of CO_2, temperature and hydrostatic pressure) at the organism-water interface, and the rate of removal of dissolved calcium, which affects the solubility equilibrium of the carbonates.

Calcium carbonate solubility increases with both total hydrostatic pressure and CO_2 partial pressure (Pytkowicz, 1967), and this mineral is more easily dissolved in deep than in shallow waters. It has long been recognized that dissolution on the deep-sea floor is more effective in those sediments which have been deposited below a certain depth of water. This depth is termed the *compensation depth*, and below it solution apparently exceeds supply and there is no net accumulation of calcium carbonate. The compensation depth occurs at ca 3500 m in the Pacific and at ca 4500 m in the Atlantic, and its existence poses a number of problems. One of the most important of these involves the solubility equilibria of calcium carbonate in sea water. Originally, it was considered that sea water was super-saturated with respect to calcite at depths above the compensation depth (see for example, Turekian, 1964). However, Berner (1965) and Pytkowicz (1965) have concluded that both Pacific and Atlantic waters are undersaturated with respect to calcite at depths below ca 500 m. This precludes the use of the model in which supersaturation was presumed to exist to depths of ca 3500–4500 m, and a number of alterna-tives have been put forward. For example, Turekian (1965) has sug-gested that some organisms actually accumulate, rather than lose, calcium carbonate from sea water on their descent to the sea floor. Olausson (1967) has postulated that the compensation depth corresponds to a transition from Deep Water to cold CO_2-rich Bottom Water, which because it is more rapidly renewed, dissolves calcium carbonate faster.

However, the results of a number of recent studies are at variance with the conclusion reached by Berner (1965) and Pytkowicz (1965), i.e. that sea water is undersaturated with respect to calcite below ca 500 m. Both Peterson (1966) and Berger (1967) have shown that the rate of dissolution of calcite in Pacific Ocean water increases sharply below a depth of ca 3000 m. Their results can be explained in terms of solution kinetics, but Li et al., (1969) offer a simpler explanation for this phenomenon, and for the position of the compensation depth. Using measured values of the partial pressure of CO_2 and of the total content of dissolved inorganic CO_2, they have calculated the degree of saturation of calcite and aragonite in ocean waters. They found that: (i) Atlantic waters only become undersaturated with respect to *calcite* at depths of ca 4000–5000 m, and Pacific waters at ca 1,500–3,000 m;

(ii) Atlantic waters become undersaturated with respect to *aragonite* at ca 1,000–2,500 m, and Pacific waters at ca 300 m. The authors concluded that the compensation depth in Atlantic sediments (ca 4,500 m) and in Pacific sediments (ca 3,500 m) reflects the transition from saturation to undersaturation with respect to calcite in the overlying waters; an explanation similar to that tentatively suggested by Turekian (1964). According to Li *et al.* (1969) the difference in the depths at which the waters of the Pacific and Atlantic oceans become undersaturated with respect to the carbonate minerals is a result of differences in the CO₂ contents of the two water masses, Pacific deep-waters having a higher content of dissolved CO_2 (see also p. 139).

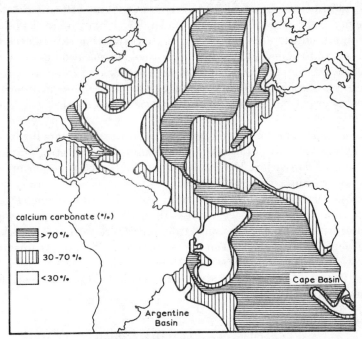

Fig. 12.11. The distribution of calcium carbonate at the sediment surface of the Atlantic Ocean (modified from Turekian, 1965).

The distribution of carbonates and the position and effect of the compensation depth can be modified by a number of processes, some of which can be illustrated by reference to the distribution of calcium carbonate in the deep-sea sediments of the Atlantic Ocean—see Figure 12.11. (i) The concentration of calcium carbonate in deep-sea sediments reflects the relative rates of accumulation of the "clay" and carbonate components. For example, the carbonate content of deep-sea sediments deposited on topographic highs such as the Mid-Atlantic Ridge system, is, in general, high because of the low rate of deposition of lithogenous solids. (ii) In the Cape Basin, high productivity in the overlying waters outweighs solution effects, and carbonate accumulation is high at all water depths (Turekian, 1965). (iii) In the Argentine Basin the CO_2-rich Antarctic deep water is renewed so rapidly that carbonate dissolution is very effective, and as a consequence the carbonate contents of the sediments is low at all depths.

The idea that the compensation depth is the only major control on the dissolution of carbonates in deep-sea sediments has been questioned by Smith et al. (1968). These authors carried out a statistical study of the relationship between the calcium carbonate content of many deep-sea sediments and the depth of water under which they were deposited; they found that there was only a moderate correlation between the two variables, and concluded that although the compensation depth effect does exist, the simple relationship between carbonate content and depth of water is complicated by other factors. One such factor may be that calcareous organisms adsorb organic molecules as a coating which may inhibit dissolution.

The organisms preserved in deep-sea carbonate sediments have been used extensively to interpret palaeo-ecological and stratigraphical conditions of deposition. For example, many species of organisms will grow only within a narrow temperature range, and past variations in surface water temperatures have been studied by examining species distribution in sedimentary cores. Information on surface water temperature has also been obtained from the $^{18}O/^{16}O$ ratios of the carbonates of foraminiferal shells. The application of carbonate organisms to palaeo-temperature determination has been reviewed by Emilliani and Flint (1963), and their use as stratigraphical indicators has been summarized by Riedel (1963).

B. SILICEOUS SKELETAL MATERIAL

Silica in the form of opal, an amorphous mater al with the composition $SiO_2.nH_2O$, is secreted by a number of marine organisms. These include diatoms, silicoflagellates, radiolaria and sponges; of which diatoms and radiolaria are quantitatively the most important.

Diatoms are plants which have frustules of a variety of shapes, the most usual types found in deep-sea sediments being discoidal or elliptical with sizes between 10–100 μm (Riedel, 1963)—see Figure 12.12a. Sediments rich in the remains of diatoms—diatomaceous oozes—are found in high latitudes such as the Antarctic region (Goodell, 1965), in

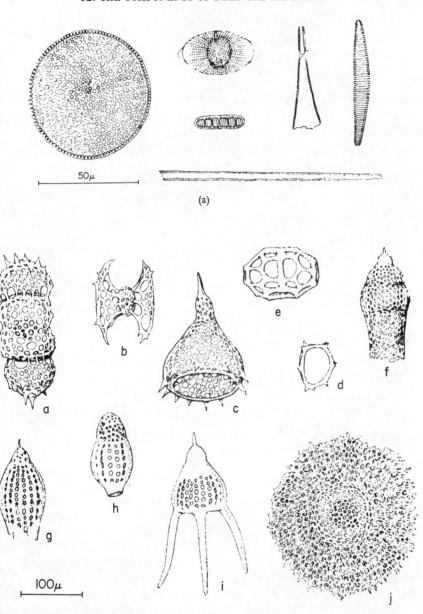

Fig. 12.12. Siliceous organisms found in deep-sea sediments (Riedel, 1963).
(a) Some Quaternary planktonic diatoms. (b) Radiolaria (Spumellaria and
Nassellaria) from Pacific sediments. a-d, Quaternary; e, f, Lower Miocene; g-j,
Eocene. (c) Tripylean Radiolaria from a Quaternary Pacific sediment.

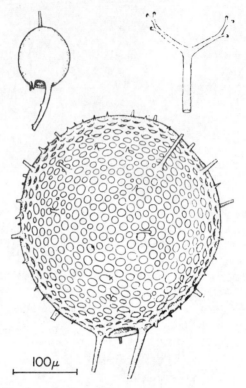

100μ

Fig. 12.12 (c). For legend see p. 377.

a band in the North Pacific Ocean and under certain areas of coastal upwelling along the western coasts of America and Africa.

Radiolaria are animals which occur in all parts of the oceans, and which contribute significant amounts of opaline silica to sediments, particularly in tropical and temperate regions (Riedel, 1963). Radiolarian tests are often spherical or helmet-shaped, and range in diameter from 50–400 μm—see Figure 12.12b and c. Radiolarian oozes occur only in significant amounts in the equatorial regions of the Pacific Ocean.

Sponges and silicoflagellates are widespread in the oceans, and although their remains are common in deep-sea sediments, they usually make up only a minor part of the biogenous silica.

According to Riedel (1959), three main factors control the contribution made by opal-secreting organisms to marine sediments. These are: (i) the rate of production of siliceous organisms in the overlying waters. Areas of high productivity associated with siliceous oozes include the Sub-arctic Convergence, the Equatorial Divergence and the divergences along the western coasts of the continents (Arrhenius, 1963); (ii) the degree of dilution of siliceous remains by lithogenous components (including volcanic materials), and by calcareous organisms; (iii) the extent of dissolution of the siliceous tests. Opaline material is very soluble in sea water, and according to Arrhenius (1963) much of it is dissolved or peptized soon after the death of the organisms. Apparently, depth of water has little effect on the dissolution equilibria, and there is apparently no "compensation depth" such as that found for calcareous tests (Riedel, 1959). Dissolution of the shells also continues after burial while they are in contact with interstitial waters undersaturated with respect to silica. Factors which may decrease dissolution include the introduction of dissolved silica into the interstitial water from the sediments themselves, e.g. from the weathering of pyroclastic material. According to Goldberg (1967), the presence of organic phases in opal can provide some protection against dissolution in slightly basic solutions (see also p. 177).

Opaline material is not as important quantitatively as calcareous shell debris in marine sediments. However, opaline phases can constitute as much as ca 70% of some sediments (Calvert, 1966; Chester and Elderfield, 1968), and the biogenous removal of silica plays an important part in the marine silica budget. The factors controlling the silica budget have been discussed by Calvert (1968), Burton and Liss (1969) and Schutz and Turekian (1965) (see also p. 175).

C. PHOSPHATES

The major host mineral for biogenous phosphate is skeletal apatite, which can constitute a few per cent (by weight) of some deep-sea sediments. Biogenous apatite undergoes dissolution on the deep-ocean floor, and in slowly accumulating pelagic sediments only the most resistant phosphate structures, such as sharks' teeth and the ear bones of whales, are preserved. Arrhenius *et al.* (1957) have presented analyses of skeletal fish debris from pelagic sediments, and have shown that the apatite phases of the debris contain a relatively high content of uranium, thorium and the rare earth elements. The concentration of these elements apparently occurs by adsorption from sea water after the death of the organisms. These rare earth phosphates may be preserved in slowly deposited sediments as microscopic crystals, following the dissolution of skeletal apatite. If the rare earths, uranium and thorium are released into solution they may be incorporated into hydrogenous phases, e.g. ferro-manganese nodules.

According to Arrhenius *et al.* (1957), there is a relationship between the size of the apatite crystals and their rare earth content. Bathypelagic fish debris were found to contain apatite crystals of 30–40 μm (along the z-axis) with a rare earth content of a few per cent; whale bone fragments have apatite crystals of 60–120 μm, and ca 0·1% rare earths; hydrogenous apatite has crystal sizes >2 mm, and a rare earth content of <0.05%.

D. SULPHATES

Barite is the principal biogenous sulphate in deep-sea sediments and has been described in Chapter 12 Section III-A (iv).

E. ORGANIC MATERIAL

This material consists of decomposable organic C and N compounds, and may be produced in the terrestrial or marine environments. It is composed of a number of substances which include: humic acids, amino acids, bitumens, hydrocarbons, lipids, porphyrins and lignin. Near-shore sediments may contain several per cent of organic matter. In contrast, in slowly deposited deep-sea sediments most of the organic matter is oxidized soon after deposition, and so escapes preservation. For this reason the nature of the organic components in marine sediments is not discussed in detail in the present text, which is confined to describing deep-sea sediments.

However, organic matter is extremely important because it exerts a strong control over the kind of diagenetic change which occurs in marine sediments subsequent to their deposition. Because of this it is necessary to understand the general relationship between the organic matter in marine sediments and their environment of deposition. This relationship may be summarized as follows: (i) the total content of organic matter in the sediments decreases in the order, near-shore > deep-sea marginal > deep-sea pelagic sediments; (ii) the extent to which organic matter is preserved in a sediment depends largely on the rate of deposition of the total sediment components; (iii) the kind of organic matter found in marine sediments varies from one oceanic area to another. For example, in near-shore sediments humic acids can constitute >50% of the total organic matter, compared to <5% in pelagic clays. In these clays much of the organic matter (>70%) is composed of substances with a high molecular weight, which are in part polymerized (Palacas *et al.*, 1966); (iv) the composition of the organic matter found in pelagic sediments is apparently invariant, regardless of location, except in anoxic environments.

The effect of organic matter on the geochemistry of deep-sea sediments is discussed in Chapter 13 Section II-B (ii).

V. COSMOGENOUS COMPONENTS

Cosmogenous components of deep-sea sediments are those derived from extra-terrestrial sources.

Small black magnetic spherules have been found in deep-sea sediments from all the major oceans (Brunn *et al.*, 1955; Crozier, 1960), and were first reported by Murray and Renard (1891) who considered them to have an extra-terrestrial origin and suggested that they were in fact micro-meteorites. This type of spherule has a metallic nucleus surrounded by a shell, usually consisting of magnetite, which Murray and Renard (1891) postulated was formed by oxidation of the outer layers during passage through the atmosphere following the ablation of molten droplets from a meteorite.

Brunn *et al.* (1955) collected magnetic particles directly from the ocean bed by means of a magnetic rake. Most of the particles obtained were spheres with diameters <0·5 mm, and consisted either wholly of magnetite, or of a silicate groundmass loaded with magnetite crystallites. While agreeing that these particles were meteoritic in origin, these authors suggested that those particles containing silicates originated from stony meteorites, whereas those having a metallic nucleus were derived from iron meteorites. Hunter and Parkin (1960) have shown that the magnetic spherules separated from Pacific and Atlantic deep-sea sediments are essentially similar, and can be divided into two principal groups; the iron spherules (density ca 6), and the stony spherules (density ca 3). The iron spherules, which range in diameter from ca 10–90 μm are blacker, smoother and more rounded than the stony spherules and always possess an eccentrically placed spherical metallic nucleus, the presence of which can often be detected by a circular step-like pattern surrounding that portion nearest to the surface—see Figure 12.13. This nucleus is surrounded by a shell of magnetite. Hunter and Parkin concluded that these particles have an extra-terrestrial origin—see also Fredriksson and Martin (1963). The stony spherules occur in much larger sizes (ranging from ca 15–250 μm) than the iron spherules, and some of them (ca 15–20%) depart from the spherical shape. Often the stony spherules have craters in the surface (a and b, in Figure 12.13b) which may have originated from bubbles of gas emerging from the heated fragment. Occasionally, a creamy-white coloration is observed at the base of the crater, and when broken open, the spherule is found to consist of an aggregation of minute translucent pale yellow-green crystals, surrounded by an outer jacket of harder material. However, the majority of the stony spherules are grey-black throughout and exhibit a dendritic structure in cross-section—see Figure 12.13b—indicating that they have cooled very rapidly from a liquid

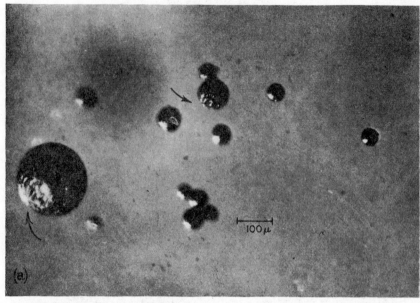

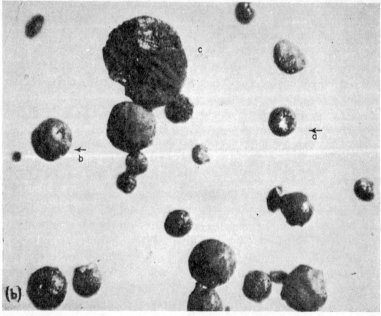

Fig. 12.13. Cosmic spherules found in pelagic sediments. (a) Iron spherules. (b) Stony spherules (from Hunter and Parkin, 1960, by courtesy of the Royal Society).

state. According to Hunter and Parkin (1960) almost all the stony spherules have a similar composition and consist essentially of a fine grained olivine (approximating in composition to forsterite) which contains a small amount of nickel. The origin of these stony spherules has not been established. In addition to these two major groups of spherules, i.e. "irons" and "stones", Hunter and Parkin (1960) have tentatively distinguished a third group, the "stony irons" (density ca 3·5–4·9).

In a study of the magnetic spherules separated from a Pacific deep-sea core Brownlow et al. (1966) divided them into three types; metal centred (i.e. having a core), hollow magnetite and solid magnetite. The metal centred spherules were assumed to have an extra-terrestrial origin, and since the hollow spherules had a similar distribution they were also tentatively assigned the same origin. However, Fredricksson and Martin (1963) have pointed out that although the nickel-iron (metal centred) spherules are definitely extra-terrestrial, the other magnetic spherules, with or without hollows, may have a terrestrial volcanic origin.

TABLE 12.7

The accretion rate of cosmic material to the earth's surface (data from Parkin et al., 1967 and Parkin and Tilles, 1968)

Method of determination	Accretion rate (tons per day)
Ni in deep-sea sediments	10,000
^{36}Cl in deep-sea sediments	5,000
Spherules in polar ice	
(i) Spherules >5 μm dia.	550
(ii) Spherules >15 μm dia.	490
Aircraft filter collection in stratosphere	550
Penetration satellite experiments	500
Os in deep-sea sediments	300
Atmospheric collections on meshes	
(a) Scilly Isles (U.K.)	
(i) Spherules >20 μm dia.	10·0
(ii) Spherules >10 μm dia.	2·0
(b) Barbados	
(i) Spherules >20 μm dia.	1·0
(ii) Spherules >10 μm dia.	0·3

Attempts have also been made to collect extra-terrestrial material from snow and ice, and from the atmosphere and space. These collections, and those made from deep-sea sediments, were directed towards making a positive identification of extra-terrestrial material, to establishing the rate at which it accreted on the earth's surface, and to understanding its origin. Comparisons between the various studies reveal very large differences in the estimated accretion rates. However, Grjebine (1967) has pointed out that such comparisons are difficult to evaluate because in most studies dust measurements were made over only a limited range of sizes, whereas the cosmic dusts range in size over several orders of magnitude. The various estimates of the influx of cosmic material to the earth's surface per day are listed in Table 12.7.

The question of the accretion rate is closely bound to that of the origin of the extra-terrestrial material reaching the earth. However, the latter question has still not been resolved, owing in part to the difficulty of actually identifying such material. From various early studies of the cosmic spherules found in marine sediments, it was concluded that they had originated as recently generated Zodiacal dust produced by meteoritic collisions (Hunter and Parkin, 1960; Brownlow et al., 1966). The particles were thought to be derived from the Zodiacal Cloud as it shifted past the earth, and it was suggested that the variation in spherules with depth in deep-sea cores reflected changes in the spatial density of this cloud.

However, recent detailed work on the amount of spherules actually arriving on the earth's surface at the present time has put these earlier conclusions in doubt. Parkin et al. (1967) collected airborne dust at Barbados and found, somewhat surprisingly, that most spherules collected were contaminants and that, in fact, less than ca 1 ton per day of cosmic material was reaching the earth. Even this figure may be an upper limit, and when contamination is fully allowed for, the rate could approach zero. These findings give rise to a paradox. Some of the spherules found in deep-sea sediments are without doubt extra-terrestrial in origin and yet their accretion rates, averaged over a long time period, exceeded those found at Barbados by several orders of magnitude. Parkin et al. (1967) argued that this means that the arrival of spherules is spasmodic, and since the rate of arrival can approach zero, i.e. no deposition, the spherules do not originate in the Zodiacal Cloud. These authors have further concluded that an origin from the surface ablation of a meteorite is more reasonable, and that a suitable meteorite is overhead about every 100 years or so. Proof that surface ablation of this kind does occur has been given by Krinov (1959), who found many spherules in the soils around the fallout area of an iron meteorite.

However, the problem of making a positive identification of cosmic material still remains. Criteria used in the past as indicators of a cosmic origin have been applied to separated and collected solids, and include chemical parameters, size distributions, density, physical form, etc. Recently, attempts have been made to measure cosmogenic nuclides, i.e. cosmic-ray produced radioactive or stable nuclides, which could give information on whether the host material contains extra-terrestrial debris. Much of the work has been carried out on bulk samples of deep-sea sediments, or on separated magnetic concentrates. The nuclides, and nuclide ratios sought have included ^{36}Cl (Schaeffer et al., 1964), $^{26}Al : {}^{10}Be$ ratios (Lal and Venkatavaradan, 1966; Amin et al., 1966), ^{64}Fe (in Arrhenius, 1967), and those of the noble gases, e.g. $^{4}He : {}^{3}He$ and $^{40}Ar : {}^{36}Ar$ ratios (Merrihue, 1964).

REFERENCES

Amin, B. S., Karkar, D. P. and Lal, D. (1966). *Deep-Sea Res.* **13**, 805.

Armstrong, F. A. J. (1965). *In* "Chemical Oceanography" (J. P. Riley and G. Skirrow, eds). Vol. 1, Academic Press, London. p. 323.

Arrhenius, G. O. S. (1963). *In* "The Sea" (M. N. Hill, ed.), Vol. 3. Interscience, New York, p. 655.

Arrhenius, G. O. S. (1965). *In* "Advances in Earth Science" (P. M. Hurley, ed.) M.I.T., Press, Cambridge, Mass. p. 155.

Arrhenius, G. O. S. (1967). Deep-Sea Sedimentation. Report, Fourteenth Gen. Assem. of the Intern. Union of Geodesy and Geophysics.

Arrhenius, G. O. S., Bramlette, M. and Picciotto, E. (1957). *Nature, Lond.* **180**, 85.

Arrhenius, G. O. S. and Korkisch, J. (1959). *Int. Oceanogr. Congs. Preprints*, **497**, (M. Sears, ed.), American Association for the Advancement of Science, Washington.

Arrhenius, G. O. S. and Bonatti, E. (1965). *In* "Progress in Oceanography" (M. Sears, ed.), Vol. 3. Pergamon Press, New York, p. 7.

Arrhenius, G. O. S., Mero, J. and Korkisch, J. (1964). *Science, N.Y.* **144**, 170.

Barnes, S. S. (1967). *Science, N.Y.* **157**, 63.

Beltagy, A. I., and Chester, R. (1970). In preparation.

Berry, R. W. and Johns, W. D. (1966). *Bull. Geol. Soc. Am.* **77**, 183.

Berger, W. H. (1967). *Science, N.Y.* **156**, 383.

Berner, R. A. (1965) *Geochim. cosmochim. Acta*, **29**, 947.

Biscaye, P. E. (1964). *Geochemistry Technical Report*, **8**, Dept. of Geology, Yale University.

Bonatti, E. (1963). *Trans. N.Y. Acad. Sci.* **25**, 938.

Bonatti, E. (1965). *Bull. Volcanologique*, **28**, 3.

Bonatti, E. (1966). *Science, N.Y.* **153**, 534.

Bonatti, E. and Nayudu, Y. R. (1965). *Am. J. Sci.* **263**, 17.

Bonatti, E. and Joensuu, O. (1967). *Abstract Amer. Geophys. Union, Annual Meeting*, Washington, D.C. 17.

Borchert, H. (1965a). *In* "Chemical Oceanography" (J. P. Riley and G. Skirrow, eds) Vol. 2. Academic Press, London, p. 159.

Borchert, H. (1965b). *In* "Chemical Oceanography" (J. P. Riley and G. Skirrow, eds), Vol. 2. Academic Press, London, p. 205.

Böstrom, K. and Peterson, M. N. A. (1966). *Econ. Geol.* **61**, 1288.

Brownlow, A. E., Hunter, W. and Parkin, D. W. (1966). *Geophys. J. R. Astrol. Soc.* **12**, 1.

Brunn, A. F., Langer, E. and Pauly, H. (1955). *Deep-Sea Res.* **2**, 230.

Burton, J. D. and Liss, P. S. (1969). *Nature, Lond.* **220**, 905.

Buser, W. (1959). *Int. Oceanogr. Congr. Reprints, Lond.* **962** (M. Sears, ed.). American Association for the Advancement of Science, Washington.

Buser, W. and Grütter, A. (1956). *Schweiz. Min. Petrogr. Mitt.* **36**, 49.

Calvert, S. E. (1966). *Geol. Soc. Am. Bull.* **77**, 569.

Calvert, S. E. (1968). *Nature, Lond.* **219**, 919.

Chester, R. (1965). *In* "Chemical Oceanography" (J. P. Riley and G. Skirrow, eds), Vol. 2. Academic Press, London, p. 23.

Chester, R. and Elderfield, H. (1968). *Geochim. cosmochim. Acta*, **32**, 1128.

Chester, R. and Hughes, M. J. (1966). *Deep-Sea Res.* **13**, 627.

Chester, R. and Hughes, M. J. (1967). *Chem. Geol.* **2**, 249.

Chester, R. and Hughes, M. J. (1969). *Deep-Sea Res.* **16**, 639.

Chow, T. J. and Goldberg, E. D. (1960). *Geochim. cosmochim. Acta*, **20**, 192.

Coombs, D. S., Ellis, A. J., Fyfe, W. S. and Taylor, A. M. (1959). *Geochim. cosmochim. Acta*, **17**, 53.

Correns, C. W. (1954). *Deep-Sea Res.* **1**, 78.

Craig, H. (1966). *Science, N.Y.* **154**, 1544.

Cronan, D. S. (1969). *Chem. Geol.* **5**, 99.

Cronan, D. S. and Tooms, J. S. (1967). *Deep-Sea Res.* **14**, 239.

Cronan, D. S., and Tooms, J. S. (1969). *Deep-Sea Res.* **16**, 35.

Crozier, W. D. (1960). *J. Geophys. Res.* **65**, 2791.

Deer, W. A., Howie, R. A. and Zussman, J. (1962) "Rock-Forming Minerals", Vol. 5. Longmans, London.

Deer, W. A., Howie, R. A. and Zussman, J. (1962) "Rock-Forming Minerals", Vol. 3. Longmans, London.

Delany, A. C., Delany Audrey C., Parkin, D. W., Griffin, J. J., Goldberg, E. D. and Reinmann, B. E. (1967). *Geochim. cosmochim. Acta*, **31**, 885.

Dietz, R. S., Emery, K. O. and Shepard, F. P. (1942). *Bull. Geol. Soc. Amer.* **53**, 815.

El Wakeel, S. K. and Riley, J. P. (1961). *Geochim. cosmochim. Acta*, **25**, 110.

Emilliani, C. and Flint, R. F. (1963). "The Sea" (M. N. Hill, ed.), Vol. 3. Interscience, New York, p. 888.

Fredriksson, K. and Martin, L. R. (1963). *Geochim. cosmochim. Acta*, **27**, 245.

Goldberg, E. D. (1954). *Jour. Geol.* **62**, 249.

Goldberg, E. D. (1961). *In* "Physics and Chemistry of the Earth" (L. H. Ahrens, F. Press, K. Rankama and S. K. Runcorn, eds), Vol. 4. Pergamon Press, Oxford, p. 281.

Goldberg, E. D. (1964). *Trans. N.Y. Acad. Sci.* **27**, 7.

Goldberg, E. D. (1967). *In* "Submarine Geology" (F. P. Shepard, ed.). Harper, New York.

Goldberg, E. D. and Arrhenius, G. O. S. (1958). *Geochim. cosmochim. Acta*, **27**, 245.

Goldberg, E. D. and Griffin, J. J. (1964). *J. Geophys. Res.* **69**, 4293.

Goldberg, E. D., Koide, M., Griffin, J. J. and Peterson, M. N. A. (1963). *In* "Isotopic and Cosmic Chemistry" (H. Craig, S. Mullen and G. J. Wassenburg, eds). North Holland, Amsterdam, p. 211.

Goldberg, E. D., Somayajulu, B. L. K., Galloway, J., Kaplan, I. R. and Faure, G. (1969). *Geochim. cosmochim. Acta*, **33**, 287–289.

Goodell, H. G. (1965). *In* Contrib. 11, Sed. Res. Lab., Dept. of Geol., Florida State University.

Gorbunova, L. N. (1966). *Oceanology*, **6**, 215.

Gorsline, A. S. (1963). *J. Geol.* **71**, 422.

Graham, J. (1959). *Science*, **129**, 1428.

Griffin, J. J. and Goldberg, E. D. (1963). *In* "The Sea" (M. N. Hill, ed.) Vol. 3. Interscience, New York, p. 728.

Griffin, J. J., Windom, H. and Goldberg, E. D. (1968). *Deep-Sea Res.* **15**, 433.

Grim, R. E. (1968). "Clay Mineralogy." McGraw-Hill, New York.

Grjebine, T. (1967). *In* "Mantles of the Earth and Terrestrial Planets" (S. K. Runcorn, ed.), Interscience, New York.

Grütter, A. and Buser, W. (1957), *Cimia*, **11**, 132.

Hathaway, J. C. and Sachs, P. L. (1965). *Am. Mineral* **50**, 852.

Heezen, B. C., Nesteroopf, W. D., Oberling, A. and Sabatiei, G. (1965). *C. r. hebd. Séanc. Acad. Sci.*, *Paris*, **260**, 5821.

Holland, H. D. (1965). *Proc. natn. Acad. Sci. U.S.A.* **53**, 1173.

Hunter, W. and Parkin ,D. W. (1960). *Proc. R. Soc. A*, **225**.

Hurley, P. M., Heezen, B. C., Pinson, W. H. and Fairbairn, H. W. (1963). *Geochim. cosmochim. Acta*, **27**, 393.

Kazakov, A. V. (1937). *U.S.S.R. Sci. Inst. Fertilisers and Insectfungicides Trans.* **142**, 95.

Kazakov, A. V. (1950). *Bull. Acad. Sci. U.S.S.R.*, Ser Geol. **5**, 42.

Kolodny, Y. and Kaplan, I. R. (1970). *Geochim. cosmochim. Acta*, **34**, 3.

Kramer, J. D. (1964). *Science, N.Y.* **146**, 637.

Kramer, J. R. (1966). *Abs. Second. Internat. Oceanographic Cong. Moscow.* **205**.

Krinov, E. L. (1959). *Sky. Telesc.* **18**, 617.

Krynine, P. D. (1948). *J. Geol.* **56**, 130.

Lal, D. and Venkatavaradan, V. S. (1966). *Science, N.Y.* **151**, 1381.

Li, Y. H., Takahashi, T. and Broecker, W. S. (1969). *J. Geophys. Res.* **74**, 5507.

Lynn, D. C. and Bonatti, E. (1965). *Mar. Geol.* **3**, 457.

McBirney, A. R. (1963). *Bull. Volcanologique*, **26**, 455.

McConnell, D. (1965). *Econ. Geol.* **60**, 1059.

Manheim, F. T. (1965). *In* "Symposium on Marine Geochemistry". University of Rhode Island, Occ. Pub. No. 3–1965.

Mason, B. (1952). "Principles of Geochemistry". John Wiley, New York.

Menard, H. W. and Shipek, C. J. (1958), *Nature, Lond.* **182**, 1156.

Mero, J. L. (1965). "The Mineral Resources of the Sea." Elsevier, New York.

Merrihue, C. M. (1964). *Ann. N.Y. Acad. Sci.* **119**, 351.

Miller, A. R., Densmore, C. D., Degens, E. T., Hathaway, J. C., Manheim, F. T., McFarlin, P. F., Pocklington, R. and Jokela, A. (1966). *Geochim. cosmochim. Acta.* **30**, 341.

Mumpton, F. A. (1960). *Am. Mineralogist*, **45**, 351.

Murray, J. and Renard, A. F. (1891). *Scient. Rep. Challenger Exped.* **3**.

Nayudu, Y. R. (1962). *Am. Geophys. Un. Monog.* **6**.

Ninkovich, D., Heezen, B. C., Conally, J. R. and Burckle, L. H. (1964). *Deep-Sea Res.* **2**, 605.

Olausson, E. (1967). *In* "Progress in Oceanography" (M. Sears, ed.), Vol. 5. Pergamon Press, Oxford, p. 245.

Palacas, J. G., Swanson, V. E. and Moore, G. W. (1966). *U.S. Geol. Survey Prof. Paper* **550-C**. C 102.

Parkin, D. W. and Tilles, D. (1968). *Science, N.Y.* **159**, 936.
Parkin, D. W., Delany, A. C. and Delany Audrey C. (1967). *Geochim. cosmochim. Acta*, **31**, 1311.
Peterson, M. N. A. (1966). *Science, N.Y.* **154**, 1542.
Peterson, M. N. A. and Goldberg, E. D. (1962). *J. Geophys. Res.* **67**, 3477.
Pettersson, H. H. (1945). *Medd. Oceanogr. Inst. Goteborg*, **213**, no. 5, I.
Pratt, R. M. and McFarlin, P. F. (1966). *Science, N.Y.* **151**, 1080.
Presley, B. J., Brooks, R. R. and Kaplan, I. R. (1967). *Science, N.Y.* **158**, 906.
Price, N. B. (1967). *Marine Geol.* **5**, 511.
Price, N. B. and Calvert, S. E. (1970). *Marine Geol.* In press.
Pytkowicz, R. M. (1967). *Geochim. cosmochim. Acta*, **31**, 63.
Pytkowicz, R. M. (1965). *Limnol. Oceanog.* **10**, 220.
Pytkowicz, R. M. and Kester, D. R. (1967). *Limnol. Oceanog.* **12**, 714.
Radczewski, O. E. (1939). *In* "Recent Marine Sediments" (P. O. Trask, ed.), American Association Petroleum Geologists, Tulsa, 496.
Rex, R. W. and Goldberg, E. D. (1958). *Tellus*, **10**, 153.
Revelle, R. and Emery, K. O. (1951). *Bull. Geol. Soc. Am.* **62**, 707.
Riedel, W. R. (1963). *In* "The Sea" (M. N. Hill, ed.), Vol. 3. Interscience, New York, p. 866.
Riedel, W. R. (1959). *In* "Silica in Sediments", American Association Petroleum Geologists, p. 80.
Riley, J. P. and Sinhaseni, P. (1958). *J. Mar. Res.* **17**, 466.
Rona, E., Hood, D. W., Mure, L. and Buglio, B. (1962). *Limnol Oceanogr.* **7**, 201.
Schaeffer, O. A., Megrue, G. H. and Thompson, S. O. (1964). *Ann. N.Y. Acad. Sci.* **119**, 347.
Schutz, D. F. and Turekian, K. K. (1965). *J. Geophys. Res.* **70**, 5519.
Sillén, L. G. (1961). *In* "Oceanography (M. Sears, ed.), International Oceanography Congress, New York, American Association for the Advancement of Science, Publ. **67**, 549.
Skornyakova, N. S., Andrushchienko, P. E. and Formina, L. S. (1962). Abs. in *Deep Sea Res.* **11** (1964) 93. Original source, *Okeanologiya*, **2** (2).
Smith, S. V., Dygas, J. A. and Chave, K. E. (1968). *Mar. Geol.* **6**, 391.
Sudo, T. and Matsuoka, M. (1959). *Geochim. cosmochim. Acta*, **17**, 1.
Summerhayes, C. P. (1967). *N. Z. Jl Mar. Freshwat. Res.* **1**, (3), 261.
Thompson, G., Borner, V. T., Nelson, W. G. and Cifelli, R. (1968). *J. Sediment. Petrol.* **38**, 1035.
Tooms, J. S. and Summerhayes, C. P. (1968). *Nature, Lond.* **218**, 1241.
Tooms, J. S., Summerhayes, C. P. and Cronan, D. S. (1969). *Oceanogr. Mar. Biol. Ann. Rev.* **7**, 49.
Turekian, K. K. (1964). *Trans. N.Y. Acad. Sci.* **26**, 312.
Turekian, K. K. (1965). *In* "Chemical Oceanography", Vol. 2 (J. P. Riley and G. Skirrow, eds). Academic Press, London, p. 81.
Turekian, K. K. (1968). *Geochim. cosmochim. Acta*, **32**, 603.
Whitehouse, V. G. and McCarter, R. S. (1958–59). *Texas A and M College, Contrib. in Ocean. and Meteorol.* **4**, 193.
Willis, J. P. and Ahrens, L. H. (1963). *Geochim. cosmochim. Acta*, **26**, 751.
Yeroshchev-Shak, V. A. (1961). *Dokl. Akad. Nauk. S.S.S.R.*, **137**, 695.
Yeroshchev-Shak, V. A. (1964). *Sov. Oceanog.* **2**, 90.
Yoder, H. S. Jr. (1957). *Proc. 6th Nat. Conf. Clays and Clay Minerals*, **42**.
Zen, E–an. (1959). *J. Sediment. Petrol.* **29**, 513.

Chapter 13

The Geochemistry of Deep-sea Sediments

I. Introduction

All the elements found in crustal rocks are probably present in deep-sea sediments, although not all of them have yet been identified. With the exception of Mn and Fe, most of the major elements are present in similar concentrations in deep-sea and continental sediments. However, some deep-sea sediments have trace element assemblages which are characteristically different from those of continental and near-shore sediments. The concentrations of a number of trace elements in deep-sea sediments are given in Table 13.1. In this table the sediments are divided into "clays" and "carbonates". This is a useful division because the principal components in the two types of sediments have different origins. Deep-sea clays consist largely of fine grained lithogenous, or secondary hydrogenous, particles of a continental or submarine volcanic origin (see also Chapter 12). The clays are deposited at a slow rate, and often contain primary hydrogenous phases such as ferro-manganese nodules, which are rich in certain trace elements. In contrast, deep-sea carbonates, which are deposited at a relatively fast rate, contain a significant amount of calcareous shell material. This material is produced within the marine environment, and has only a relatively low content of most trace elements—see Table 13.9. Because of this, the clays are enriched in trace elements compared to the carbonates; strontium is an exception to this—see Table 13.1.

One of the most important features of the geochemistry of the deep-sea clays is that they are also enriched in certain trace elements relative to near-shore and continental sediments. This is illustrated in Figure 13.1, which shows the distribution of nickel (on a carbonate-free basis) in the sediments of the Atlantic Ocean. It is evident from this figure that the highest contents of nickel are found in those sediments farthest from the continents. In addition, some of the trace elements are more strongly enriched in deep-sea clays of the Pacific than in those of the Atlantic. This can be seen from Table 13.2, which lists the average

TABLE 13.1

The concentrations of some trace elements in deep-sea sediments* (in ppm)

Element	Deep-sea carbonate	Deep-sea clay	Element	Deep-sea carbonate	Deep-sea clay
Li	5	57	Cs	0·4	6
Be	0·X	2·6	Ba	190	2300
B	55	230	La	10	115
Sc	2	19	Ce	35	345
V	20	120	Pr	3·3	33
Cr	11	90	Nd	14	140
Mn	1000	6700	Sm	3·8	38
Fe	9000	65000	Eu	0·6	6
Co	7	74	Gd	3·8	38
Ni	30	225	Tb	0·6	6
Cu	30	250	Dy	2·7	27
Zn	35	165	Ho	0·8	7·5
Ga	13	20	Er	1·5	15
Ge	0·2	2	Tm	0·1	1·2
As	1	13	Yb	1·5	15
Se	0·17	0·17	Lu	0·5	4·5
Rb	10	110	Hf	0·41	4·1
Sr	2000	180	Re	0·004	0·001
Y	42	90	Au	0·00X	0·00X
Zr	20	150	Hg	0·0X	0·X
Nb	4·6	14	Tl	0·16	0·8
Mo	3	27	Pb	9	80
Ag	0·0X	0·11	Th		5
Cd	0·0X	0·42	U		1
In	0·02	0·08			
Sn	0·X	1·5			
Sb	0·15	1·0			

* Data from Turekian and Wedepohl (1961); data for In and Re from A. D. Matthews (unpublished). Two types of deep-sea sediments are listed: (i) pelagic clay, which is essentially free from calcium carbonate; (ii) carbonates, which in the purest sampled form contained ca 10% clay. Some trace elements are concentrated in Pacific relative to Atlantic deep-sea clays (see text), and for these elements an average of the two oceanic clays was used for the abundance table. For some elements, only order of magnitude estimates could be made; these are indicated by the symbol X.

contents of a number of trace elements in both near-shore sediments and Atlantic and Pacific deep-sea clays. It is such variations which have given marine sediments a particular appeal to many geochemists.

In recent years, a great deal of research has been carried out with the aim of establishing the distribution and original sources of those trace elements which are enriched in deep-sea sediments. At present, our knowledge of the processes involved in the formation of the sediments does not permit their geochemistry to be fully understood. Nevertheless, sufficient information is available to enable a general description of the major processes which affect their chemical composition to be given.

TABLE 13.2

The distribution of some trace elements in marine sediments (in ppm)

Trace element	Near-shore sediments*	Atlantic deep-sea clay†	Pacific deep-sea clay‡	Ferro-manganese nodules§	Ratio, Pacific: Atlantic clay
Cr	100	80	77	<10	0·96
V	130	140	130	590	1·1
Cu	48	115	570	3300	5·0
Pb	20	52	162	1500	3·1
Ni	55	79	293	5700	3·7
Co	13	39	116	3400	3·0
Mn	850	3982	12500	220000	3·1

* Data from Wedepohl (1960).
† Data from Turekian and Imbrie (1966), and Wedepohl (1960).
‡ Data from El Wakeel and Riley (1961), Goldberg and Arrhenius (1958) and Landergren (1964).
§ Data from Table 12.6.

These processes are reviewed below in terms of the general framework which has been used in the three preceding chapters. Attention is focused on the geochemistry of those elements which are enriched in deep-sea sediments because it is through them that the processes controlling the overall geochemistry of the sediments can be characterized. Major elements are mentioned only briefly, and for a more detailed description of their geochemistry the reader is referred to Chester (1965a).

II. FACTORS CONTROLLING THE CHEMICAL COMPOSITION OF DEEP-SEA SEDIMENTS

The chemical compositions of deep-sea sediments are governed by a number of factors, the most important of which are: A, the relative proportions of the minerals constituting the sediments; B, the paths by which the elements are introduced into the marine environment; C, the mechanisms by which the elements are incorporated into the sediments; and D, the overall pattern of sedimentation in a particular oceanic area. These

factors are discussed in the following sections, and an attempt is made
to assess the extent to which they combine to influence the overall
distribution of certain elements in deep-sea sediments.

A. THE RELATIVE PROPORTIONS OF THE MINERALS CONSTITUTING DEEP-SEA SEDIMENTS

The relative proportions of the component minerals of deep-sea
sediments will obviously exert a fundamental control on their chemical
compositions, particularly with respect to their content of major

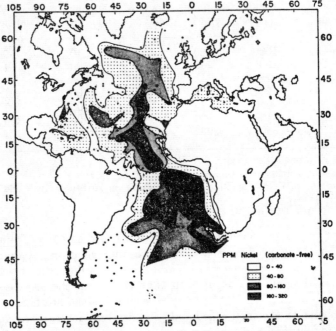

Fig. 13.1. The distribution of nickel at the sediment surface of the Atlantic
Ocean (ppm, on a carbonate-free basis; modified from Turekian and Imbrie, 1967).

elements. This is evident from Table 13.3, which gives the major
element composition of three of the principal types of deep-sea pelagic
sediments. These analyses serve to illustrate the overall trends in the
distribution of major elements which are largely a function of the
mineralogy of the sediments. The mineral components have been described in detail in Chapter 12, and for this reason individual major
elements are not discussed in the present section. However, their
distribution among the minerals of deep-sea sediments is summarized
in Table 13.4.

TABLE 13.3

Average chemical composition of pelagic sediments* (weight % oxides)

	Original compositions				Compositions on carbonate, water and organic C-free basis			
	Calcareous	Lithogenous	Siliceous	Oceanic average	Calcareous	Lithogenous	Siliceous	Oceanic average†
SiO_2	26·96	55·34	63·91	42·72	59·86	60·44	70·61	61·52
TiO_2	0·38	0·84	0·65	0·59	0·94	0·92	0·72	0·90
Al_2O_3	7·97	17·84	13·30	12·29	18·34	19·06	14·72	18·12
Fe_2O_3	3·00	7·04	5·66	4·89	9·77	8·93	7·08	9·09
FeO	0·87	1·13	0·67	0·94	—	—	—	—
MnO	0·33	0·48	0·50	0·41	0·75	0·52	0·55	0·64
CaO	0·30	0·93	0·75	0·60	1·04	1·01	0·83	1·00
MgO	1·29	3·42	1·95	2·18	3·05	3·73	2·17	3·19
Na_2O	0·80	1·53	0·94	1·10	2·50	1·67	0·93	1·98
K_2O	1·48	3·26	1·90	2·10	3·29	3·56	2·09	3·23
P_2O_5	0·15	0·14	0·27	0·16	0·46	0·16	0·30	0·33
H_2O	3·91	6·54	7·13	5·35				
$CaCO_3$	50·09	0·79	1·09	24·87				
$MgCO_3$	2·16	0·83	1·04	1·51				
Org. C.	0·31	0·24	0·22	0·27				
Org. N.	—	0·016	0·016	0·015				
Total	100·0	100·0	100·0	100·0	100·0	100·0	100·0	100·0
Total Fe_2O_3	3·89	8·23	6·42	—	—	—	—	—

* Data based on the analyses of 25 pelagic sediments from all the major oceans (El Wakeel and Riley, 1961).

† Weighted mean calculated on an areal basis. Percentage of deep-ocean floor covered by sediments, 48·7% calcareous, 37·8% lithogenous, 13·5% siliceous.

TABLE 13.4

The distribution of major elements among the minerals of deep-sea sediments*

Element	Lithogenous minerals			Hydrogenous minerals		Biogenous minerals	Cosmogenous minerals	Average content in deep-sea sediments (weight %, oxide)†
	Resistant minerals	Weathering residues	Igneous, metamorphic and sedimentary minerals; and pyroclastic material	Primary	Secondary			
Si	Quartz	Clay Minerals	Micas, Feldspars, Amphiboles, Pyroxenes, Pyroclastic material	—	Zeolites, Montmorillonite	Opal	—	SiO_2, 42·7
Al	—	Clay Minerals	Micas, Feldspars, Amphiboles, Pyroxenes, Pyroclastic material	—	Zeolites, Montmorillonite	—	—	Al_2O_3, 12·3
Ti	Rutile, Anatase	Clay Minerals	Micas, Feldspars, Amphiboles, Pyroxenes, Pyroclastic material	Anatase (?) Ferromanganese nodules	Zeolites, Montmorillonite	—	—	TiO_2, 0·59
Na, K	—	Clay Minerals	Micas, Feldspars, Amphiboles, Pyroxenes, Pyroclastic material	—	Zeolites, Montmorillonite	—	—	$\begin{cases} Na_2O, & 1·1 \\ K_2O, & 2·1 \end{cases}$
P	—	Clay Minerals	Micas, Feldspars, Amphiboles, Pyroxenes, Pyroclastic material	Phosphates	Zeolites, Montmorillonite	Phosphates	—	P_2O_5, 0·16

		Clay Minerals	(detrital)	(authigenic)	Zeolites, Montmorillonite	Carbonates		
Ca	—	Clay Minerals	Micas, Feldspars, Amphiboles, Pyroxenes, Calcite, Dolomite, Pyroclastic material	Carbonates, Phosphates	Zeolites, Montmorillonite	Carbonates	—	CaO 0·75
Mg	—	Clay Minerals	Micas, Feldspars, Amphiboles, Pyroxenes, Calcite, Dolomite, Pyroclastic material	Carbonates	Zeolites, Montmorillonite	Carbonates	—	MgO 2·18
Fe	Haematite, Goethite	Clay Minerals	Micas, Feldspars, Amphiboles, Pyroxenes, Pyroclastic material, Iron oxides (as coatings on other minerals, and as discrete grains)	Ferro-manganese nodules, Goethite	Zeolites, Montmorillonite	Associated with calcareous shell debris as oxide coatings.	Cosmic spherules	Fe_2O_3, 24·89
Mn	—	Clay Minerals	Micas, Feldspars, Amphiboles, Pyroxenes, Pyroclastic material, Manganese oxides (as coatings on other minerals, and as descrete grains)	Ferro-manganese nodules, Manganese oxides	Zeolites, Montmorillonite	Associated with calcareous shell debris as oxide coatings.	—	MnO, 0·41

* In this table the minerals are classified in terms of a number of sedimentary components according to the scheme outlined in Chapter 12. Some of these components, e.g. ferro-manganese nodules, contain more than one mineral, and for these the individual minerals are not specified. Only the principal host minerals are listed and their quantitative importance is indicated by the following typescripts; e.g. **Quartz** > CLAY MINERALS > Carbonates. † From Table 13.3.

The relative proportions of the minerals in deep-sea sediments also affect the distribution of the trace elements, although the relationship is more complex than that involving the major elements. Because of this, some prior knowledge of the geochemical history of the trace elements is necessary in order to understand their distribution in the sediments.

B. THE PATHS BY WHICH ELEMENTS ARE INTRODUCED INTO THE MARINE ENVIRONMENT

Elements are introduced into the marine environment either in solution, or associated with solid and colloidal material. Elements in solid source materials include those located within the lattice structures of lithogenous minerals, and those incorporated into surface (and inter-sheet) positions by adsorption and ion-exchange processes. Some elements are also brought to the sea in association with organic particulate matter.

(i) Elements associated with solid material

Solids introduced into the oceans result from the weathering of the continents, from volcanic activity (both terrestrial and submarine), and from extra-terrestrial sources.

Lithogenous material originating from continental weathering makes by far the most important contribution of solids to deep-sea sediments. This material can be transported to deep-sea areas by a variety of agencies, e.g. water, wind and ice transport, and it makes up the "background" input of suspended material which is present in all deep-sea waters. This "background" material is extremely important because its chemical composition represents a fundamental unit upon which those elements removed from the sea water itself are superimposed.

When the trace element composition of this lithogenous material is considered, a fundamental distinction must be made between those elements held in lattice positions within lithogenous minerals, and those held in surface and inter-sheet positions i.e. elements which have been removed from solution. This distinction is necessary because although the former elements have a continental origin, the latter can be acquired from solution in both the continental and marine environments. In order to establish the distribution of the trace elements which are held in lattice positions in lithogenous material Chester and Hughes (1967) have outlined a chemical technique designed to separate these elements from others present in deep-sea sediments. The technique, which involves leaching the sediments with a combined acid-reducing agent solution, can be applied to those deep-sea sediments in which the sole non-lithogenous components are ferro-manganese minerals, carbonate minerals and adsorbed trace elements. To date, the technique has been used to investigate trace element partition patterns in deep-sea sediments from the North Pacific and North Atlantic Oceans. From the results of these studies it is possible to ascertain whether variations in lattice-held lithogenous trace elements can account for the overall variations reported in the trace element compositions of deep-sea sediments.

The non-lithogenous material of the deep-sea sediments of the North Atlantic consists principally of ferro-manganese nodules and carbonate

TABLE 13.5

Average lattice-held lithogenous trace element contents of deep-sea sediments (*in ppm*)

Trace element	Average of 38 North Atlantic deep-sea sediments* (Lithogenous fractions)	Average of a North Pacific deep-sea clay core† (Lithogenous fractions)	Average of near-shore sediments‡ (Total sediment)
Ni	63	46	55
Co	12	16	13
Ga	19	—	19
Cr	72	91	100
V	120	92	130
Ba	421	—	750
Sr	141	—	250
Mn	582	743	850

* Data from Chester and Messiha-Hanna (1970); † data from Chester and Hughes (1969); ‡ data from Wedepohl (1960).

minerals, together with only minor amounts of authigenic montmorillonite, opal and the zeolites. Because of this, the residues obtained by leaching the sediments with the acid-reducing agent will consist largely of continentally-derived minerals in which the trace elements are held in lattice positions. Chester and Messiha-Hanna (1970) have used this technique to study the partition of the elements Fe, Mn, Ni, Co, Cr, V, Ga, Sr and Ba between the lithogenous and non-lithogenous fractions of a number of deep-sea and near-shore North Atlantic sediments. The average trace element contents of the lattice-held lithogenous fractions of these sediments are given in Table 13.5, together with those of a North Pacific deep-sea clay. The trace element analyses of a number of near-shore sediments are also included in this table. Near-shore sediments may be regarded as representing an early stage in the adjustment of lithogenous material to the marine environment. For this reason the compositions of near-shore sediments should give an indication of those of deep-sea clays, in the absence of processes which produce trace element enrichment. It can be seen from Table 13.5 that the concentrations of Cr, V, Mn, Ni, Co, Ga and Sr in the lattice-held lithogenous fractions of deep-sea sediments are similar to those in the total samples of near-shore sediments. This implies that the trace elements which have a similar abundance in near-shore sediments and deep-sea clays (see Table 13.2) are largely lithogenous, i.e. detrital, in origin. This has been confirmed by Chester and Messiha-Hanna (1970), who have shown that ca 70% of the Cr and V in North Atlantic deep-sea and near-shore sediments is held in lattice positions within lithogenous minerals. It is evident that the distribution of these two elements in the sediments is controlled by the input of detrital material to the Atlantic Ocean.

In contrast, Chester and Messiha-Hanna (1970) found that, on average, $<50\%$ of the Co, Ni, and Mn in the sediments had a lattice-held lithogenous origin, and that the contributions from this source to the total amount of each element in the sediments varied geographically. This variation is illustrated in Figure 13.2 which shows that the fraction of lattice-held lithogenous nickel decreases from values of $>\frac{2}{3}$ near to the continents to $<\frac{1}{3}$ in mid-ocean areas. Nickel is one of the trace elements which are concentrated in deep-sea clays (see Table 13.2), and the mid-ocean sediments which contain the minimum lattice-held lithogenous nickel are in fact those which have high total nickel contents (see Figure 13.1).

Chester and Messiha-Hanna (1970) have shown that there are local variations in the trace element composition of the lattice-held lithogenous material in Atlantic deep-sea sediments. These variations are the result of a number of factors of which the most important are: (i) variations in the chemical and mineralogical compositions of the continental source rocks; (ii) regional variations in weathering conditions which result in the concentration of specific minerals, e.g. the clay minerals, in certain areas (see also Chapter 12, Section IIB(ii)(c)); (iii) dilution of the lithogenous material by minerals which have only a small trace element content, e.g. quartz (see also p. 417). For those trace elements, e.g. Cr and V, which have a similar abundance in near-shore sediments and deep-sea clays, the variations in the composition of the lattice-held lithogenous material are of the same order of magnitude as those in the total sediments. However, for trace elements such as Mn, Ni and Co, which are concentrated in deep-sea clays, the variations in the composition of this lithogenous material are small compared to those in the total sediments, and they cannot account for the enrichment of these elements in deep-sea clays.

From the data given in Table 13.5 it seems likely that the composition of the lattice-held lithogenous material in deep-sea sediments of the Atlantic and North Pacific Oceans is similar, although local variations do

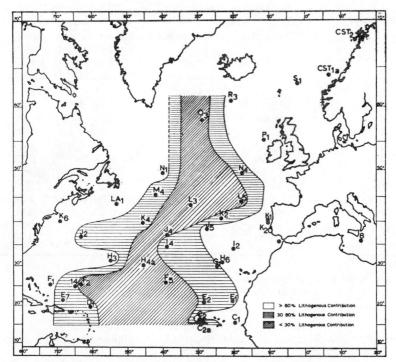

Fig. 13.2. The distribution of lattice-held lithogenous nickel at the sediment surface of the North Atlantic (the lithogenous contributions are expressed as a percentage of the total nickel in the sediments).

occur. This type of lithogenous material is particularly important in those oceanic areas which receive a large input of land-derived solids; these areas include the Atlantic and North Pacific Oceans. In other areas, e.g. the South Pacific, much of the solid material is introduced by submarine volcanic activity and is subsequently altered by submarine weathering processes. The trace element composition of this material is probably different from the average composition of the lithogenous material given in Table 13.5, and its geochemistry is discussed below.

The lithogenous solids introduced into the oceans from terrestrial volcanic activity are an ubiquitous component of deep-sea sediments,

TABLE 13.6 *The chemical composition of oceanic basalts;*

A. *Major element composition (wt. % oxides; data from Nicholls, 1965a)*

| | Oceanic island basalts | | Mid-Atlantic Ridge basalts | |
	Average tholeiitic basalt	Average alkalic basalt	Tholeiitic basalt	High alumina basalt
SiO_2	49·36	46·46	50·47	48·13
TiO_2	2·50	3·01	1·04	0·72
Al_2O_3	13·94	14·64	15·93	17·07
Fe_2O_3	3·03	3·37	0·95	1·17
FeO	8·53	9·11	7·88	8·65
MnO	0·16	0·14	0·13	0·13
MgO	8·44	8·19	8·75	10·29
CaO	10·30	10·33	11·38	11·26
Na_2O	2·13	2·92	2·60	2·39
K_2O	0·38	0·84	0·10	0·09
H_2O (total)	—	—	0·59	0·29
P_2O_5	0·26	0·37	0·11	0·10

B. *Trace element composition (ppm; data from Engel et al., 1965)*

Trace element	Average oceanic tholeiitic basalt	Average oceanic alkalic basalt
Cr	297	67
V	292	252
Cu	77	36
Ni	97	51
Co	32	25
Ga	17	22
Li	9	11
Rb	10	33
Ba	14	498
Sr	130	815
Zr	95	54

but are usually present in large amounts only in restricted areas. The larger sized pyroclastic material originating from terrestrial volcanic activity is ejected for only relatively short distances from the source, and most of that present in deep-sea sediments consists of pumice and volcanic ash (see also p. 328).

Basalts are the predominant type of volcanic rock found on the ocean floor, although other types, e.g. trachytes, occur in subordinate amounts (Nicholls, 1965a). Basalts have cooled quickly from a molten state, and often have a glassy appearance. They consist primarily of pyroxenes and plagioclases, and may also contain olivine and/or nepheline. Oceanic basalts are divided into two major groups, or suites; the tholeiitic basalts (which do not contain nepheline), and the alkalic basalts (which contain normative nepheline). Each of these suites is divided into a number of sub-groups (see Harris, 1969), although the classification is somewhat arbitrary. Samples of oceanic basalts have been obtained from outcrops on mid-ocean islands and by dredging the deep-sea floor, particularly in the region of the submerged portions of the mid-ocean ridges. Both tholeiitic and alkalic basalts are common on oceanic islands, but basalts dredged from the deep-sea floor are largely tholeiitic in character. The latter range in composition from the more typical tholeiitic basalts (Table 13.6, col. 3) which are not obviously associated with the mid-ocean ridges, to high alumina basalts (Table 13.6, col. 4) which appear to have a definite association with the ridge systems (Nicholls, 1965b). Both of these types of basalts are low in potassium compared to the tholeiitic basalts associated with oceanic islands (Table 13.6, col. 1). The analyses of the major types of oceanic basalts are shown in Table 13.6. These basalts have been studied in great detail in recent years because they may give an indication of the composition of the earth's mantle and its relationship to sea-floor spreading. A voluminous literature on them now exists, and reviews have been published by Nicholls (1965a and b), Melson et al. (1968), and Harris (1969).

In addition to basalts, ultrabasic rocks have been found associated with the mid-ocean ridge system in the Atlantic. In a recent study, Melson et al., (1968) have divided the rocks of the islands, mountains and ridges of the Mid-Atlantic Ridge system into two major petrographic types; thick basaltic volcanic piles, and partially serpentized ultrabasic intrusions. The ultrabasic intrusions appear to be common in the fracture and transverse fault zones of the Mid-Atlantic Ridge, but are rarely exposed elsewhere in the Atlantic. Low grade metamorphic derivatives of the basaltic and ultrabasic rocks are also found associated with the ridge system.

The original chemical composition of volcanic material in deep-sea areas will depend on the composition of the lava from which it was formed. However, its composition can be extensively modified by contact with sea water. For example, Nicholls (1963) has published analyses (Table 13.7) of basaltic glass from the floor of the Atlantic and of the residual hydroxide-rich material which is formed from it by submarine weathering. It can be seen that this process has resulted in marked increases in the ferric-ferrous iron ratio and in the water

TABLE 13.7

Analyses of basaltic glasses from the floor of the Atlantic Ocean and residual hydroxide-rich material resulting from sea water attack on basaltic glass (Results in weight percentages; from Nicholls, 1965)

	1	2	3
SiO_2	50·47	48·13	25·81
Al_2O_3	15·93	17·07	19·35
TiO_2	1·04	0·72	2·02
Fe_2O_3	0·95	1·17	15·05
FeO	7·88	8·65	0·63
MnO	0·13	0·13	0·15
MgO	8·75	10·29	1·29
CaO	11·38	11·26	1·25
Na_2O	2·60	2·39	1·68
K_2O	0·10	0·09	0·89
H_2O^+	0·53	0·27 ⎫	31·02
H_2O^-	0·06	0·02 ⎭	
P_2O_5	0·11	0·10	0·08
Cl	—	—	0·97
Uncorrected total	—	—	100·19
Less O for Cl	—	—	0·22
Total	99·93	100·29	99·97

1. Basaltic glass, 50° 44′ N., 29° 52′ W.
2. Basaltic glass, 28° 54′ N., 43° 19′ W.
3. Residual hydroxide-rich material, 19° 23′ N., 40° 53′ W. (see Nicholls, G. D. and Bowen, V. T. (1961), *Nature*, **192**, 156).

content, a smaller increase in the total iron content, a slight increase in the alumina content, marked decreases in the calcium and magnesium contents, and a lesser decrease in the content of silica in the residual weathered material.

The only material in deep-sea sediments which has been shown definitely to have an extra-terrestrial origin are the iron spherules (see p. 381). Chemically, the iron spherules consist largely of iron oxide with smaller amounts of nickel. Although cosmic spherules comprise only a very minor proportion of the total solids in deep-sea sediments, their importance has been stressed by various authors. For example, Pettersson (1959) has invoked a cosmic origin to explain the relatively high content of nickel in some deep-sea sediments. However, Chester and Hughes (1966) have examined the distribution of cosmic spherules in deep-sea sediments, and have concluded that even in those sediments which contained the maximum number of spherules reported in the literature, ca 5000/kg dry sediment, nickel of cosmic origin would amount to only ca 2% of the total nickel (see also Laevastu and Mellis, 1955).

It was shown above that variations in lattice-held lithogenous trace elements cannot account for the overall variations in the distribution of those trace elements which are concentrated in deep-sea sediments. However, elements can also be associated with lithogenous material in non-lattice positions. For example, river-borne detritus will contain elements which are held in surface positions by adsorption reactions. Few estimates of the chemical composition of river-borne detritus are available, and these only give the composition of the total detritus. Turekian and Scott (1967) have analysed detritus from 17 U.S. rivers on the Eastern and Gulf of Mexico seaboards, and have reported average Ni and Co contents of 126 ppm and 36 ppm respectively in the detritus. According to Chester and Messiha-Hanna (1970) the average Ni and Co contents of the lattice-held lithogenous fraction of Atlantic deep-sea sediments are 63 ppm and 12 ppm respectively, which are similar to the total concentrations in near-shore sediments. It would appear, therefore, that river-borne detritus can contain considerable amounts of adsorbed Ni and Co. These adsorbed elements may behave in one of two ways when they enter the sea; they may be transported and deposited with the detritus, or they may undergo ion-exchange or desorption reactions on introduction to sea water.

This problem has been considered by Kharkar et al. (1968), who have studied the distribution of a number of trace elements in river waters and river detritus. In addition, these authors carried out laboratory studies on the adsorption and desorption of the trace elements on, and from, minerals commonly occurring in river detritus. They concluded that trace elements which have been adsorbed from solution by clay mineral species in the river environment are released, to a greater or lesser extent, on contact with sea water. This release is apparently brought about by ion-exchange processes involving the ions of elements such as sodium and magnesium. The effect of desorption varies from one trace element to another. For example, there is little desorption of chromium from river-borne detritus; about 10% of the dissolved silver supplied to the sea by rivers has been desorbed at the fresh water-saline water boundary; the supply of cobalt desorbed from river-borne detritus exceeds that originally in solution in river water by a factor of two. However, the authors found that there was considerably less desorption of certain trace elements, e.g. cobalt, from iron and manganese oxides than from clay minerals on contact with sea water.

To summarize; it would appear that variations in the lattice-held trace elements contents of both continental and submarine volcanic solids are too small to account for the observed variations in the distribution of those trace elements which are concentrated in deep-sea sediments. Some of the trace elements associated with lithogenous material in non-lattice positions are desorbed in contact with sea water, but others may be transported to deep-sea areas by fine grained lithogenous particles such as manganese and iron oxides. The effect of this process on the overall geochemistry of deep-sea sediments is discussed in section 13-D above.

(ii) Elements in solution

Elements which are incorporated into deep-sea sediments from solution may have originated from a number of sources. These include: river run-off, mobilization and migration in interstitial waters (i.e. diagenetic transport), volcanic emanations, and submarine weathering.

River run-off provides by far the most important single source of the dissolved material supplied to sea water. Estimates of the extent of river run-off on a worldwide basis are very difficult to make, and those which are available are at best rough approximations. Turekian (1969) has estimated the total river discharge into the World Ocean to be ca $3\cdot6\times10^{16}$l/year, and the total dissolved load to be ca $3\cdot6\times10^{15}$g/year. However, the overall composition of the dissolved material in rivers cannot be evaluated with any certainty because regional variations in the drainage area petrology of different rivers, and seasonal variations in drainage characteristics in the same river, can produce large differences in the amount and composition of the dissolved material. According to Turekian (1969), man-made contamination is another factor which must be considered because, in some rivers, it may no longer be regarded as trivial. An average major and trace element composition of river water is given in Table 4.1 (pp. 64-67), but these data must be considered as only tentative. The distribution of this dissolved material in sea water has been discussed in Chapter 4.

Elements present in interstitial waters may originate in two ways:

(i) from sea water which was trapped within the accumulating sediment;
(ii) by liberation into solution from the sediments through diagenetic
mobilization processes such as solution, ion-exchange, desorption, etc.
On a major scale, element mobilization and subsequent upward migration in the interstitial water may provide an important source of some
of the elements incorporated into certain deep-sea sediments and
their surface components such as ferro-manganese nodules.

In recent years a large number of studies have been made of the
composition of interstitial waters; these have been reviewed by Brooks
et al. (1968), and by Price and Calvert (1970). The results of these
studies are difficult to evaluate. Some major elements, e.g. potassium,
appear to be appreciably enriched in interstitial waters relative to sea
water, and others do not. However, interstitial waters do seem to be
considerably enriched in trace elements, some of which may be present
at concentrations more than 1,000 times greater than those found in sea
water—see Table 13.8.

The post-depositional mobilization and migration of elements occurs
in the interstitial water-sediment complex. Little is known of the
reactions occurring during these processes, and much more work needs
to be done before they can be described in detail, although the general
processes operative in element mobilization and migration may be
summarized as follows. Many of the elements which are brought to
the sea-floor in association with solid material are in an oxidized form.
If the conditions in the sediment environment are reducing, either at
depth or at the surface, some elements are released into solution if they
can be reduced to a lower valency state. If this occurs at depth within
the sediment, the elements will be released into the interstitial waters
where they may remain in solution until upward migration or precipitation occurs. Under reducing conditions the sediment-interstitial
water complexes are rich in organic matter, and this can affect the
subsequent fate of the elements. For example, some elements may be
adsorbed onto organic components or may form chelates, such as amino
acid complexes. If these complexes migrate into an environment which
is oxidizing in character the metals in them will tend to precipitate
from solution to an extent which depends on the stability of their
complexes (Price, 1967). Ahrens (1966) has predicted the stability of some
amino-acid chelates to be in the order $Mn < Fe < Co < Zn < Ni < Cu$, and
as a result of this, chelated manganese should be released, and so become
available for precipitation more readily than iron, and iron more
readily than cobalt etc. Reduction is important in post-deposition
reactions and will affect Mn and Fe, but is unlikely to be responsible
for the direct release of Co, Zn, Ni and Cu. Other reactions which

TABLE 13.8

The composition of marine sediment interstitial waters

A. Major element composition (data from Siever et al., 1965)

	Cl (g/l)	Ca (g/l)	Mg (g/l)	Na (g/l)	K (g/l)	SiO$_2$ (mg/l)
Average composition of interstitial water from 22 near-shore and deep-sea sediments	19·8	0·43	1·20	10·8	0·54	50
Sea water	19·3	0·42	1·30	10·7	0·37	1

B. Trace element composition (μg/l; A. data from Presley et al., 1967; B. data from Brooks et al., 1968)

		Mn	Fe	Ni	Co	Cu	Zn	Eh(mv)	pH
Near-shore sediments, Southern California Borderland	Grey-green mud[A]	179	7·9	3·5	0·3	—	—	+400	8·2
	Black mud[A]	1·6	33	5·2	1·3	—	—	−90	7·8
	Sediment, Santa Cruz Basin[B]	—	7·2	3·3	0·3	8·1	40	+400	8·2
	Sediment, Santa Barbara Basin[B]	—	37	6·4	1·3	3·1	17	−90	7·8
	Grey-green clay[A]	3803	8·0	317	14·5	—	—	+380	8·0
	Brown clay[A]	810	14·7	18	4·7	—	—	+320	7·7
Deep-sea sediments, East Pacific rise area	Green-grey clay[A]	1187	7·5	248	12	—	—	+370	7·9
	Calcareous ooze[A]	176	3·9	24	3·0	—	—	+410	7·7
	Red clay[A]	5598	4·0	14	4·1	—	—	+420	7·7
	Brown clay[A]	10	3·2	3·2	1·7	—	—	+290	7·7
	Sea water, off the coast of southern California	2[A]	2[A]	2[A]	0·1[A]	2·1[B]	4·4[B]	—	—

The trace element concentrations and the Eh and pH values are averages based on a number of samples from each core.

occur in the sediment-interstitial water complex include desorption, ion-exchange and solution, all of which can release elements into interstitial waters.

Once an element is released into interstitial water, in complexed, molecular or ionic form, upward migration may occur. Two main mechanisms are involved in this process: (i) compaction of sediments at depth which results in the upward transfer of entrapped water; (ii) ionic and/or molecular diffusion (Lynn and Bonatti, 1965). Diffusion results mainly from interstitial waters having a higher content of a particular element than the sea water above them. This concentration gradient gives rise to an upward flux of the element.

The quantitative importance of diagenetic (or post-depositional) migration as a source of the elements incorporated into the upper layers of deep-sea sediments is difficult to assess. The extent to which elements are mobilized prior to migration is governed largely by the redox potential and pH of the environment, which, in turn, is controlled by the organic content of the sediment. In general, the marginal areas of the deep-sea regions have a higher organic content than the central regions (see also p. 367). For example, Lynn and Bonatti (1965) have shown that in the Pacific Ocean the top layer of the sediments is in an oxidized condition, but that it is underlain by a layer of reduced sediments. The thickness of the oxidized layer increases seaward from <1 cm in near-shore areas to between ca 8 and 15 cm in marginal areas. In the central region of the Pacific the oxidized layer extends below the penetration depth of normal coring techniques, although the reduced layer might be present at greater depths. Lynn and Bonatti (1965) found a high content of manganese in the top portions of several deep-sea cores from the Pacific. In these cores the sediments underlying the manganese-rich zones showed evidence of reducing conditions. The authors concluded that the manganese had been dissolved from the reduced layers and had migrated upwards through the interstitial water by ionic or molecular diffusion, and had precipitated in the surface oxidized layers.

More direct evidence of the post-depositional migration of manganese has been provided by Li et al. (1969). These authors have studied the distribution of manganese in both the sediments and the interstitial waters of an Arctic deep-sea core. In the upper layer of the core the conditions were oxidizing, but below a depth of ca 20 cm the sediments were in a reduced condition. The manganese content of the sediments decreased with depth, being highest in the upper oxidizing layer. In contrast, the manganese content of the interstitial waters showed a constant increase with depth to about one metre from the top of the core—

see Figure 13.3. Li *et al.* (1969) concluded that the distribution of manganese in the interstitial water-sediment complex resulted from the
reduction of manganese oxides at depth. This process released manganese into the interstitial water through which it migrated upwards
until precipitation occurred in the oxidizing layers of the sediment
column—a conclusion similar to that reached by Lynn and Bonatti
(1965). Further evidence for the post-depositional migration of manganese and other metals in Pacific Ocean sediments has been given by
Price and Calvert (1970).

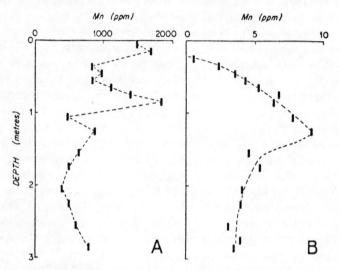

Fig. 13.3. The distribution of manganese in the sediment and interstitial water
of an Arctic deep-sea core. A. Distribution of manganese in the sediment; B.
distribution of manganese in the interstitial water. (Modified from Li *et al.*, 1969).

It was shown above that the distribution of manganese in deep-sea
sediments has provided evidence for its post-depositional migration.
For other elements, such as potassium and sodium, the only evidence
available at present for the existence of a diffusion flux is given by their
distribution in interstitial water. Various studies have shown that
potassium may be enriched in some interstitial waters (Shiskina, 1964;
Siever *et al.*, 1965; Brooks *et al.*, 1968). However, Mangelsdorf *et al.* (1969)
have thrown doubt on the validity of results obtained from interstitial
waters squeezed out of marine sediments. These authors showed that
spurious potassium enrichments can be obtained if sampling procedures
are not carefully standardized, particularly with respect to the temperatures at which the samples are stored and squeezed.

Many authors have suggested that some of the trace elements which are enriched in deep-sea sediments have a volcanic origin—see for example, Pettersson (1945), Goldberg and Arrhenius (1958), Wedepohl (1960), Arrhenius *et al.* (1964). However, theories regarding the importance of volcanic mechanisms in the supply of trace elements have been the subject of much controversy, and the problem is by no means resolved yet (see also p. 364).

The most important submarine process involved in volcanic supply mechanisms is probably the leaching of elements from hot effusive lava as it comes into contact with cold sea water. A further process is the attack by sea water on solidified volcanic rocks (see also p. 402). Bonatti (1965) has distinguished between two types of eruptive mechanisms in the emplacement of basaltic volcanic rocks on the floor of the Southeast Pacific Ocean. In one of these, the lava has effused quietly onto the sea floor, and any significant reaction with sea water has been prevented by the instantaneous formation of a thin insulating crust of glass at the surface of the lava flow. The lavas effused in this kind of eruption often have a pillow-like structure, and apparently issue from fissure-type volcanic vents. Submarine vulcanism of this type does not have a notable effect on the chemistry of sea water except in the immediate area covered by the lava. In the second type of eruption, which is referred to as hyaloclastic, there is evidence of extensive physical and chemical interaction between the hot lava and sea water during effusion. Much of the lava is shattered and pulverized because of rapid cooling on contact with cold sea water. The resulting debris (termed hyaloclastites) is rapidly hydrated at the prevailing high temperatures to form palagonite (see also p. 329). During this process sea water leaches various elements from the lava, particularly the transition elements such as manganese and iron. According to Bonatti (1965), this kind of vulcanism can affect both sediments and sea water over large areas adjacent to the activity; an important effect being the formation of montmorillonite, nontronite and the zeolites from fine grained hyaloclastic material. The type of effusion which occurs in submarine eruptions is apparently controlled by the viscosity of the lava; the more viscous lavas giving rise to hyaloclastic eruptions.

In a recent study, Böstrom *et al.* (1969), have shown that the non-carbonate fractions of deep-sea sediments from the crests of active ocean ridges, i.e. centres for ocean-floor spreading, have higher manganese and iron contents than those from inactive ridges and other volcanic regions. The active ridges have a high heat flow, and Böstrom *et al.* (1969) have concluded that the sediments on them have acquired these two elements as a result of volcanic processes associated with ocean-floor spreading. During vulcanism they have probably been leached from hot magma as it came into contact with sea water. The phenomenon may be directly related to the out-gassing of the earth's mantle through emanations which ascend the sub-oceanic part of the mid-ocean ridge system*.

There is no doubt that in some oceanic areas, e.g. the South Pacific, submarine vulcanism has had an important influence on the mineralogical and chemical composition of deep-sea sediments. This is particularly relevant in the South Pacific because it is an area which receives only a small input of land-derived detritus. The overall effect of this is that the sediment components are dominated by minerals of volcanic origin, such as montmorillonite and the zeolites (see also p. 349). In other parts of the oceans, the influence of submarine vulcanism is restricted to regions immediately associated with eruptive processes, e.g. the crests of active oceanic ridges. However, the question which has not so far been resolved is the importance of submarine vulcanism as an oceanwide mechanism for the supply of trace elements to deep-sea sediments.

C. THE MECHANISMS BY WHICH ELEMENTS ARE INCORPORATED INTO DEEP-SEA SEDIMENTS

Some elements are incorporated into deep-sea sediments in association with solid material. Others are removed directly from sea water, by both hydrogenous and biogenous processes.

The hydrogenous processes include co-precipitation, ion-exchange and adsorption reactions. Of these reactions, co-precipitation is particularly important, and the incorporation of elements into phases which are formed directly from sea water is one of the major factors influencing the trace element geochemistry of deep-sea sediments. The oxides of iron and manganese, and ferro-manganese nodules are the most common hydrogenous phases in the sediments. The nodules, which have been described in detail in Chapter 12, contain high concentrations of those trace elements which are enriched in deep-sea clays relative to near-shore sediments—see Table 13.2. Micro-nodules are particularly common in Pacific pelagic deep-sea clays, and their presence is probably responsible for the unique trace element composition of these sediments. The influence exerted by the nodules on the geochemistry of pelagic sediments has been stressed by many authors. For example, Goldberg (1954) has shown that there is a correlation between manganese and nickel, and between iron and cobalt in pelagic clays, and has concluded that much of the nickel and cobalt has been removed from sea water

---

* Recent research (see for example: Böstrom, K., (1970). *Earth Plan. Sci. Lett.* **9**, 348, and *Nature, Lond.* **227**, 1041; also Hörowitz, K., (1970). *Marine Geol.* **9**, 41) has confirmed the importance of active ocean ridges in the supply of trace elements to sea water. The supply is restricted, but in certain local areas it can contribute up to 10% of the total content of some trace elements, e.g. Fe, Mn, Pb, Zn, Hg in the sediments.

during the formation of ferro-manganese nodules (see also Landergren, 1964). Goldberg and Arrhenius (1958) and El Wakeel and Riley (1961) have also suggested that much of the Pb, Cu and Zn in pelagic sediments is present in the micro-nodules. Chester and Hughes (1969) have examined the distribution of trace elements in a North Pacific lithogenous clay which contained an abundance of micro-nodules. They found that the hydrogenous trace elements, i.e. those removed from sea water, were partitioned between micro-nodules and iron oxide phases. The nodule-associated trace elements varied between 1 and 90% of the total sediment trace element content, and decreased in the order Mn >Co >Ni >Cu >Pb = V >Cr. For the elements Mn, Co and Ni over half of the total content in the sediment was held in the micro-nodules. Some elements, e.g. vanadium, were concentrated in the iron oxide phases relative to micro-nodules, but it was not possible to distinguish between adsorption and co-precipitation as the mechanisms by which they had been removed from sea water.

Phosphates are important hydrogenous phases, and their trace element geochemistry has been reviewed by Tooms *et al.* (1969). However, large-scale phosphate deposits are confined mainly to near-shore areas, and in deep-sea sediments francolite is the only phosphate mineral which has a widespread occurrence, particularly in deposits on seamounts and other topographic highs. This mineral is usually present in the sediments in only small amounts, but since it may contain relatively high concentrations of zirconium and the rare earth elements it can make a significant contribution to the trace element content of the sediments with which it is associated. Co-precipitation within sulphide minerals may also be a significant process in the removal of some trace elements from sea water, but the process is restricted to near shore strongly reducing environments.

Ion-exchange between the clay minerals and the major elements in sea water is an important mechanism in marine geochemistry. Weaver (1967) has reviewed this subject, and has concluded that under present day conditions clay minerals acquire more sodium and magnesium than potassium from sea water. Weaver has also stressed that a major proportion of the potassium removed from sea water is held initially in interstitial waters and is only fixed in clay minerals, mainly illites, after deep burial of the sediments. The amounts of potassium removed from sea water by clay minerals appears to have changed during geological time, and the possibility exists that clay minerals may now be contributing this element to sea water. Much work still remains to be done on this topic, particularly with respect to the distribution of potassium between lattice and non-lattice positions in illites, and the changes that can affect this element in both these sites when the mineral is introduced into sea water.

Various authors, e.g. Landergren (1954) have stressed the importance of adsorption reactions in the geochemistry of boron. Landergren has shown that there is a relationship between the boron content of a sediment and the salinity of the water under which it was deposited; the more saline the water, the higher the content of boron. Recently, this simple relationship has been questioned, and it is now considered that other factors such as the rate of deposition, chemical composition of the clay minerals and the organic content may also affect the boron content of deep-sea sediments. It was originally thought that boron was incorporated into the sediments by adsorption from sea water on to illites. However, Goldberg and Arrhenius (1958) have shown that much of the boron in deep-sea sediments is held in lattice structures of the clay minerals, and the actual mechanism by which boron is removed from sea water is not known (see also Thompson and Melson, 1970).

A number of authors, e.g. Krauskopf (1956), have shown that various minerals are capable of adsorbing trace elements from sea water. Chester (1965b) has discussed the mechanisms by which these elements are adsorbed onto clay mineral surfaces, and has shown that the clay mineral illite is capable of adsorbing >50% of the cobalt and >90% of the zinc from sea water containing the metals at their natural concentrations. However, Kharkar et al. (1968) have shown that river-borne detritus will desorb considerable amounts of cobalt on contact with sea water. This is difficult to reconcile with the experimental results reported by Krauskopf (1956) and Chester (1965b), who have both found that cobalt is adsorbed from sea water onto clay minerals, despite the presence of major ions such as Na^+ and K^+ in the water. At present, the overall importance of the adsorption of trace elements from sea water onto clay minerals cannot be evaluated. However, it is likely that clay minerals brought to the sea surface by eolian transport, i.e. clays which have not been in river water, will adsorb trace elements from sea water.

Oxide coatings on mineral surfaces are important in adsorption reactions. It was shown above that a significant result of the study made by Kharkar et al., (1968) was that there was much less desorption of certain trace elements, e.g. cobalt, from iron and manganese oxides than from clay minerals on contact with sea water. It is possible, therefore, that adsorption of trace elements from sea water by river-borne iron and manganese oxides may be one of the major pathways by which some trace elements are incorporated into deep-sea sediments. Oxides of this type differ from those of a marine hydrogenous origin because they will contain trace elements acquired in the river environment, in addition to those removed directly from sea water.

Organic matter, either in the form of discrete particles, or in association with mineral species (e.g. the clay minerals) is an extremely efficient adsorbent for many trace elements. In those sediments which contain a relatively high content of organic matter, this material can significantly affect their geochemistry (see also p. 407). However, the surface layers of most deep-sea pelagic sediments are in an oxidized condition, and it is unlikely that organic matter at the sediment surface will be in contact with sea water for sufficient time for it to be an active trace element adsorbent. It is possible that organic matter will adsorb trace elements as it descends the water column, and that these elements will be released at the sediment surface, and so become available for incorporation into the sediment by other mechanisms.

Biogenous processes involving living organisms are extremely important in the removal of some elements from sea water. Elements may be associated with organisms in a number of ways which include: (i) incorporation into hard skeletal parts; (ii) incorporation into soft body parts; (iii) association with bodily processes, e.g. digestive and excretory functions. Elements associated with bodily processes may be retained within the organism during its life, or may be expelled by excretion, e.g. in faecal pellets.

The principal skeletal materials in deep-sea sediments are calcium carbonate, opal and apatite. Opal-secreting organisms remove large quantities of silica from sea water, and their sedimentation results in the concentration of silicon in some deep-sea sediments. However, siliceous sediments have a restricted distribution. In contrast, calcareous sediments are found in many areas, and cover ca 50% of the deep-sea floor. Because of this, the sea floor accumulation of carbonate secreting organisms is an important sedimentary process, and a large proportion of the calcium in deep-sea sediments is contained in calcareous shell debris. The shells also contain other elements, and the concentrations of some of these are given in Table 13.9, together with their concentrations in deep-sea sediments. It is evident from this table that calcareous shells contain a relatively high content of strontium, and Chester and Messiha-Hanna (1970) have shown that >80% of the total content of this element in deep-sea sediments is held in calcareous shell material. According to Nicholls et al. (1959), the sedimentation of pteropod shells may be one of the major pathways by which V and Pb are removed from sea water. In addition to the elements incorporated within the shell material, calcareous shells may have a coating of manganese or iron oxide. However, these oxide coatings do not appear to concentrate trace elements from sea water, and on average calcareous shells have only a small content of the trace elements, such as Mn, Ni, Co and Cu, which are concentrated in deep-sea clays—see Table 13.9. According to Goldberg (1962), these elements, which are probably present in sea water chiefly as cationic species, can form strong organic complexes and tend to be concentrated in the soft parts of organisms. These soft body parts are not preserved in deep-sea sediments. Because

TABLE 13.9

The concentrations of some elements in carbonate shell material and deep-sea sediments (ppm)

Element	Average of foraminiferal shells from deep-sea sediments*	Pteropod shell†	Cocco-lith ooze‡	Average of the >30 μm fraction of a deep-sea carbonate core§	Average, deep-sea carbonates‖	Average, deep-sea clays¶
Sr	1,112	—	1,468	—	2,000	180
Ba	10–30	—	175	—	190	2,300
Mg	1,417	—	1,100	—	4,000	21,000
Fe	1,213	—	—	—	9,000	65,000
Mn	335	—	263	—	1,000	6,700
Cr	—	1	5	41	11	90
V	15	85	—	20	20	120
Ni	21	2	4	4	30	225
Co	—	20	4	<1	7	74
Cu	23	30	13	15	30	250
Pb	138	200	—	—	9	80
Ti	525	5	113	—	770	4,600
Al	2522	—	—	—	20,000	84,000
Si	7536	—	—	—	32,000	250,000

* Data from Turekian (1965b); † data from Nicholls *et al.* (1959); ‡ data from Thompson and Bowen (1969); § data from H. Elderfield (unpublished). The >30 μm fraction of the sediment consisted almost exclusively of shell material; ‖ data from Turekian and Wedepohl (1961); ¶ data from Turekian and Wedepohl (1961).

of this, the elements associated with them will be released on decomposition of the organisms, although some of these elements may be permanently incorporated into the sediments by other processes. For example, Revelle *et al.*, (1955) have shown that some of the copper in pelagic clays has a non-skeletal biogenous origin, although the mechanism by which it is incorporated into the clays is not known. Barium is another example of an element which may be incorporated into deep-sea sediments by biogenous mechanisms which involve non-skeletal material (see also p. 357).

It will be evident from the various factors discussed above that trace elements can be incorporated into deep-sea sediments by a number of mechanisms. The importance of each mechanism varies from element to element. This is shown in Table 13.10 which summarizes the distribution of a number of geochemically dissimilar trace elements among the components of deep-sea sediments.

TABLE 13.10

The distribution of a number of trace elements among the principal components of deep-sea sediments

Trace Element	Lithogenous Components *		Hydrogenous Components		Biogenous Components	
	Lattice positions	Non-lattice positions	Primary †	Secondary ‡	Skeletal processes	Non-skeletal processes
Cr	▓			▓		
V	▓					
Ni	▓	▓		▓		
Co	▓			▓		
Cu	▓			▓	▓	
Ba	▓			▓	▓	
Sr	▓			▓		

▓ Components indicated in this way are important in the incorporation of the trace elements into deep-sea sediments

▢ Components indicated in this way do not contribute significant amounts of the trace elements to deep-sea sediments

* These refer to the clay minerals, quartz, feldspars, etc., with minor amounts of iron and manganese oxides.
  † These refer to ferro-manganese nodules, and iron and manganese oxides.
  ‡ These refer to montmorillonite and the zeolites.
  The skeletal processes refer to those of carbonate-secreting organisms.

The mechanisms by which trace elements are removed from sea water have been discussed above with respect to deep-sea sediments. Trace element removal mechanisms in the oceans as a whole have been discussed by Krauskopf (1956) and Turekian (1965a). The results of these studies indicate that some trace elements are more effectively removed from sea water than others, and, more important, that some elements are more efficiently removed in *particular parts* of the ocean. This inhomogenous removal of trace elements from sea water is reflected in their distribution in marine sediments (Turekian, 1968). Some elements, e.g. silver, are removed largely into near-shore sediments by processes such as sulphide precipitation. In contrast, the barium which reaches the sea in a dissolved form is concentrated in deep-sea sediments, to which it is brought by the biogenous precipitation of barite which dissolves as it descends the water column. The solubility increases with pressure, and results in the preferential preservation of barite on topographic highs such as the mid-oceanic ridge and flank systems—see also p. 357.

D. THE OVERALL PATTERN OF SEDIMENTATION

It is convenient to divide the trace elements in marine sediments into two broad groups: (i) those which have a similar abundance in near-shore sediments and deep-sea clays, e.g. Cr and V; (ii) those which are enriched in Atlantic deep-sea clays relative to near-shore sediments, and which are even more enriched in Pacific deep-sea clays, e.g. Ni, Co, Pb, Zn, Cu and Mn (see Table 13.2).

It was shown above that the distribution of Cr and V in North Atlantic deep-sea sediments is largely controlled by the input of detrital material from the continents, i.e. there is little Cr and V in the sediments in excess of that held in the lattice structures of lithogenous minerals. However, in mid-ocean areas of the Atlantic the deep-sea sediments are enriched in elements such as Mn, Ni and Co; an enrichment which cannot be accounted for by variations in lattice-held lithogenous trace elements (see also p. 398). It is evident, therefore, that other explanations must be sought to account for the "excess" content of those elements which are enriched in deep-sea clays. In recent years a number of such explanations have been advanced which relate trace element enrichment patterns to the overall conditions of sedimentation in the oceans.

One explanation involves the concept that the deep-sea sediments receive an independent constant supply of the "excess" trace elements from sea water*. In this theory, the trace elements are thought to be removed homogeneously from sea-water, i.e. equally in all parts of the ocean, and are superimposed on deep-sea "clay" which accumulates at varying-rates in different parts of the World Ocean. Thus, the deep-sea clays of the Pacific are thought to be more strongly enriched in "excess" trace elements than those of the Atlantic because they are deposited at a slower rate. Wedepohl (1960) has used the trace element composition of near-shore sediments, and both Atlantic and Pacific deep-sea clays, to show that, in the light of this theory, it would appear

---

* The trace elements in solution are supplied to sea water by processes such as river run-off, volcanic emanations and post-depositional migration—see Chapter 13, II B, *ii*.

that the average deep-sea clay accumulates ca 3·4 times faster in the Atlantic than in the Pacific Ocean. However, Turekian (1967) has criticized this theory on the grounds that the "excess" trace elements have an inhomogenous distribution in the oceans, and that their sources of supply, and mechanisms of removal, probably differ in the Atlantic and Pacific Oceans.

Because of this, Turekian (1965a, 1967) has proposed an alternative theory to explain the "excess" contents of Ni, Co and Mn in deep-sea clays. This involves the differential transport of pelagic and non-pelagic lithogenous material in the oceans. The major feature in the distribution of Ni, Co and Mn in Atlantic sediments is their relatively high concentrations in mid-ocean areas. Sediments deposited in these central areas are pelagic in character, i.e. their lithogenous material is composed of fine-grained particles which have been deposited slowly from the upper water layers (see also p. 284). These fine grained particles are composed mainly of clay minerals (which may be coated with iron and manganese oxides), together with discrete grains of manganese and iron oxides (see also p. 352). The particles have a large surface area and will have acquired a high content of trace elements in the river environment. On contact with sea water some of the trace elements previously adsorbed by the clay minerals will be desorbed (see also p. 403), but most of those held by manganese and iron oxides will be retained and subsequently transported to deep-sea areas. In contrast, the sediments of the abyssal plains which fringe the continents are mainly non-pelagic, and contain relatively large particles which have been deposited by bottom processes (see also p. 285). These particles, which are rich in quartz, have only low contents of trace elements such as Ni, Co and Mn, elements which in non-pelagic sediments are held largely in lattice positions in the lithogenous minerals.

Turekian's theory relates the distribution of trace elements to the classification of deep-sea sediments into pelagic and non-pelagic types which has been discussed in Chapter 10. It is the *pelagic* sediments which contain the "excess" trace elements, and Turekian has attempted to relate the differences between Pacific and Atlantic deep-sea clays to the greater areal extent of pelagic sedimentation in the latter ocean. However, it is difficult to understand why this should result in Pacific *pelagic* clays having a higher content of trace elements than Atlantic *pelagic* clays. This question throws doubt on a fundamental assumption involved in the "differential transport" theory, i.e. that the removal of trace elements such as Ni, Co and Mn from deep-sea waters is of secondary importance to their removal in river and near-shore environments. This assumption does not take sufficient account of the influence exerted on the geochemistry of pelagic sediments by ferro-manganese micro-nodules—phases which have grown directly from elements dissolved in sea water. These nodules are particularly abundant in Pacific pelagic clays, and may contain over half of the Mn, Ni and Co in the sediments.

The enrichment of certain trace elements in the Pacific relative to Atlantic deep-sea clays is intimately related to the trace element supply mechanisms in the two oceans. It is now thought that dissolved trace elements supplied to the Atlantic Ocean are largely continental in origin, and are introduced by river run-off (Turekian, 1968). However. a number of authors have suggested that the Pacific Ocean receives an additional supply of dissolved trace elements from volcanic processes (see for example: Goldberg and Arrhenius, 1958; Wedepohl, 1960; Turekian, 1967; see also p. 409). Trace elements introduced into sea water in this way will be incorporated into deep-sea clays largely by hydrogenous processes such as the formation of ferro-manganese nodules. These nodules have their highest abundance in Pacific pelagic clays, and offer an explanation for the concentration of "excess" trace elements in these clays relative to those of the Atlantic. However, the removal of trace elements from sea water into deep-sea sediments cannot be neglected even in the Atlantic Ocean, e.g. it is now believed that much of the land-derived fractions of the deep-sea sediments in the Atlantic between ca 10° N. and ca 35° N. have an eolian origin (see also p. 334). Much of this eolian dust originates from the desert areas of Africa, and so will not have any river-acquired trace elements. Some of this dust will be transported directly to deep-sea areas, and since the sediments in these areas have a high content of trace elements, it must be assumed that considerable proportions of them have been derived from sea water*.

In the discussion given in the preceding sections the trace element geochemistry of deep-sea sediments has been related to the distribution of a small number of elements in Atlantic and Pacific sediments. The elements were selected because their geochemistry is known in some detail, and because they served to illustrate some of the general principles which govern trace element distribution in deep-sea sediments. To provide a useful framework which offers at least a general explanation for the observed trace element distribution patterns these principles should be applicable to all deep-sea sediments, and not simply those of the Atlantic and Pacific Oceans. Evidence which indicates the oceanwide similarity in trace element distribution patterns has been provided by Bender and Schultz (1969), who have studied the distribution of Mn, Ni, Co and Cu in deep-sea sediments from the Indian Ocean. These authors found that the concentration of the elements

---

* This has been confirmed by recent work. Chester and Johnson (unpublished data) have shown that the eolian dusts in this region do not have a sufficiently high content of Mn, Ni and Co to account for the enrichment of these trace elements in mid-ocean sediments. Further, it has been shown that the dusts are capable of adsorbing trace elements from sea water.

increased from near-continental to deep-sea areas in a manner similar to that reported in the Atlantic and Pacific Oceans.

The essential problems involved in developing a satisfactory explanation for the distribution of trace elements in deep-sea sediments are complex. One of the most critical problems relates to the supply of dissolved trace elements to sea water by river run-off, and in particular to whether or not the deep-sea areas act as a gigantic sink for these elements. According to Turekian (1965) the removal of dissolved trace elements from sea water occurs inhomogeneously, and he has concluded that rivers do not supply important amounts of elements such as Co and Ni to deep-sea areas in a dissolved form. In his view, therefore, the deep-sea does not act as the ultimate sink for these elements which, although transported within the oceans, are removed more effectively in near-shore areas. This contrasts with the conclusion reached by Manheim (1965), that the Pacific and Indian Oceans, and to a much lesser extent the Atlantic Ocean, do act as a sink for continentally derived manganese and associated trace elements such as Co and Ni.

Until questions such as these are resolved, it will not be possible to postulate a single theory which can adequately explain the distribution of trace elements in deep-sea sediments.

III. SUMMARY

The most important features of the distribution of trace elements in deep-sea sediments are summarized below.

(i) The major sources of the trace elements in deep-sea sediments are: river run-off which introduces dissolved, desorbed and detrital trace elements; volcanic emanations which introduce dissolved and detrital trace elements; post-depositional migration which introduces only dissolved trace elements; and eolian transport, which introduces largely detrital trace elements.

(ii) Dissolved trace elements, and some of those associated with detrital material, undergo inhomogeneous removal from sea water; some are incorporated predominantly into near-shore sediments, and others into deep-sea sediments.

(iii) Dissolved trace elements are brought to the sediment surfaces by oceanic mixing processes, post-depositional migration, and vertical transport by biological processes in which the organisms act only as an intermediate reservoir and subsequently release their trace elements into solution.

(iv) Trace elements are incorporated into the sediments associated with detrital material, or within hydrogenous and biogenous solids.

(v) The importance of post-depositional migration is difficult to assess, but in some deep-sea sediments, e.g. those of the Pacific marginal areas, it must be considered as a source of the trace elements incorporated into the upper layers of the sediments.

(vi) Some trace elements have a similar abundance in near-shore sediments and deep-sea clays. These elements, which include Cr and V, are introduced into the sediments largely within the lattice structures of lithogenous minerals.

(vii) Certain trace elements are enriched in Atlantic deep-sea clays relative to near-shore sediments, and are even more strongly enriched in Pacific deep-sea clays. This enrichment is found in pelagic deep-sea clays, and in the Atlantic it can be largely accounted for by trace elements supplied from the continents. In the Pacific Ocean there is an additional supply of trace elements from volcanic emanations.

(viii) Two major processes contribute to the enrichment of trace elements in pelagic sediments; the differential transport of lithogenous material, and the removal of trace elements from sea water by processes such as the formation of ferro-manganese nodules.

(ix) The enrichment of trace elements in Pacific relative to Atlantic deep-sea clays can be related to four major factors: (i) the deep-sea clays of the Pacific have a lower rate of accumulation than those of the Atlantic; (ii) a higher percentage of the Pacific deep-sea floor is composed of areas remote from the land masses; (iii) in addition to river run-off Pacific ocean waters receive an independent supply of trace elements from volcanic sources which are located mainly in the Central Pacific; (iv) Pacific deep-sea clays contain more micro-ferro-manganese nodules than those of the Atlantic; this is the result of a combination of various factors, including (i), (ii) and (iii) above, which have been discussed in detail in Chapter 12.

REFERENCES

Ahrens, L. H. (1966). *Geochim. cosmochim. Acta*, **30**, 1111.

Arrhenius, G. O. S., Mero, J. and Korkisch, J. (1964). *Science, N.Y.* **144**, 170.

Bender, M. L. and Schultz, C. (1969). *Geochim. cosmochim. Acta*, **33**, 292.

Bonatti, E. (1965). *Second International Oceanographic Congress, Abstracts of papers*, **55**, Moscow.

Böstrom, K., Peterson, M. N. A., Joensuu, O. and Fisher, D. E. (1969). *J. Geophys. Res.* **74**, 3261.

Brooks, R. R., Presley, B. J. and Kaplan, I. R. (1968). *Geochim. cosmochim. Acta*, **32**, 397.

Chester, R. (1965a). *In* "Chemical Oceanography", Vol. 2 (J. P. Riley and G. Skirrow, eds). Academic Press, London. p. 23.

Chester, R. (1965b). *Nature, Lond.* **206**, 884.

Chester, R. and Hughes, M. J. (1966). *Deep-Sea Res.* **13**, 627.

Chester, R. and Hughes, M. J. (1967). *Chem. Geol.* **2**, 249.

Chester, R. and Hughes, M. J. (1969). *Deep-Sea Res.* **16**, 639.

Chester, R. and Messiha-Hanna, R. G. (1970). *Geochim. cosmochim Acta.* In press.

El Wakeel, S. K. and Riley, J. P. (1961). *Geochim. cosmochim. Acta,* **25**, 110.

Engel, A. E. J., Engel, C. and Haven, R. G. (1965). *Bull. Geol. Soc. Am.* **76**, 719.

Goldberg, E. D. (1954). *J. Geol.* **62**, 249.

Goldberg, E. D. (1962). *Liminal Oceanogr.* Supplement to Vol. 7, LXXVI.

Goldberg, E. D., and Arrhenius, G. O. S. (1958). *Geochim. cosmochim. Acta,* **27**, 245.

Harris, P. G. (1967). *In* "International Dictionary of Geophysics", Vol. 2. (S. K. Runcorn, ed. in chief). Pergamon Press, Oxford. p. 1043.

Kharkar, D. P., Turekian, K. K. and Bertine, K. K. (1968). *Geochim. cosmochim. Acta,* **32**, 285.

Krauskopf, K. B. (1956). *Geochim. cosmochim. Acta,* **9**,

Laevastu, T., and Mellis, O. (1965). *Trans. Am. Geophys. Un.,* **36**, 385.

Landergren, S. (1954). *Rep. Swed. Deep-Sea Exped.* **7**, Goteburg.

Landergren, S. (1964). *Rep. Swed. Deep-Sea Exped.* **10**, Goteburg.

Li, Y. H., Bischoff, J. and Mathieu, G. (1969). *Earth Plan. Sci. Lett.* **7**, 265.

Lynn, D. C. and Bonatti, E. (1965). *Marine Geol.* **3**, 457.

Mangelsdorf, P. C. Jr., Wilson, T. R. S. and Daniell, E. (1969). *Science, N.Y.* **165**, 171.

Manheim, F. T. (1965). *In* "Symposium on Marine Geochemistry". University of Rhode Island, Occ. Publ. no. 3–1965.

Melson, W. G., Thompson, G. and Van Andel, T. H. (1968). *J. Geophys. Res.* **73**, 5925.

Nicholls, G. D. (1963). *Sci. Prog.* **1**, 12.

Nicholls, G. D. (1965a). *Phil. Trans. R. Soc.* **258**, 168.

Nicholls, G. D. (1965b). *Min. Mag.* **34**, 373.

Nicholls, G. D., Curl, H. and Bowen, V. T. (1959). *Limnol. Oceanogr.* **4**, 472.

Pettersson, H. (1945). *Medd. Oceanogr. Inst. Goteburg,* **2B**, No. 5, 1.

Pettersson, H. (1959). *Geochim. cosmochim. Acta.* **17**, 209,

Presley, B. J., Brooks, R. R. and Kaplan, I. R. (1967). *Science, N.Y.* **158**, 906.

Price, N. B. (1967). *Marine Geol.* **5**, 511.

Price, N. B. and Calvert, S. E. (1970). *Marine Geol.* In press.

Revelle, R., Bramlette, M. M., Arrhenius, G. O. S. and Goldberg, E. D. (1955). *Geol. Am. Spec. Paper.* **62**, 221.

Shiskina, O. V. (1964). *Geokhimiya,* **6**, 564.

Siever, R., Beck, K. C. and Berner, R. A. (1965). *J. Geol.* **73**, 39.

Thompson, G., and Bowen, V. T. (1969). *J. mar. Res.,* **27**, 32.

Thompson, G. and Melson, W. G. (1970). *Earth. Plan. Sci. Lett,* **8**, 61.

Tooms, J. S., Summerhayes, C. P. and Cronan, D. S. (1969). *Oceanogr. Mar. Biol. Ann. Rev.* **7**, 49.

Turekian, K. K. (1965a). *In* "Symposium on Marine Geochemistry". University Rhode Island, Occ. Pub. No. 3–1965.

Turekian, K. K. (1965b). *In* "Chemical Oceanography" (J. P. Riley and G. Skirrow, eds), Vol. 2. Academic Press, London. p. 81.

Turekian, K. K. (1967). *In* "Progress in Oceanography" (M. Sears, ed.), Vol. 4. Pergamon Press, Oxford.

Turekian, K. K. (1968). *Geochim. cosmochim. Acta,* **32**, 603.

Turekian, K. K. (1969). *In* "Handbook of Geochemistry" (K. H. Wedepohl, ed.) Vol. 1. Springer-Verlag, Berlin. p. 296.

Turekian, K. K., and Imbrie, J. (1967). *Earth Plan. Sci. Lett.* **1**, 161.

Turekian, K. K. and Scott, M. R. (1967). *Environ. Sci. Tech.* **1**, 940.

Turekian, K. K. and Wedepohl, L. H. (1961). *Bull. Geol. Soc. Am.,* **72**, 195.

Weaver, C. A. (1967). *Geochim. cosmochim. Acta,* **31**, 258.

Wedepohl, K. H. (1960). *Geochim. cosmochim. Acta,* **18**, 200.

Appendix

TABLE 1

Preparation of artificial sea water ($S = 35 \cdot 00^{o}/_{oo}$) having the composition shown in Table 4.5 (after Kester *et al., 1967, Limnol. Oceanog.* **12**, 176).

	g
NaCl	23·926
Na_2SO_4	4·008
KCl	0·677
$NaHCO_3$	0·196
KBr	0·098
H_3BO_3	0·026
NaF	0·003

Dissolve in ca 750 ml of distilled water and then add
53·27 ml of 1·0 M $MgCl_2$. $6H_2O$ solution*
10·33 ml of 1·0 M $CaCl_2$. $2H_2O$ solution*
0·90 ml of 0·1 M $SrCl_26H_2O$ solution*
and distilled water to 1 kg

* Standardized by titration with silver nitrate solution using Mohr's method.

TABLE 2

Collected conversion factors

Conversion	Factor	Reciprocal
$\mu g\ NO_3^- \rightarrow \mu g\ N$	0·2259	4·427
$\mu g\ NO_2^- \rightarrow \mu g\ N$	0·3045	3·286
$\mu g\ NH_3 \rightarrow \mu g\ N$	0·8225	1·216
$\mu g\ NH_4^+ \rightarrow \mu g\ N$	0·7764	1·287
$\mu g\ PO_4^{3-} \rightarrow \mu g\ P$	0·3261	3·066
$\mu g\ P_2O_5 \rightarrow \mu g\ P$	0·4364	2·291
$\mu g\ SiO_2 \rightarrow \mu g\ Si$	0·4675	2·139
$\mu g\ SiO_4^{4-} \rightarrow \mu g\ Si$	0·3050	3·278
$\mu g\ N \rightarrow \mu g$-at. N	0·07138	14·008
$\mu g\ P \rightarrow \mu g$-at. P	0·03228	30·975
$\mu g\ Si \rightarrow \mu g$-at. Si	0·03560	28·09

TABLE 3

Table for conversion of weights of nitrogen, phosphorus and silicon expressed in terms of $\mu g/l$. into $\mu g\text{-}at./l$.

μg N, P, or Si/l.	μg-at. N/l.	μg-at. P/l.	μg-at. Si/l.
1	0·071	0·032	0·036
2	0·143	0·065	0·071
3	0·214	0·097	0·107
4	0·286	0·129	0·142
5	0·357	0·161	0·178
6	0·428	0·194	0·214
7	0·500	0·226	0·249
8	0·571	0·258	0·284
9	0·643	0·291	0·320
10	0·714	0·323	0·356
20	1·428	0·646	0·712
30	2·142	0·968	1·068
40	2·856	1·291	1·424
50	3·569	1·614	1·780
60	4·283	1·937	2·136
70	4·997	2·260	2·492
80	5·711	2·582	2·848
90	6·425	2·905	3·204
100	7·139	3·228	3·560

Author Index

Subject Index

(Numbers in bold type indicate the page on which a subject is treated most fully.)

17